AF344314

From Source to Seep: Geochemical Applications
in Hydrocarbon Systems

The Geological Society of London
Books Editorial Committee

Chief Editor
RICK LAW (USA)

Society Books Editors
JIM GRIFFITHS (UK)
DAVE HODGSON (UK)
PHIL LEAT (UK)
TERESA SABATO CERALDI (UK)
DANIELA SCHMIDT (UK)
RANDELL STEPHENSON (UK)
ROB STRACHAN (UK)
MARK WHITEMAN (UK)

Society Books Advisors
GHULAM BHAT (India)
MARIE-FRANÇOISE BRUNET (France)
ANNE-CHRISTINE DA SILVA (Belgium)
JASPER KNIGHT (South Africa)
SPENCER LUCAS (USA)
DOLORES PEREIRA (Spain)
VIRGINIA TOY (New Zealand)
GEORG ZELLMER (New Zealand)

Geological Society books refereeing procedures

The Society makes every effort to ensure that the scientific and production quality of its books matches that of its journals. Since 1997, all book proposals have been refereed by specialist reviewers as well as by the Society's Books Editorial Committee. If the referees identify weaknesses in the proposal, these must be addressed before the proposal is accepted.

Once the book is accepted, the Society Book Editors ensure that the volume editors follow strict guidelines on refereeing and quality control. We insist that individual papers can only be accepted after satisfactory review by two independent referees. The questions on the review forms are similar to those for *Journal of the Geological Society*. The referees' forms and comments must be available to the Society's Book Editors on request.

Although many of the books result from meetings, the editors are expected to commission papers that were not presented at the meeting to ensure that the book provides a balanced coverage of the subject. Being accepted for presentation at the meeting does not guarantee inclusion in the book.

More information about submitting a proposal and producing a book for the Society can be found on its website: www.geolsoc.org.uk.

It is recommended that reference to all or part of this book should be made in one of the following ways:

LAWSON, M., FORMOLO, M. J. & EILER, J. M. (eds) 2018. *From Source to Seep: Geochemical Applications in Hydrocarbon Systems*. Geological Society, London, Special Publications, **468**.

BYRNE, D. J., BARRY, P. H., LAWSON, M. & BALLENTINE, C. J. 2018. Noble gases in conventional and unconventional petroleum systems. *In*: LAWSON, M., FORMOLO, M. J. & EILER, J. M. (eds) *From Source to Seep: Geochemical Applications in Hydrocarbon Systems*. Geological Society, London, Special Publications, **468**, 127–149. First published online December 14, 2017, https://doi.org/10.1144/SP468.5

GEOLOGICAL SOCIETY SPECIAL PUBLICATION NO. 468

From Source to Seep: Geochemical Applications in Hydrocarbon Systems

EDITED BY

M. LAWSON
ExxonMobil Upstream Research Company, USA

M. J. FORMOLO
ExxonMobil Upstream Research Company, USA

and

J. M. EILER
California Institute of Technology, USA

2018
Published by
The Geological Society
London

THE GEOLOGICAL SOCIETY

The Geological Society of London (GSL) was founded in 1807. It is the oldest national geological society in the world and the largest in Europe. It was incorporated under Royal Charter in 1825 and is Registered Charity 210161.

The Society is the UK national learned and professional society for geology with a worldwide Fellowship (FGS) of over 10 000. The Society has the power to confer Chartered status on suitably qualified Fellows, and about 2000 of the Fellowship carry the title (CGeol). Chartered Geologists may also obtain the equivalent European title, European Geologist (EurGeol). One fifth of the Society's fellowship resides outside the UK. To find out more about the Society, log on to www.geolsoc.org.uk.

The Geological Society Publishing House (Bath, UK) produces the Society's international journals and books, and acts as European distributor for selected publications of the American Association of Petroleum Geologists (AAPG), the Indonesian Petroleum Association (IPA), the Geological Society of America (GSA), the Society for Sedimentary Geology (SEPM) and the Geologists' Association (GA). Joint marketing agreements ensure that GSL Fellows may purchase these societies' publications at a discount. The Society's online bookshop (accessible from www.geolsoc.org.uk) offers secure book purchasing with your credit or debit card.

To find out about joining the Society and benefiting from substantial discounts on publications of GSL and other societies worldwide, consult www.geolsoc.org.uk, or contact the Fellowship Department at: The Geological Society, Burlington House, Piccadilly, London W1J 0BG: Tel. +44 (0)20 7434 9944; Fax +44 (0)20 7439 8975; E-mail: enquiries@geolsoc.org.uk.

For information about the Society's meetings, consult *Events* on www.geolsoc.org.uk. To find out more about the Society's Corporate Affiliates Scheme, write to enquiries@geolsoc.org.uk.

Published by The Geological Society from:
The Geological Society Publishing House, Unit 7, Brassmill Enterprise Centre, Brassmill Lane, Bath BA1 3JN, UK

The Lyell Collection: www.lyellcollection.org
Online bookshop: www.geolsoc.org.uk/bookshop
Orders: Tel. +44 (0)1225 445046, Fax +44 (0)1225 442836

The publishers make no representation, express or implied, with regard to the accuracy of the information contained in this book and cannot accept any legal responsibility for any errors or omissions that may be made.

© The Geological Society of London 2018. No reproduction, copy or transmission of all or part of this publication may be made without the prior written permission of the publisher. In the UK, users may clear copying permissions and make payment to The Copyright Licensing Agency Ltd, Saffron House, 6–10 Kirby Street, London EC1N 8TS UK, and in the USA to the Copyright Clearance Center, 222 Rosewood Drive, Danvers, MA 01923, USA. Other countries may have a local reproduction rights agency for such payments. Full information on the Society's permissions policy can be found at: www.geolsoc.org.uk/permissions

British Library Cataloguing in Publication Data

A catalogue record for this book is available from the British Library.
ISBN 978-1-78620-366-3
ISSN 0305-8719

Distributors
For details of international agents and distributors see:
www.geolsoc.org.uk/agentsdistributors

Typeset by Nova Techset Private Limited, Bengaluru & Chennai, India
Printed and bound by CPI Group (UK) Ltd, Croydon CR0 4YY

Contents

LAWSON, M., FORMOLO, M. J., SUMMA, L. & EILER, J. M. Geochemical applications in petroleum systems analysis: new constraints and the power of integration — 1

Source-rock identification and the temperature/timing of hydrocarbon generation

STOLPER, D. A., LAWSON, M., FORMOLO, M. J., DAVIS, C. L., DOUGLAS, P. M. J. & EILER, J. M. The utility of methane clumped isotopes to constrain the origins of methane in natural gas accumulations — 23

EILER, J. M., CLOG, M., LAWSON, M., LLOYD, M., PIASECKI, A., PONTON, C. & XIE, H. The isotopic structures of geological organic compounds — 53

GAO, Y., CASEY, J. F., BERNARDO, L. M., YANG, W. & BISSADA, K. K. (A.) Vanadium isotope composition of crude oil: effects of source, maturation and biodegradation — 83

PEDENTCHOUK, N. & TURICH, C. Carbon and hydrogen isotopic compositions of *n*-alkanes as a tool in petroleum exploration — 105

Mechanisms and time-scales associated with hydrocarbon migration, trapping, storage and alteration

BYRNE, D. J., BARRY, P. H., LAWSON, M. & BALLENTINE, C. J. Noble gases in conventional and unconventional petroleum systems — 127

MOORE, M. T., VINSON, D. S., WHYTE, C. J., EYMOLD, W. K., WALSH, T. B. & DARRAH, T. H. Differentiating between biogenic and thermogenic sources of natural gas in coalbed methane reservoirs from the Illinois Basin using noble gas and hydrocarbon geochemistry — 151

The impact of fluid flow on reservoir properties

MacDONALD, J. M., JOHN, C. M. & GIRARD, J.-P. Testing clumped isotopes as a reservoir characterization tool: a comparison with fluid inclusions in a dolomitized sedimentary carbonate reservoir buried to 2–4 km — 189

Index — 203

Geochemical applications in petroleum systems analysis: new constraints and the power of integration

M. LAWSON[1]*, M. J. FORMOLO[1], L. SUMMA[1,2] & J. M. EILER[3]

[1]*ExxonMobil Upstream Research Company, 22777 Springwoods Village Pkwy, Spring, TX 77389, USA*

[2]*Present address: Department of Earth, Environmental and Planetary Sciences, Rice University, 6100 Main Street, Houston, TX 77005, USA*

[3]*California Institute of Technology, 1200 E California Blvd, Pasadena, CA 91125, USA*

**Correspondence: Michael.lawson@exxonmobil.com*

Abstract: This paper provides an overview of the role that geochemistry plays in petroleum systems analysis, and how this can be used to derive constraints on the key elements and processes that give rise to a successful petroleum system. We discuss the history of petroleum geochemistry before reflecting on the next frontier in geochemical applications in hydrocarbon systems. We then review the individual contributions to this Special Publication. These papers present new geochemical techniques that allow us to develop a more systematic understanding of critical petroleum system elements; including the temperature and timing of source-rock deposition and maturation, the mechanisms and timescales associated with hydrocarbon migration, trapping, storage and alteration, and the impact of fluid flow on reservoir properties. Finally, we provide a practical example of how these different geochemical techniques can be integrated to constrain and generate a robust understanding of the prolific Paleozoic petroleum system of the Bighorn Basin.

Gold Open Access: This article is published under the terms of the CC-BY 3.0 license.

By 2035, global energy demand is anticipated to increase by 33% as the population continues to grow (WEO 2011; Khatib 2012). While renewable energy sources are expanding rapidly, hydrocarbons are predicted to supply 80% of the global demand until at least 2040 (Khatib 2012; ExxonMobil 2016). Historically, increases in energy demand have driven oil and gas companies to transition hydrocarbon exploration towards increasingly complex environments. Beginning in the 1900s and through to the 1950s, oil and gas exploration was largely focused towards onshore assets. A move to shallow offshore environments in the 1960s preceded a shift to ultra-deep-water (>1500 m) environments in the 1980s in regions such as West Africa and the Gulf of Mexico (May *et al.* 2010). To meet future energy demand, exploration is again moving into increasingly challenging geological settings, such as tight/unconventional resources, deep targets in ultra-deep-water basins, salt basins where subsurface imaging is extremely difficult, and harsh and environmentally sensitive drilling environments such as the Arctic (Summa 2015). Exploration companies must also contend with technical and drilling challenges associated with shallow gas in deep-water settings, heavy oil and potentially corrosive non-hydrocarbon gases, such as CO_2 and H_2S.

While exploration environments and technical challenges have evolved, the key questions asked of the petroleum systems analyst have remained broadly the same since Colonel Edwin Drake struck oil at a depth of 69 ft in Titusville, Pennsylvania, USA in 1859 (Link 1952). This is true both in frontier basins with no prior history of exploration, and in mature basins with thousands of wells. The first and most important question asked of the exploration geologist is whether a petroleum system (thermogenic or biogenic) exists. In other words, is there an organic-rich rock buried to sufficient temperature and pressure to generate hydrocarbons? Or is there a robust biogenic gas system?

If evidence suggests hydrocarbons were generated, the petroleum systems analyst must determine whether they were trapped and, if so, is the accumulation dominantly oil or gas. In systems where a robust petroleum system has been demonstrated and subsurface accumulations have been discovered, a more challenging question arises when a drilled well yields negative results: why was this well dry? This becomes even more challenging when one considers the potential for purely biogenic gas occurrences that are not associated with a thermogenic hydrocarbon system.

Biogenic gas accumulations represent some of the largest recent gas discoveries (e.g. the Zohr

From: LAWSON, M., FORMOLO, M. J. & EILER, J. M. (eds) 2018. *From Source to Seep: Geochemical Applications in Hydrocarbon Systems*. Geological Society, London, Special Publications, **468**, 1–21.
First published online February 19, 2018, https://doi.org/10.1144/SP468.6

© 2018 ExxonMobil Upstream Research Company. Published by The Geological Society of London.
Publishing disclaimer: www.geolsoc.org.uk/pub_ethics

Field, Egypt: Esestime *et al.* 2016) and globally may represent up to 20–30% of all natural gas resources (Grunau 1984; Whiticar *et al.* 1986; Rice 1993). Odedra *et al.* (2005) estimated that if only 10% of the microbially produced methane is trapped, it would provide a resource base of 20 000 trillion cubic ft (TCF). These estimates are conservative as they are dominated by primary microbial gas and do not include potentially vast volumes of secondary microbial gas as proposed in the West Siberian Basin (Milkov 2010). Overall, these estimates are poorly constrained due to the difficult nature of predicting the generation and accumulation of microbial gas over geological timescales. However, even with the inherit complexities in understanding biogenic gas systems, the volumes of microbial gas can be of great economic value and the key question that remains is: What are the principal elements that give rise to large biogenic gas accumulations, and how can we develop an effective exploration strategy to discover them? Regardless of thermogenic or biogenic origin, the petroleum systems analyst needs a framework and set of tools to evaluate exploration opportunities and risk of the presence of a working hydrocarbon system.

It has been over 20 years since Magoon & Dow (1994) laid the foundation for petroleum systems analysis as an integrated science. They defined the petroleum system as:

> [A] natural system that encompasses a pod of active source rock and all related oil and gas and which includes all the geologic elements and processes that are essential if a hydrocarbon accumulation is to exist. The term *system* describes the interdependent elements and processes that form the functional unit that creates hydrocarbon accumulations. The *essential elements* include a petroleum source rock, reservoir rock, seal rock, and overburden rock, and the *processes* are trap formation and the generation-migration-accumulation of petroleum. These essential elements and processes must occur in time and space so that organic matter included in a source rock can be converted to a petroleum accumulation. A petroleum system exists wherever the essential elements and processes occur.

Geochemical techniques have contributed significantly to our understanding of petroleum systems globally. However, to meet the increasing challenges of discovering new hydrocarbon resources, it is essential we continue to advance our understanding of these systems through new geochemical approaches and analytical developments. Such development requires that academic and industry-led research efforts converge in ways that are unique to the geosciences. This Special Publication arose from the session of the same name at Goldschmidt 2015, and presents contributions that incorporate non-traditional geochemical and isotopic techniques, specifically recent advancements in carbonate and hydrocarbon clumped and position-specific isotope geochemistry, noble gas geochemistry, trace element geochemistry, isotope chemistry of heavy elements, and molecular geochemistry.

The aim of this volume is not to simply present new techniques, but to showcase how recent advances and state-of-the-art approaches in geochemistry from both industry and academia can generate new ways of thinking about old problems. Techniques and approaches presented herein focus on addressing three key hydrocarbon systems issues: (1) source-rock identification and the temperature/timing of hydrocarbon generation (**Eiler *et al.* 2017**; **Gao *et al.* 2017**; **Pedentchouk & Turich 2017**; **Stolper *et al.* 2017**); (2) the mechanisms and timescales associated with hydrocarbon migration, trapping, storage and alteration (**Byrne *et al.* 2017**; **Moore *et al.* 2018**); and (3) the impact of fluid flow on reservoir properties (**MacDonald *et al.* 2017**).

While this volume focuses on how new geochemical techniques can be used in petroleum systems analysis, we also recognize the potential uplift that these same techniques could offer in the environmental sciences. Increased environmental awareness means that new and increasingly sophisticated approaches are required to ensure that human activities, such as resource exploration and extraction, have a minimal impact on the natural environment. Two of the most obvious challenges in this respect is the reduction of greenhouse gas emissions that contribute to global climate change and the protection of potable water resources. Almost all studies addressing environmental questions contain an element of forensic fingerprinting and reconstruction, and the geochemical tools discussed in this Special Publication are likely to play an important role in this evolving landscape. For example, the carbonate clumped isotope technique applied in **MacDonald *et al.* (2017)** could also be used to study the cementation history of aquifers targeted for managed aquifer recharge (MAR), a strategy of using aquifers to store and improve the quality of water for future use (Maliva *et al.* 2015). The noble gases discussed in **Byrne *et al.* (2017)**, and applied in **Moore *et al.* (2018)**, are already a well-established technique studying geological sequestration of carbon dioxide and other gases (e.g. Gilfillan *et al.* 2009). Finally, the isotopic approaches described in **Eiler *et al.* (2017)**, **Gao *et al.* (2017)**, **Pedentchouk & Turich (2017)** and **Stolper *et al.* (2017)** all have the capability to elucidate the origin of hydrocarbons in the environment.

In this introductory chapter, we first provide a brief history of petroleum geochemistry and its application in petroleum system analysis. We then introduce a new generation of geochemical techniques that are discussed in different chapters of this Special Publication. Finally, we provide a case study from the Bighorn Basin that integrates traditional and new geochemical exploration techniques to generate new petroleum systems concepts for the prolific Paleozoic petroleum system.

The first wave of geochemical tools for petroleum systems analysis

Petroleum geochemistry has played an important role in hydrocarbon exploration for over 50 years (Hedberg 1964). However, even before petroleum geochemistry became established as a field, and before characterization of oils and gases was possible, scientists pondered the origin of hydrocarbons. Perhaps, the first person to propose the now accepted concept that economic accumulations of hydrocarbons have an organic origin was Thomas S. Hunt. He suggested that organic matter in North American Paleozoic rocks, which he believed was the precursor of local hydrocarbon accumulations, was likely to be derived from marine vegetation or the remains of marine animals (Hunt 1861). Decades later, Wallace Pratt, the former Chief Geologist of Humble Oil (which ultimately became ExxonMobil) was an early adopter of petroleum geochemistry principles. He popularized the concept that small hydrocarbon molecules were formed from larger parent molecules as a result of thermal cracking at depth (Pratt 1943). This is one of the guiding principles from which predictions of commodity type (oil v. gas) are now based. However, it was not until almost 100 years after Hunt introduced the organic origin of hydrocarbons hypothesis that it became widely accepted, when geochemical (Treibs 1936; Eglinton & Calvin 1967) and isotopic evidence (Craig 1953) provided compelling evidence in support of the theory. These early days of proto-petroleum geochemistry laid the foundations that future developments in analytical capabilities would allow us to investigate.

Prior to the 1980s, analytical and technological limitations restricted our ability to appreciate the many ways in which geochemistry could contribute to petroleum systems analysis. This changed in the 1980s and 1990s, when a series of geochemical tools were developed that provided important information about petroleum systems. To list and describe each of these contributions is beyond the scope of this review. We highlight here a small selection of the developments that we consider to have contributed significantly to petroleum systems analysis. A more detailed history of petroleum geochemistry can be found in Hunt *et al.* (2002) and Kvenvolden (2006, 2008). Similarly, we refer the interested reader to Tissot & Welte (1984) and Hunt (1996), seminal publications that describe the breadth and scope of petroleum geochemistry.

Advances in analytical organic geochemistry, specifically high-resolution gas chromatography and mass spectrometry, and newly recognized biological markers provided the first step-change in our ability to constrain the origin and post-formational history of liquid hydrocarbons through empirical observations of source rocks and oils. Through these advances, it became possible to: (i) determine the organic matter type and depositional environment of a source rock (e.g. Mackenzie *et al.* 1984; Tissot & Welte 1984; Moldowan *et al.* 1985; Peters *et al.* 1986, 2005); (ii) quantify the thermal maturity of a source rock (e.g. Tissot *et al.* 1987; Waples 1994); (iii) constrain the age of the source rock from which oil has been generated (e.g. Moldowan *et al.* 1994; Holba *et al.* 1998); (iv) identify and qualitatively estimate the extent of post-generation alteration, such as biodegradation (e.g. Wenger *et al.* 2002) and thermochemical sulphate reduction (Zhang *et al.* 2007); and (v) genetically link reservoired or seeped oils with other oils or to the source rocks from which they were generated (e.g. Curiale 1994). These developments also prompted comprehensive experimental studies to understand the processes, timescales and temperatures associated with source-rock maturation, hydrocarbon generation and expulsion from source rocks (e.g. Lewan 1985, 1994; Vandenbroucke *et al.* 1999). A central concept to many of these studies is that the generation of hydrocarbons is a unidirectional, kinetically controlled process. These advances in our ability to model hydrocarbon generation and characterize fluids and rocks to constrain their origin and depositional history have significantly improved our understanding of the petroleum system. One of the broader implications of this is that we are now able to calibrate geological models away from the well in which the sample was obtained to predict the potential distribution of oil v. gas within a basin, constrain the relative timing of maturation to other key petroleum systems elements such as reservoir deposition and structure timing, and, ultimately, to develop predictions of the location and hydrocarbon properties of potential accumulations.

At the same time, the use of chemical and isotopic signatures of natural gases also became well established in hydrocarbon exploration. Of particular importance in these early years of gas geochemistry applications in the petroleum industry were two specific capabilities: (i) being able to distinguish thermogenic gases from those that are generated by microbial processes (e.g. Bernard *et al.* 1976; Schoell 1980; Rice & Claypool 1981); and (ii) the ability to estimate the thermal maturity of a source rock from the carbon (δ^{13}C) and hydrogen (δD) isotopic signature of hydrocarbon gases (e.g. Whiticar 1994).

The 1980s and 1990s also saw significant advances in our ability to constrain the thermal history of sedimentary basins. This included the development of: (i) fluid-inclusion microthermometry (e.g. Goldstein & Reynolds 1994), which provides key constraints on the maximum burial temperature, and can be used to constrain the timing of hydrocarbon charge; (ii) vitrinite reflectance (e.g. Barker & Pawlewicz 1986), which can help constrain both the maximum temperature and time-integrated maturity of any given stratigraphic interval; and (iii) thermochronology techniques, such as illite age analysis (e.g. Pevear 1992, 1998) and U–Th/He dating (e.g. Farley *et al.* 1996; Wolf *et al.* 1996, 1997) that can be used to constrain both the burial and uplift phases of a basin history, respectively. This information provides previously unavailable calibration parameters that can now be utilized by advanced computing tools to calibrate more realistic models of basin thermal evolution. The most advanced tool in this regard is four-dimensional (4D) basin modelling, which models the evolution of a sedimentary basin in 3D space through time. These models are often the major platform for integrating information provided by not only geochemistry, but also geophysical techniques. When integrated with an understanding of the regional geology, such models can be used to make improved predictions of the spatial and temporal properties of key hydrocarbon systems elements and processes. These elements include regional source-rock maturation, hydrocarbon generation timing, hydrocarbon migration pathways and the preservation or deterioration of reservoir quality. A detailed discussion of such models is beyond the scope of this chapter, and as such we refer the interested reader to Hantschel & Kauerauf (2009) and Nemčok (2016).

The new frontier in geochemical applications in petroleum systems analysis

One of the significant challenges for the next decade is the requirement to work with increasingly small and limited datasets in frontier environments. As such, there is a need to generate a greater depth of information from what few samples are available. As is often the case, such advances require a significant improvement in analytical capabilities and more sophisticated approaches to obtaining and understanding data. Developments over the past decade in high-resolution Fourier transform (e.g. FT-ICR-MS: Oldenburg *et al.* 2015) and multicollector mass spectrometry (e.g. high-resolution isotope ratio mass spectrometry: Eiler *et al.* 2013), and new approaches to mining increasingly large datasets have risen to meet this challenge. These techniques now allow the geochemist to probe samples with greater precision and specificity down to

the atomic level, and have triggered a renaissance of innovation in petroleum geochemistry. The following section introduces some of the techniques that have developed during this period and how these can be applied in petroleum systems analysis.

Clumped isotope thermometry of carbonates represents a step-change from the traditional thermometry techniques described above, such as fluid-inclusion microthermometry, vitrinite reflectance and thermochronology. The development of high-resolution isotope ratio mass spectrometers has, for the first time, provided an opportunity to investigate the temperature dependence of multiple isotopic substitutions in both inorganic and organic molecules (Ghosh *et al.* 2006; Eiler 2007; Eiler *et al.* 2013; Stolper *et al.* 2014*a*, *b*). The carbonate clumped isotope thermometer was the first of the new suite of thermometers, and has been demonstrated to be applicable in the 50–300°C range, which is therefore relevant to processes such as dolomitization (Ferry *et al.* 2011) and burial diagenesis (Huntington *et al.* 2011; Shenton *et al.* 2015; Lawson *et al.* 2017). Information from this thermometer helps constrain the thermal history of sedimentary basins, and predictions of the preservation or destruction of porosity and permeability in carbonate reservoirs. In this volume, **MacDonald *et al.* (2017)** present a detailed clumped isotope study from a carbonate reservoir offshore Angola to show that carbonate clumped isotope geochemistry can yield new insights on the temperature and timing of dolomitization.

In addition to the carbonate thermometer, further developments in high-resolution multicollector mass spectrometry (e.g. Eiler *et al.* 2013) have resulted in the development of the first hydrocarbon clumped isotope geothermometer (Stolper *et al.* 2014*a*). The $^{13}CH_3D$ isotopologue of methane records and preserves the generation temperature of methane in both biogenic and thermogenic systems (Ono *et al.* 2014; Stolper *et al.* 2014*a*, *b*, 2015; Wang *et al.* 2015). Also in this volume, **Stolper *et al.* (2017)** review the current state-of-the-art in methane thermometry and describe potential new applications in hydrocarbon exploration. While the methane body of work represents the most significant to date in hydrocarbon clumped isotope and position-specific geochemistry, significant steps forward are being made in our understanding of the history of larger hydrocarbons. For example, it is now possible to measure $^{13}C-^{13}C$ clumping in ethane, and a preliminary dataset has now been reported in Clog *et al.* (2018). Similarly, three independent techniques have been developed to measure the ^{13}C position-specific signatures of propane (Gao *et al.* 2016; Gilbert *et al.* 2016, Piasecki *et al.* 2016). **Eiler *et al.* (2017)**, in this volume, review published and unpublished work on the isotopic anatomy of organic molecules larger than methane, and outline potential future constraints offered by clumped isotope and site-specific analysis of organic molecules (e.g. biosynthesis, maturation, and the environmental conditions of hydrocarbon generation and storage).

Noble gas geochemistry, unlike clumped isotope geochemistry described above, has a longer history in petroleum systems analysis. However, the exploitation of unconventional resources over the past decade has provided a new natural laboratory to test some of the key hypotheses that are also relevant to conventional systems. While the other techniques discussed in this Special Publication are sensitive tracers of hydrocarbon origin, accumulation and alteration, the noble gases are chemically inert and as such do not partake in any chemical or biological reactions that may perturb the geochemical record of other tracers. Instead, they provide information on the physical interaction and mixing of fluids in the subsurface, and can yield constraints on the timescales associated with hydrocarbon storage. Prior applications of noble gas geochemistry focused on the role of groundwater flow in hydrocarbon migration (Zartman *et al.* 1961; Bosch & Mazor 1988; Ballentine *et al.* 1991; Barry *et al.* 2016) and cementation (Ballentine *et al.* 1996). Recent advances in multicollector mass spectrometry now provide increasingly sensitive measurements of noble gases with greater precision. This capability has many potential uses in petroleum systems analysis, and has recently been applied to develop accurate estimates of the residence time of fluids in Precambrian crust (Holland *et al.* 2013) and in hydrocarbon accumulations of northern Germany (Barry *et al.* 2017). In this volume, **Byrne *et al.* (2017)** take a new look at how noble gas geochemistry can be used to constrain the migration and accumulation history of hydrocarbons, with a particular focus on unconventional accumulations that have only recently begun to be studied (e.g. Hunt *et al.* 2012) despite the fact that these resources have been intensely developed over the last 10 years. Also in this volume, **Moore *et al.* (2018)** describe how noble gases can be used to determine the history of hydrocarbons and the origin of non-hydrocarbon gases, such as H_2S, in unconventional resources, with a case study from the Illinois Basin.

Molecular geochemistry is, perhaps, the most widely applied technique in petroleum systems analysis. Among other applications, it is routinely used to tie oils to specific source intervals during exploration, to determine if oils are connected laterally within reservoir intervals to guide development of discovered hydrocarbons, and to determine the composition of oils, which governs the ultimate value of any accumulation that is going to be produced. However, like all other techniques, this approach also has its limitations. For example, molecular geochemistry may be unable to tie oil to a specific source interval if

there are multiple source rocks in the basin with similar molecular signatures. Such a scenario is made more difficult by the fact that a relative paucity of biomarkers have been discovered that provide definitive and high-resolution age constraints on source rock age. This is a common challenge for the petroleum systems expert in frontier basins with multiple Jurassic or Cretaceous marine clastic source intervals that could all plausibly be the source of hydrocarbons encountered during the initial stages of exploration. Recent advances in established techniques, such as compound specific isotope analysis, may finally provide a path to constraining such systems. **Pedentchouk & Turich (2017)**, in this volume, review the history of compound-specific carbon and hydrogen isotope analysis of hydrocarbons, and how such data can be used in exploration activities. They provide several case-study examples of how this technique has been used to better understand the origin of hydrocarbons in systems where conventional molecular geochemistry alone may not provide unambiguous results.

Finally, recent developments in the analysis of the isotopic signatures of heavy metals, such as rhenium, osmium, vanadium, nickel and molybdenum (e.g. Ventura *et al.* 2015; Georgiev *et al.* 2016), now provide new constraints on the origin and history of hydrocarbons. The study by **Gao *et al.* (2017)** in this volume reports on some of the first results of vanadium isotope ($^{51}V/^{50}V$) analysis of oils. This study discusses the main controls on the isotopic system, and how this can be applied to provide new constraints on the origin and history of oils in petroleum systems studies.

Bighorn Basin petroleum systems study: the power of integration

In the preceding sections, we have described the history of geochemistry in petroleum systems analysis, and what insights can be gained from individual techniques. However, it is only when this information is integrated can one really develop a holistic understanding of all elements and processes that contribute to a successful petroleum system. We will dedicate the following section to providing a case example of how this may be achieved for a prolific hydrocarbon system. As a laboratory for this approach, we use the Bighorn Basin of north-central Wyoming (Fig. 1), one of several significant petroleum-bearing basins in the Rocky Mountain Province of the western USA. The Bighorn Basin has produced over 2 billion barrels of oil and minor gas since the early 1900s (De Bruin 1997). After more than 100 years production is on the decline and there is limited exploration activity. Nonetheless, the wealth of existing data makes this

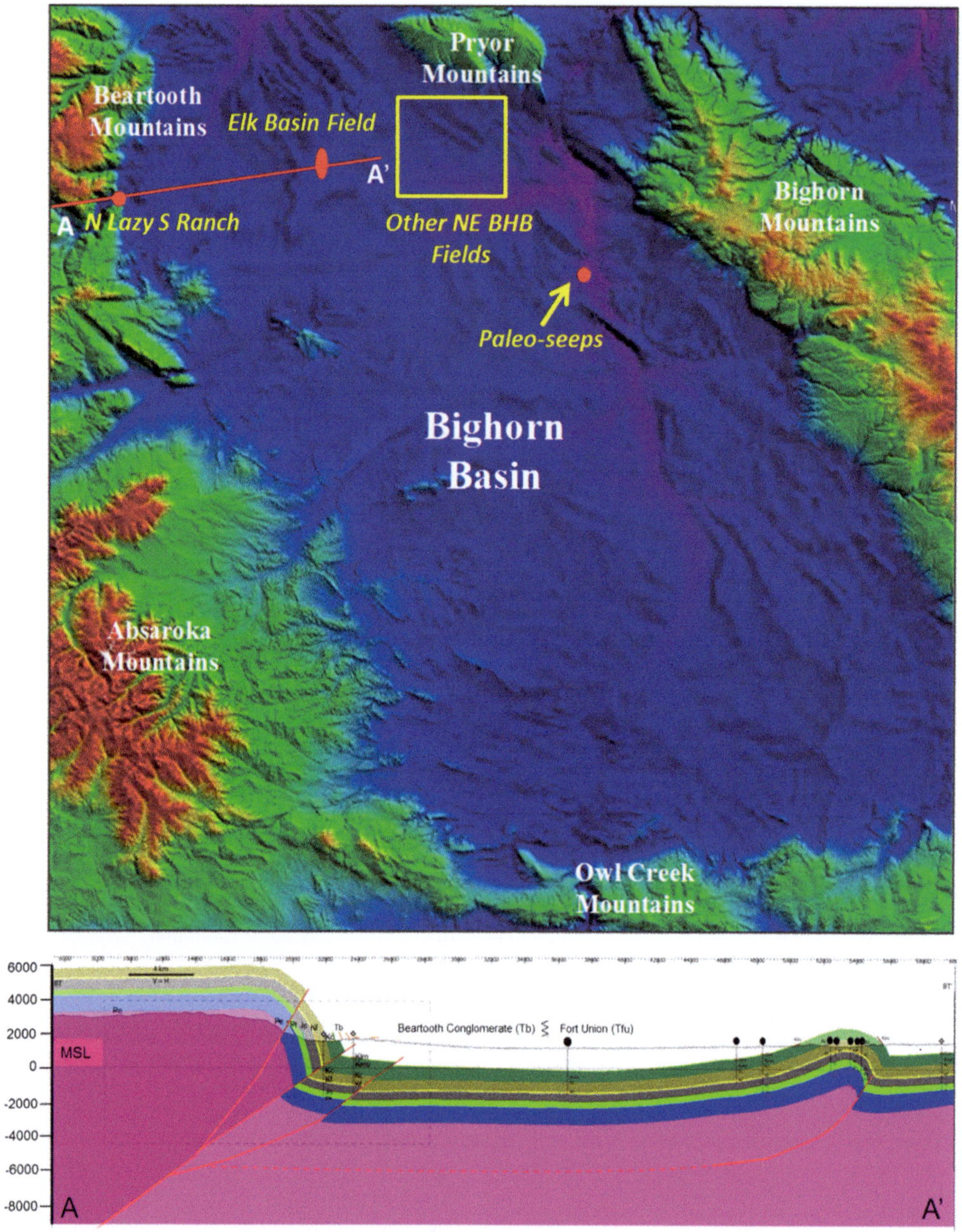

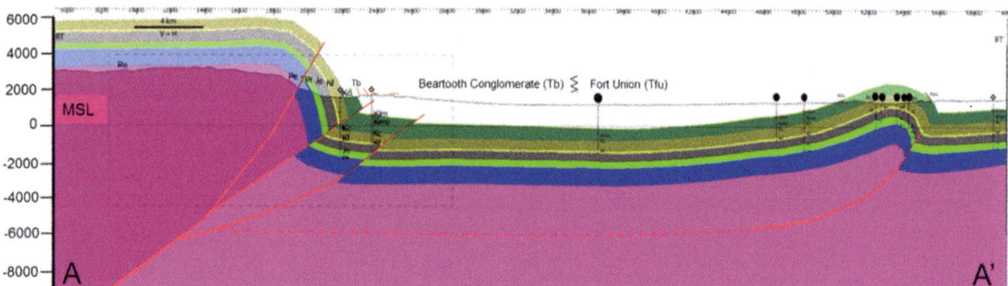

Fig. 1. Digital elevation map and representative structural cross-section for the Bighorn Basin showing sample localities describe in the text. The basin is bounded on all sides by Precambrian basement uplifts of 10 000–12 000 ft elevation (Beartooth Mountains, Pryor Mountains, Bighorn Mountains and Owl Creek Mountains), and by Tertiary volcanic rocks of the Absaroka Mountains. The basin is filled with up to 25 000 ft of sediments of Paleozoic (blue), Mesozoic (green, brown) and Tertiary (yellow) ages. Large basement-involved reverse faults define the basin edges and form anticlinal structures that serve as traps for hydrocarbon accumulation. MSL, mean sea level.

basin an ideal laboratory in which to test the power of geochemical integration to new petroleum systems concepts.

This example begins with a fundamental genetic analysis of the basin (Fig. 2) and focuses on two major tectonic events associated with long-lived plate convergence: (1) the Cretaceous Sevier Orogeny, in which convergence drove thin-skinned thrusting and development of a broad foreland basin; and (2) the Early Tertiary Laramide Orogeny, in which increased convergence rates and shallowing subduction drove thick-skinned deformation and the formation of basement-involved uplifts and intermontane basins (e.g. Sheldon 1967; Stone 1967). These events resulted in two petroleum systems in the Bighorn Basin: (1) a relatively 'simple' Mesozoic system in which local Mesozoic source rocks generate hydrocarbons that fill Mesozoic reservoirs during Early Tertiary trap formation; and (2) a more 'complex' Paleozoic system, in which hydrocarbons are likely to have migrated from the Sevier foredeep, at least partly prior to trap formation. Ironically, the relatively simple Mesozoic system has produced only approximately 10% of the basin's hydrocarbons, while the complex Paleozoic system has produced 90% (Stone 1967). We have integrated both traditional and novel geochemical approaches to better understand the success of the Paleozoic system.

We provide supporting evidence from key Paleozoic hydrocarbon systems elements of source distribution, maturation timing, and lateral migration timing and pathways in Figures 3–5. Palaeogeographical maps for key time intervals are coupled to source rock geochemistry and burial history analyses. Oils in Paleozoic reservoirs came from a clastic-starved carbonate source rock that correlates with organic-rich rocks in the Permian Phosphoria Formation (Fig. 3), a well-documented source rock in western North America with total organic carbon (TOC) values of >20% in northern Utah and eastern Idaho (e.g. Claypool *et al.* 1978). In the Bighorn Basin, the Phosphoria Formation was deposited on the shelf, and consists of a stack of shallow-marine limestones and redbeds with no source potential. This geographical position and associated lithofacies suggest that long-distance lateral migration was required to fill Bighorn Basin traps with oils sourced from the Phosphoria (Sheldon 1967; Stone 1967). The closest organic-rich Phosphoria Formation with no evidence of clastic input is *c.* 100 km west and SW of the Bighorn Basin, in the foreland of the Sevier orogenic system (Fig. 3).

The hinterland load of the Sevier system is likely to have driven these source rocks into the oil window, at which point oils began to migrate eastwards through the foreland (Cheney & Sheldon 1959; Campbell 1962; Sheldon 1967). We hypothesize that hydrocarbon migration began in the distal hinterland given its proximity to the sedimentary load from the Sevier Orogeny and progressed eastwards as the foreland advanced, from *c.* 180 to 80 Ma, based on multiple burial histories constructed for eastern Idaho and western Wyoming (e.g. Burtner

& Nigrini 1994; Maughan 1984; Stone 1967). This requirement for long-distance lateral migration is consistent with previous predictions from molecular geochemistry (e.g. Claypool *et al.* 1978). A 1D basin model was built for the footwall of the Meade Thrust, using ExxonMobil proprietary basin-modelling software, in order to capture the burial history of the Phosphoria Formation in the area where it is believed to be an effective source rock (T. Becker pers. comm. 2015) (Fig. 4). Model results suggest that the Phosphoria source rock probably reached the oil window at approximately 170 Ma at this location (Fig. 4). Becker (pers. comm. 2015) applied illite age analysis on diagenetic illite sampled from the Meade Thrust to confirm this hypothesis. Diagenetic illite forms at temperatures of approximately 90–100°C, which roughly corresponds to the temperatures required for the onset of oil and gas generation (Eslinger & Pevear 1988). K–Ar age dating of this illite yielded an age of 178.6 ± 8 Ma at this location (T. Becker pers. comm.). At this time there were several high-permeability units that could have served as lateral migration pathways for hydrocarbons from the Phosphoria Formation, and the foreland was sufficiently unstructured and simple at that time to allow long-distance migration to occur (see the crustal-scale cross-section in Fig. 4).

As noted above, the present-day Bighorn Basin and its associated anticlinal traps developed during the Early Tertiary, when the foreland was broken up in response to increased plate convergence rates. The timing of trap development is debated, but is thought to be coincident with the Laramide Orogeny at *c.* 85–55 Ma (e.g. Lillis & Selby 2013), and suggests that a significant fraction of long-distance hydrocarbon migration from the west occurred prior to trap formation. The Sevier Foreland System is thought to have been broken and disconnected by the Laramide-related basement-involved uplifts by approximately 55 Ma (e.g. Sheldon 1967; Stone 1967), as can be seen in the centre of the crustal cross-section (see Fig. 5). A 1D basin model built at the location of the Johnson Watson well in the Elk Basin (T. Becker pers. comm. 2015) (Fig. 5) captures the magnitude of this event. Apatite fission-track analysis of drill cuttings from the well confirms that at least 1 km of uplift occurred in the area beginning at *c.* 55 Ma. We note here that not all basin reconstructions predict uplift at this time in the basin history (e.g. Roberts *et al.* 2008), and so the development and distribution of these barriers to hydrocarbon migration may not be a basin-wide phenomenon. However, our model is consistent with the timing and approximate magnitude of uplift of another 1D model developed for this part of the basin presented in Hagen & Surdam (1984). Such a scenario typically presents a big risk for a successful petroleum system, even though this system works

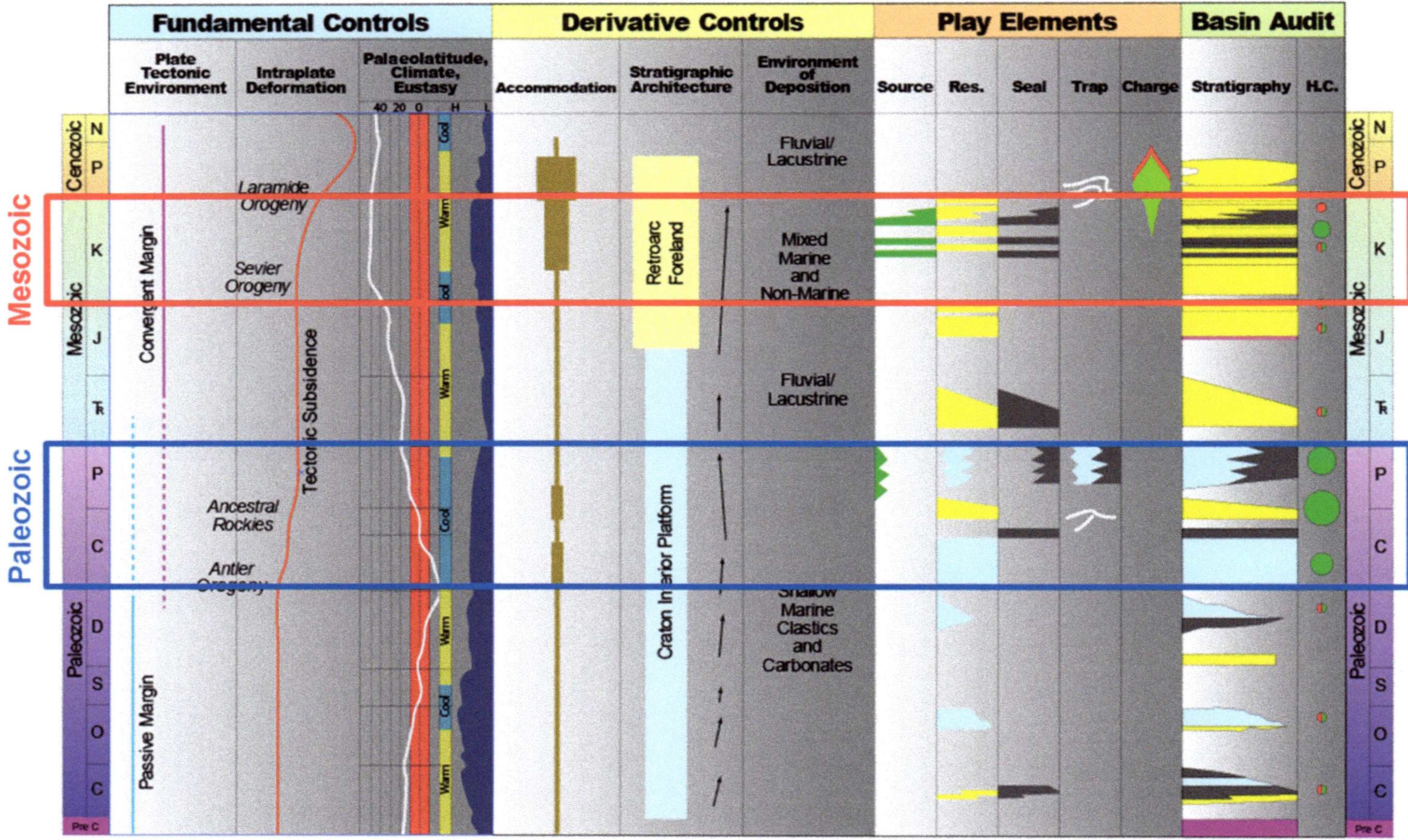

Fig. 2. Play summary chart for the Bighorn Basin. The play summary chart is used as a tool for genetic analysis of key petroleum systems elements and their relative timing. In the Bighorn Basin, there are two primary petroleum systems: (1) a Mesozoic system, outlined in red, in which local source rocks mature at the same time as local traps are forming; and (2) a Paleozoic system, outlined in blue, for which there is no apparent local source, and long-distance migration from source rocks to the west is likely to predate the formation of large structural traps, such as the one shown in Figure 1.

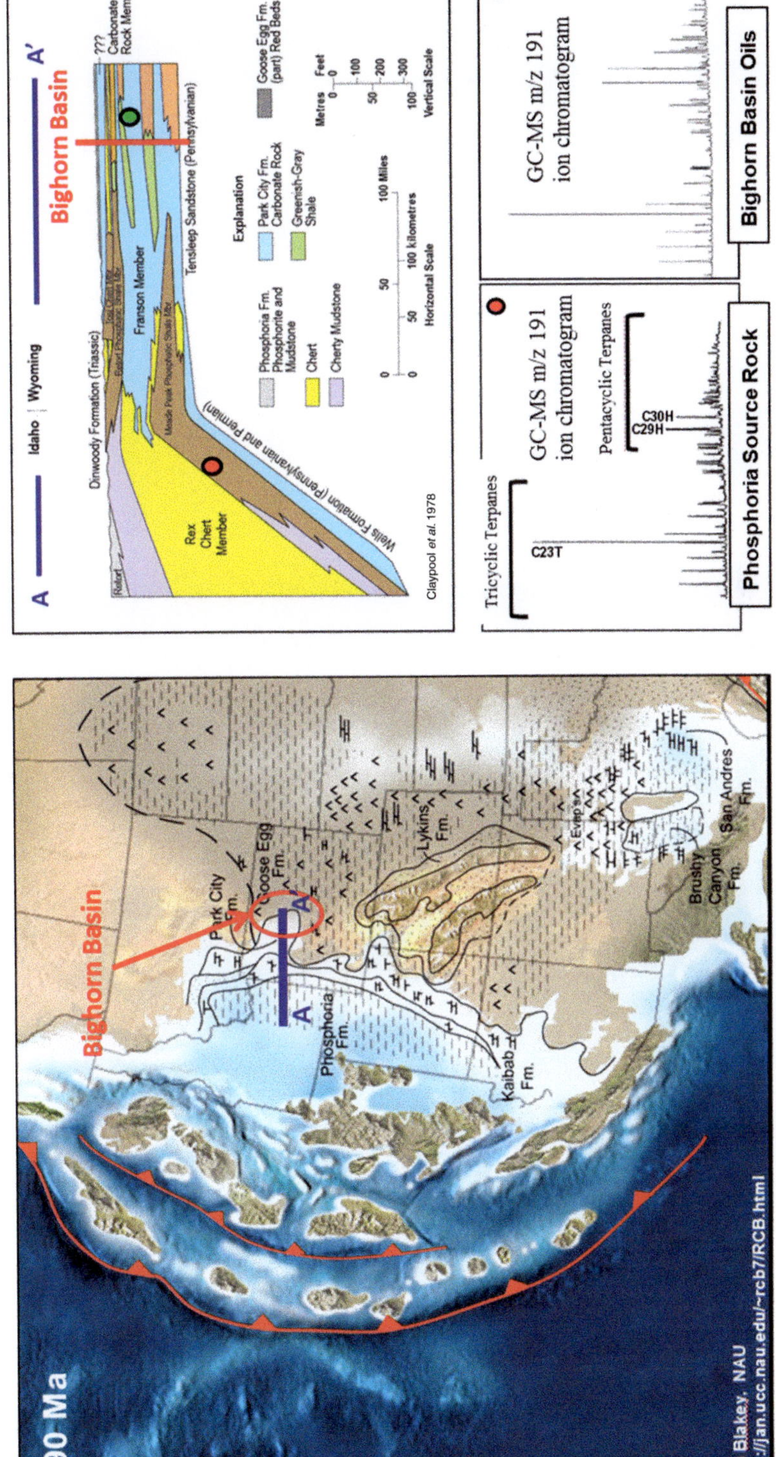

Fig. 3. Western North American palaeogeography at c. 290 Ma (Blakey 2011), at the time of deposition of key source rocks of the Phosphoria Formation. The geochemistry of oils reservoired in the Bighorn Basin correlates well with the geochemistry of Phosphoria source rocks, as illustrated by the m/z 191 chromatograms of Terpane biomarkers. The Phosphoria Formation is a largely a clastic-starved source rock that reaches TOC values of >20% along the western part of cross-section A–A'.

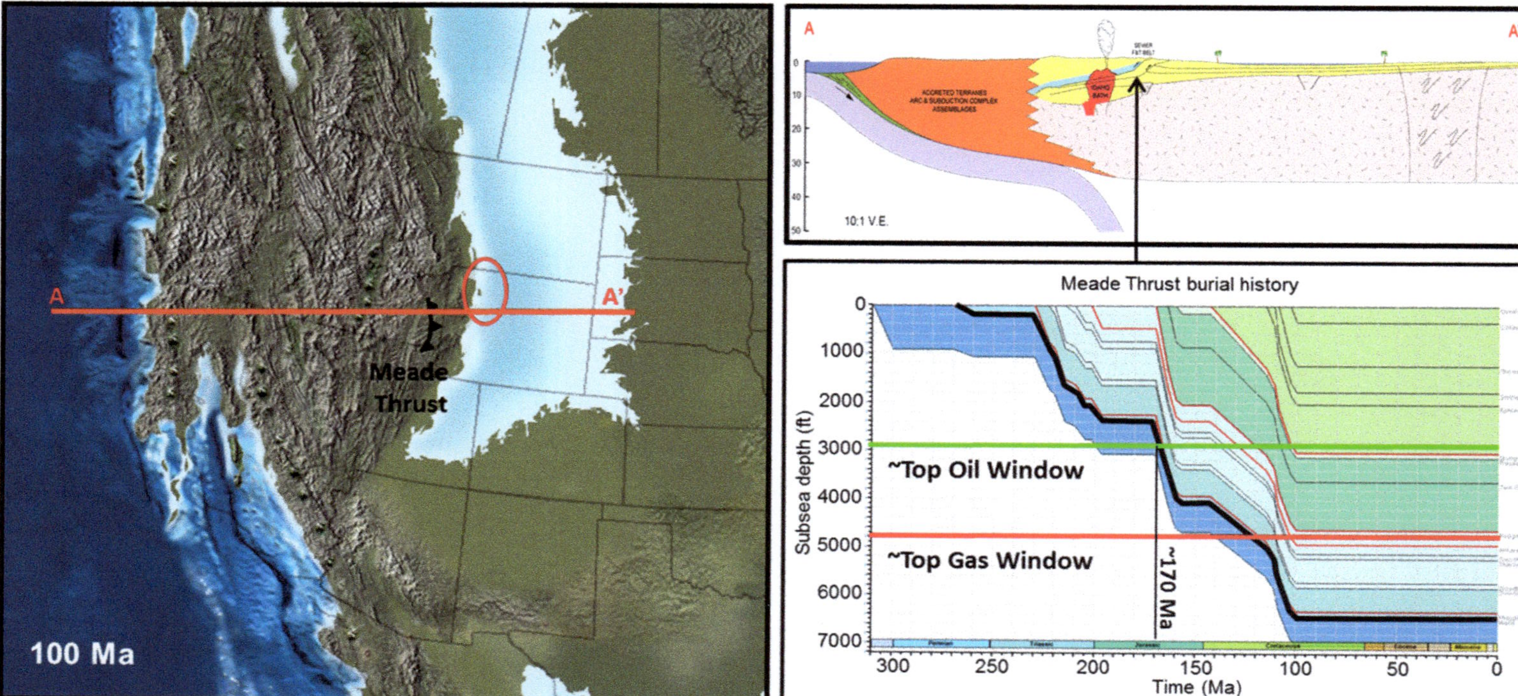

Fig. 4. Western North American palaeogeography at *c.* 100 Ma (Blakey 2011), which is a key time for the continuing maturation of Phosphoria source rocks and the long-distance migration of the resulting oils. A 1D burial history model from the footwall of the Meade Thrust shows that, at this locality, the Phosphoria (highlighted in black) is likely to have reached the oil window at *c.* 170 Ma. The model is supported by a K–Ar age of 178.6 ± 8 Ma from diagenetic illite in the Meade Thrust. V.E., vertical exaggeration.

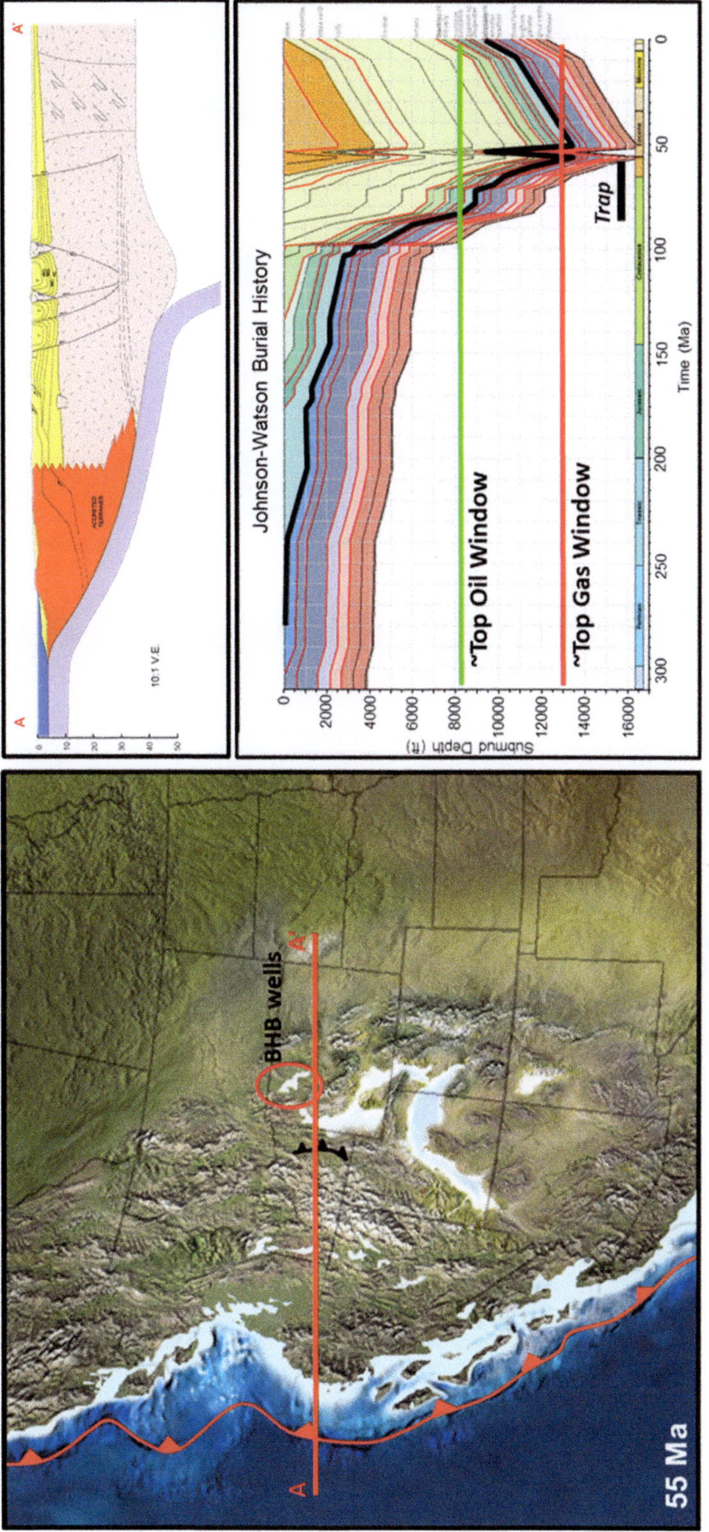

Fig. 5. Western North American palaeogeography at *c.* 55 Ma (Blakey 2011), during the development of the present-day Bighorn Basin (BHB) and its structural traps. The Sevier Foreland system was 'broken' by the basement involved uplifts, as shown in the crustal-scale cross-section. Large, basement-involved anticlinal traps developed in response to this deformation, and local Mesozoic-aged source rocks were buried and began to generate hydrocarbons, as shown by the burial history from a well in the Elk Basin Field (see Fig. 1).

well. What do geochemical tools have to offer to help us understand this complex petroleum system?

As noted in the previous section, and in other papers in this Special Publication, one of the new challenges for the next decade is the requirement to work with increasingly small and limited datasets in frontier environments. As an example, new high-resolution analytical techniques for hydrocarbon-bearing fluid inclusions now yield direct forensic evidence to constrain the source facies and maturity of petroleum source rocks, as well as the timing of migration and trapping. Paleozoic reservoirs in the Elk Basin oil field from the northern Bighorn Basin (Fig. 1) contain abundant hydrocarbon and aqueous fluid inclusions that reflect a range of fluid properties and suggest a complex charge history (Fig. 6, S. Becker, pers. comm., 2015). The oil and water inclusions display a wide range of homogenization temperatures. Bluish-white fluorescing inclusions in diagenetic calcite from one of the reservoir intervals, the Tensleep Formation, have homogenization temperatures of 60–100°C, and co-exist with water inclusions that have homogenization temperatures of 100–120°C. Brown and dull-yellow fluorescing inclusions from diagenetic anhydrite in the Phosphoria Formation have homogenization temperatures of 55–90°C, and co-exist with water inclusions that have homogenization temperatures of 105–110°C. The brown inclusions are consistent with lower API gravities associated with a biodegraded oil (Fig. 6). These trapping temperatures suggest that hydrocarbons began to be trapped as early as 80 Ma based on the thermal history for the Tensleep Formation at this location presented in Figure 4. This corresponds with the very early stages of trap development. Time-of-flight secondary ion mass spectrometry (ToFSIMS) analysis of individual inclusions (e.g. Siljeström *et al.* 2010) within these assemblages indicates that the oils in the inclusions are molecularly similar to those in the reservoir, and correlate well with a clastic-starved carbonate source rock (Siljeström *et al.* 2016).

Re–Os geochronology is one of the more significant recent developments in the analysis of the isotopic signatures of heavy metals, with applications to hydrocarbon systems analysis. This technique provide new constraints on the origin and history of hydrocarbons (Ventura *et al.* 2015; Georgiev *et al.* 2016). In this volume, **Gao *et al.* (2017)** discuss the systematics of the vanadium system, and here we describe results from the analysis of Re–Os in whole oils and asphaltenes from three wells in the NE corner of the Bighorn Basin (Fig. 1). The analyses were performed at the University of Arizona Re–Os laboratory (Kirk, pers. comm., 2017), using techniques similar to those published by Lillis & Selby (2013). They produced a well-constrained apparent isochron age of 185 Ma (Fig. 7), similar to both the age of

earliest hydrocarbon generation from the Phosphoria Formation to the west (Late Triassic–Early Jurassic: e.g. Sheldon 1967; Maughan 1984) and the Re–Os age of 211 ± 21 Ma published by Lillis & Selby (2013) for oils from the western Bighorn Basin. Both of these ages are considerably younger than the depositional age of the Phosphoria Formation (*c.* 285–250 Ma). Nonetheless, we are puzzled as to how to interpret these results. Although the oils all appear to have been generated from the same clastic-starved Phosphoria source-rock facies, they also appear to have been generated at different source-rock maturation stages (Fig. 7). Assuming that hydrocarbon migration occurred over a long period of time, in an open system, several scenarios could explain the data. Taking our data together with that of Lillis & Selby (2013), one scenario is that the Phosphoria-sourced oils were trapped and then remigrated into Laramide structures, mixing oils of different ages and maturities. Alternatively, these oils and Re–Os ages could also reflect early migration followed by bypass of later oil into updip traps, or some other process. Detailed analysis of individual fractions of the oil might help to resolve the alternative scenarios.

A relatively recent discovery of what we believe to be palaeoseeps in Campanian sandstones on the eastern side of the basin (Fig. 1) allows us to track the petroleum system from source all the way to seeps, and place additional constraints on migration timing. The mounds have a carbonate-cemented siltstone matrix with diverse macrofauna, capped with laminated algal carbonate and filled with calcite-lined fractures (Fig. 8). Assuming the features are palaeoseeps, they provide direct evidence of hydrocarbon migration in the Campanian (i.e. *c.* 80 Ma), similar to the time at which fluid inclusions were likely trapped. To find evidence for thermogenic hydrocarbons and their time of trapping, we completed preliminary isotopic analysis of released gases from matrix and fracture carbonates upon crushing. Matrix cements have a convincing contribution from thermogenic hydrocarbons, with $\delta^{13}C$ values for methane that range from −57 to −38‰. In contrast, vein cements appear to contain a greater contribution of biogenic methane, with $\delta^{13}C$ values for methane that range from −67 to −56‰. The only possible origin for thermogenic gas at this time is a distal source to the west, as local Mesozoic source rocks were still immature (Roberts *et al.* 2008). While we cannot rule out the possibility that the Campanian mounds were formed by, and charged with, biogenic methane, we believe these data provide direct support for the assertion that hydrocarbons sourced from the Phosphoria Formation continued to migrate to seeps in the Bighorn Basin during the Campanian.

While each of these techniques provide insights in to specific parts of the Paleozoic petroleum system

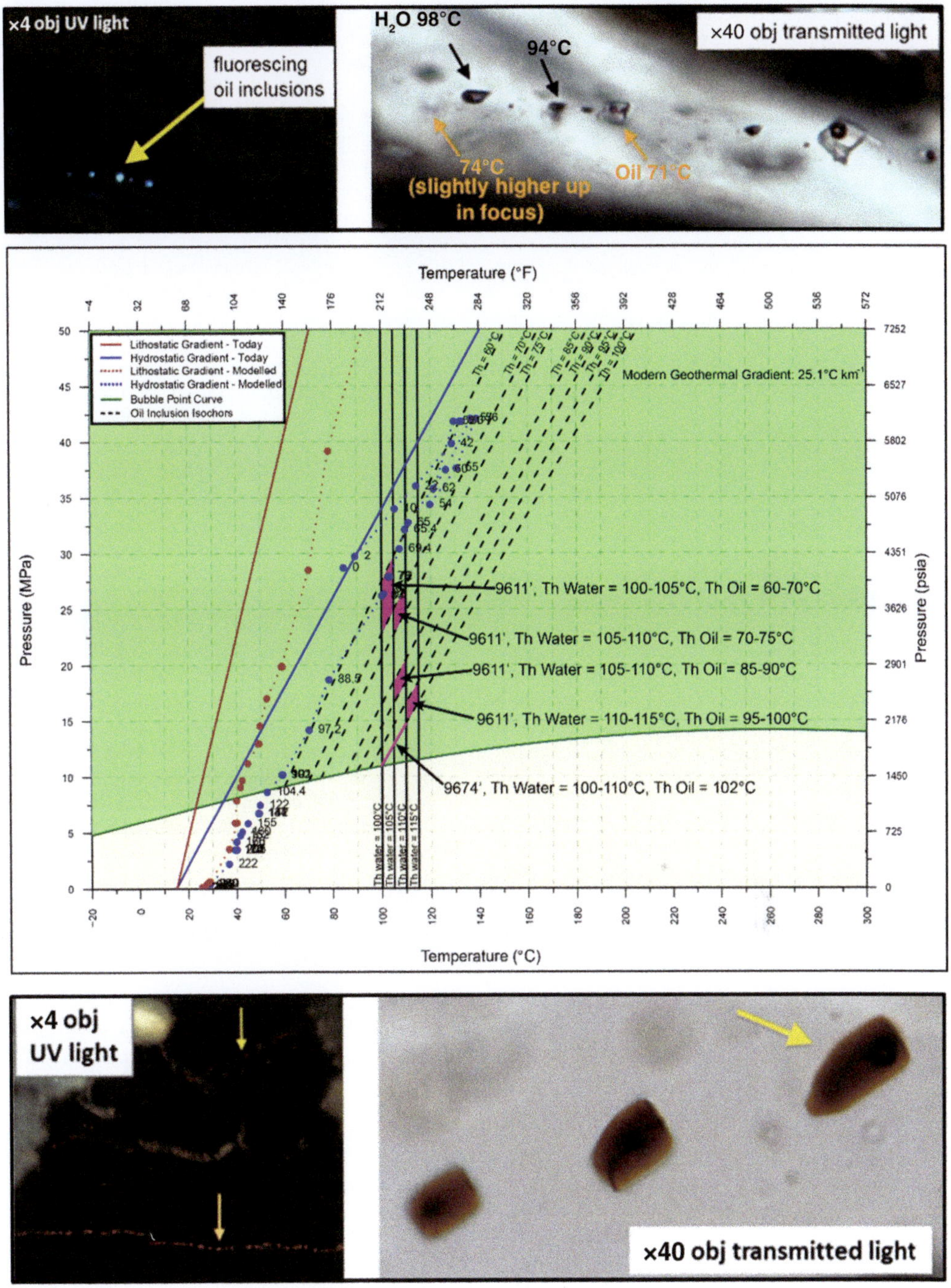

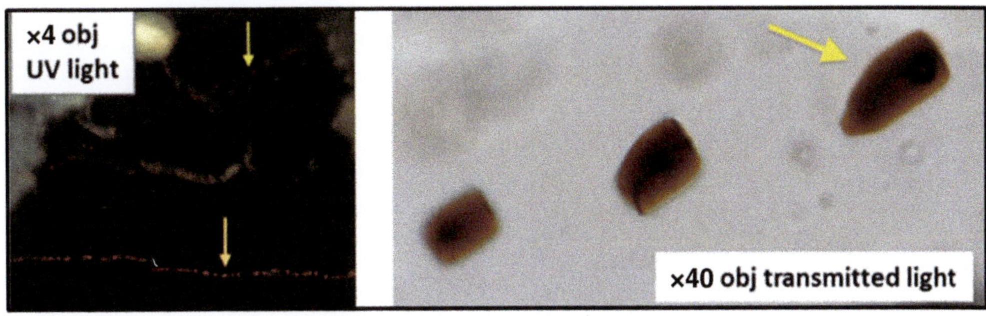

Fig. 6. Fluid-inclusion evidence for the timing of trap fill (S. Becker, pers. comm., 2015). Several populations of fluid inclusions from the Elk Basin Field are illustrated. Co-existing oil and water inclusions were used to estimate the pressure and temperature at the time of trapping, and were tied to time through a burial history model. Oil-inclusion isochores were calculated from pressure–volume–temperature (PVT) analysis of the present-day reservoired oils, and modelled hydrostatic and lithostatic pressure gradients based on the basin history in Figure 5. The results suggest that oil inclusions began to be trapped as early as 80 Ma. Trapping and post-emplacement alteration of fluid-inclusion oils is likely to have continued during the main stages of trap development, and possibly during mid-Tertiary exhumation.

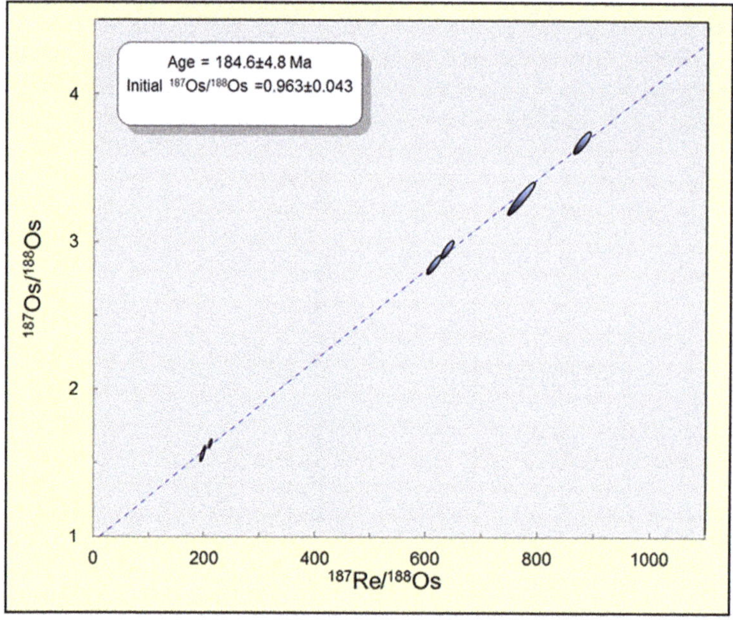

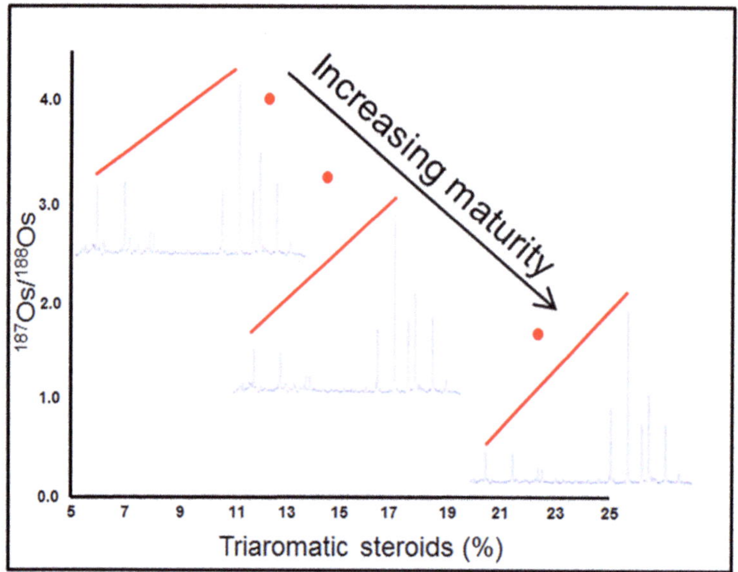

Fig. 7. Re–Os isotopic data from whole oils and asphaltenes sampled from the Byron, Garland and Little Polecat fields within the NE portion of the Bighorn Basin (Fig. 1). The data show a linear correlation on a Re–Os isochron diagram, and give a very well-constrained isochron age within error of the main array of 184.6 ± 4.8 Ma (^{187}Os/^{188}Os $= 0.963 \pm 0.043$), MSWD = 2.4. However, analysis of triaromatic steroids from the same oil samples suggests a trend of increasing maturity. Specifically, oils were generated from the same source facies, but at a different source-rock maturation stage, making it unlikely that all of the oils were trapped at the same time. The lowest maturity oils have the most radiogenic ^{187}Os/^{188}Os ratio, and the highest maturity oils have the least radiogenic ^{187}Os/^{188}Os ratio.

of the Bighorn Basin, it is only when they are all integrated that one gets a more complete understanding of the basin evolution and how it controls the distribution of hydrocarbons in the basin. Hydrocarbons in the Paleozoic reservoirs have biomarker signatures that are consistent with being sourced from

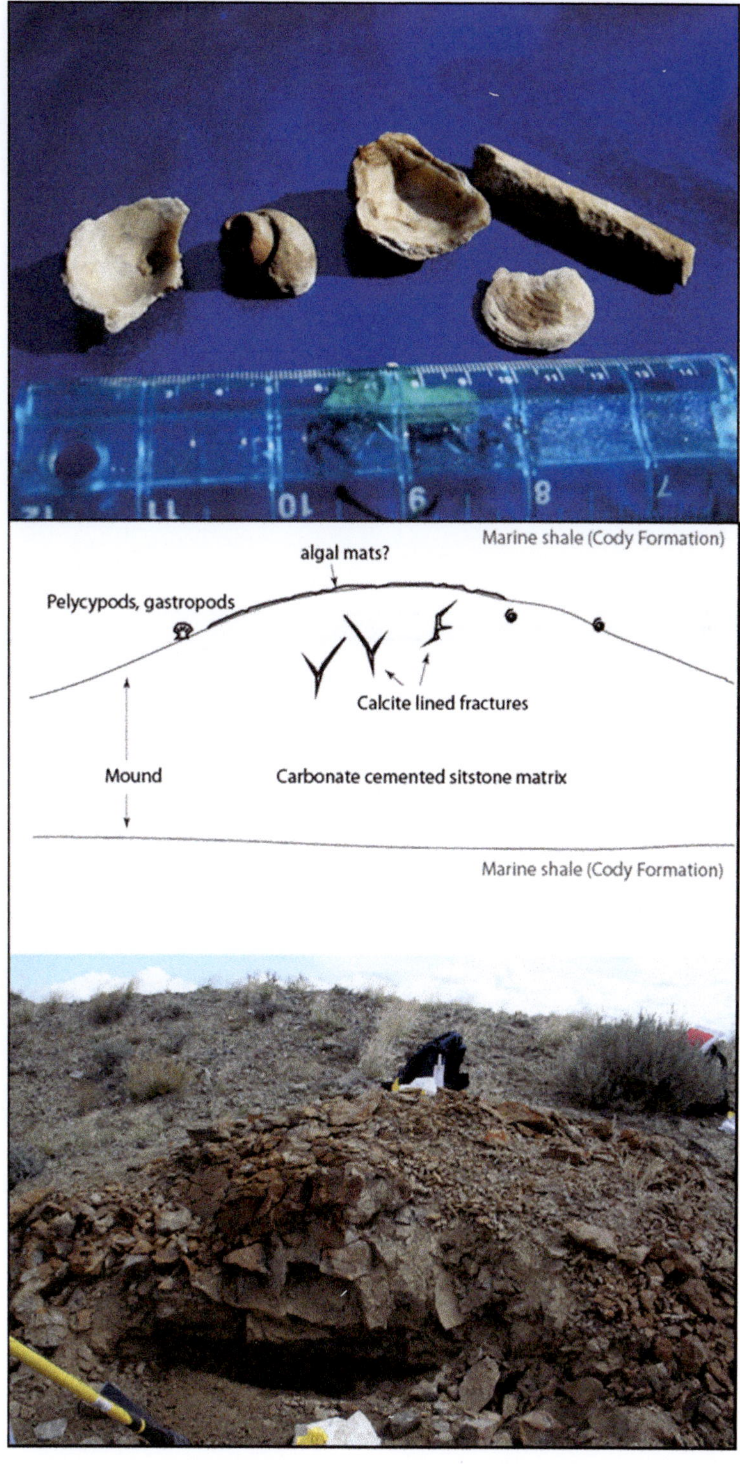

Fig. 8. Inferred palaeoseep mounds observed in Campanian mudstones (80 Ma) of the Cody Formation (Fig. 2) on the east side of the Bighorn Basin (Fig. 1). These macrofauna are consistent with, but not limited to, seep environments.

the Permian Phosphoria Formation, which requires oils to have migrated at least 100 km from the west and SW where the Phosphoria has high organic carbon contents. Basin modelling of this source kitchen to the west of the Bighorn Basin suggests that the Phosphoria Formation entered the oil window at *c.* 170 Ma. This is consistent with absolute age constraints from K–Ar dating of diagenetic illite

in this location. Furthermore, these Paleozoic oils have Re–Os ages that appear to be roughly consistent with the timing of hydrocarbon generation. However, the present-day traps that host these Phosphoria-sourced oils were not formed until after 85 Ma. One well-accepted migration model suggests that the oils were initially hosted in stratigraphic traps in the Phophoria and Tensleep formations prior to

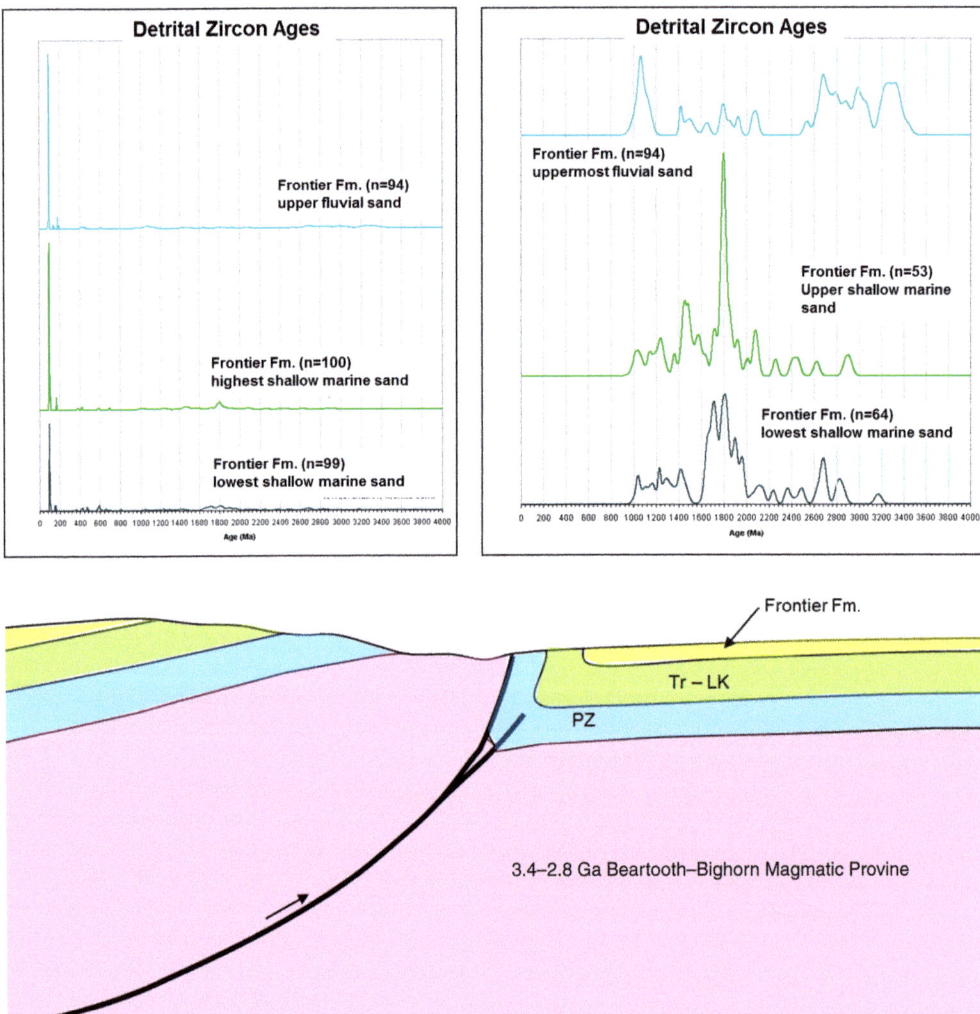

Fig. 9. Detrital zircon analysis of three shoreface/fluvial sands from the Cretaceous Frontier Formation (see Fig. 2) at the N Lazy S Ranch site (Fig. 1). Data were reproduced from May *et al.* (2013). The bottom two units are fine-grained quartzose sandstones, whereas the top unit is a coarse-grained arkosic sandstone. The zircon age probability distributions for all three samples are dominated by single clusters of young ages, consistent with the depositional age of the units. The age probability distribution for Precambrian grains in the three units is also plotted. The two lower samples display a broad spectrum of Proterozoic and Archean ages, with a prominent peak at 1800 Ma. The stratigraphically highest sample is unique in that it contains abundant Archean grains ranging from 3400 to 2600 Ma. The appearance of so many Archean grains, and the coarse-grained arkosic nature of the sandstone, suggests that these rocks were derived from Archean basement immediately to the west (Fig. 1). PZ, Paleozoic; Tr–LK, Triassic–Lower Cretaceous.

structuring and remigration (Stone 1967; Curry 2005), although potential top seals, and particularly the lateral seals for those traps, are lithologically heterogeneous (e.g. Campbell 1962; Paull & Paull 1986; Simmons & Scholle 1990). Alternatively, it is possible that incipient 'palaeostructures' enhanced the stratigraphic trapping mechanisms and improved the chances for 'catching' the early migrated hydrocarbons. Subtle Paleozoic structures have been recognized in the far NE portion of the basin (e.g. Simmons & Scholle 1990). We hypothesize that these types of subtle local features may have been more widespread than previously recognized.

To test the 'palaeostructure' hypothesis, we used yet another geochemical tool. The Frontier Formation (Fig. 2) is a major Mesozoic reservoir, deposited in the Sevier foreland at *c.* 100–90 Ma, when the volcanic arc to the west was active and first-cycle zircons were deposited with the sands. Zircons were analysed using laser ablation inductively coupled plasma mass spectrometry (LA-ICP-MS) at the Arizona LaserCron Center, using methods described by Gehrels *et al.* (2006). Detrital zircon age distributions for a site in the NW Bighorn Basin (Fig. 1), adjacent to the present-day Beartooth uplift, show a predominance of first-cycle zircons (Fig. 9) indicative of a young age, and consistent with the timing of Frontier deposition. However, these samples also contain a 3.4–2.7 Ga peak that can only be coming from the immediately adjacent basement (May *et al.* 2013), implying at least minor, local structural relief during Frontier deposition. The same sands are also poorly sorted and arkosic, indicating a local provenance. Our hypothesis is that subtle topographical highs began to form in the Middle Mesozoic on the western side of the basin, and helped to focus the pathways for hydrocarbons migrating from the west, within the underlying Phosphoria and Tensleep formations. These subtle highs would have enhanced the stratigraphic trapping potential within the Phosphoria and Tensleep formations, essentially creating combined structural–stratigraphic traps that either evolved to form the present-day structural traps and/or spilled hydrocarbons that remigrated into those traps.

In the preceding illustrative example, the combination of conventional and novel geochemical tools allowed us to better constrain the timing and pathways of hydrocarbon migration, but there are still unresolved issues. Geochemical approaches that could help address these issues include tools that constrain with high resolution: (1) the specific age of source rocks within a basin from the analysis of migrated fluids; (2) the distances that hydrocarbons have migrated from source to trap; and (3) the accumulation history of hydrocarbons, such as the timing of first charge to a trap, the identification of contributions of higher maturity fluids in mixed

oils and evidence of remigration from downdip structures.

Concluding remarks

We hope that the papers presented in this Special Publication provide those interested in petroleum systems analysis with a new perspective on how complex geological processes and signals can be understood and quantified. As hydrocarbon exploration continues to evolve and transition into increasingly complex and harsh environments, the need to glean the maximum information out of each sample continues to motivate analytical and technological developments in geochemistry and petroleum systems analysis. The current crop of technologies that are being developed and refined promise to provide a new generation of constraints that have the potential to be as important as those developed in the 1980s and 1990s. However, it is only when old and new techniques are integrated that can one hope to develop a more holistic understanding of the key elements and processes that contribute to the development of a robust petroleum system.

The integrated example of the Bighorn Basin presented here has been developed over many years at Exxon Mobil Corporation, and many people have contributed to this. We would like to thank Geoffrey Ellis for a thorough review and his suggestions to improve this manuscript. We thank Steve May, Robert Pottorf, Stephen Becker (ExxonMobil Upstream Research Company), Sebastien Dreyfus and Thomas Becker (ExxonMobil Exploration Company (a division of Exxon Mobil Corporation)) for significant contributions to this interpretation. We also thank James Kirk and Joaquin Ruiz for Re–Os analysis that was performed at the University of Arizona.

References

BALLENTINE, C.J., O'NIONS, R.K., OXBURGH, E.R., HORVATH, F. & DEAK, J. 1991. Rare gas constraints on hydrocarbon accumulation, crustal degassing and groundwater flow in the Pannonian Basin. *Earth and Planetary Science Letters*, **105**, 229–246.

BALLENTINE, C.J., O'NIONS, R.K. & COLEMAN, M.L. 1996. A Magnus opus: Helium, neon, and argon isotopes in a North Sea oilfield. *Geochimica et Cosmochimica Acta*, **60**, 831–849.

BARKER, C.E. & PAWLEWICZ, M.J. 1986. The correlation of vitrinite reflectance with maximum paleotemperature in humic organic matter. *In*: BUNTEBARTH, G. & STEGENA, L. (eds) *Paleogeothermics*. Springer, New York, 79–93.

BARRY, P.H., LAWSON, M., WARR, O., MABRY, J.C., BYRNE, D.J., MEURER, W.P. & BALLENTINE, C.J. 2016. Noble gas solubility models of hydrocarbon charge mechanism in the Sleipner Vest gas field. *Geochimica et Cosmochimica Acta*, **194**, 291–309.

BARRY, P.H., LAWSON, M., MEURER, W.P., DANABALAN, D., BYRNE, D.J., MABRY, J.C. & BALLENTINE, C.J. 2017. Determining fluid migration and isolation times in multiphase crustal domains using noble gases. *Geology*, **45**, 775–778, https://doi.org/10.1130/G38900.1

BERNARD, B.B., BROOKS, J.M. & SACKETT, W.M. 1976. Natural gas seepage in the Gulf of Mexico. *Earth and Planetary Science Letters*, **31**, 48–54.

BLAKEY, R.C. 2011. *Paleogeography and Geologic Evolution of North America*, http://jan.ucc.nau.edu/rcb7/nam.html [last accessed May 2017].

BOSCH, A. & MAZOR, E. 1988. Natural gas association with water and oil as depicted by atmospheric noble gases: case studies from the southeastern Mediterranean Coastal Plain. *Earth and Planetary Science Letters*, **87**, 338–346.

BURTNER, R.L. & NIGRINI, A. 1994. Thermochronology of the Idaho–Wyoming Thrust Belt during the Sevier Orogeny: a new, calibrated multiprocess thermal model. *AAPG Bulletin*, **78**, 1586–1612.

BYRNE, D.J., BARRY, P.H., LAWSON, M. & BALLENTINE, C.J. 2017. Noble gases in conventional and unconventional petroleum systems. *In*: LAWSON, M., FORMOLO, M.J. & EILER, J.M. (eds) *From Source to Seep: Geochemical Applications in Hydrocarbon Systems*. Geological Society, London, Special Publications, **468**. First published online December 14, 2017, https://doi.org/10.1144/SP468.5

CAMPBELL, C.V. 1962. Depositional environments of Phosphoria Formation (Permian) in southeastern Bighorn Basin, Wyoming. *AAPG Bulletin*, **46**, 478–503.

CHENEY, T.M. & SHELDON, R.P. 1959. Permian stratigraphy and oil potential, Wyoming and Utah. *In*: WILLIAMS, N.C. (ed.) *Guidebook to the Geology of the Wasatch and Uinta Mountains: Transition Area, Tenth Annual Field Conference*. Utah Geological Association, Salt Lake City, UT, 90–100.

CLAYPOOL, G.E., LOVE, A.H. & MAUGHAN, E.K. 1978. Organic geochemistry, incipient metamorphism, and oil generation in black shale members of Phosphoria Formation, western interior United States. *AAPG Bulletin*, **62**, 98–120.

CLOG, M., LAWSON, M., FERREIRA, A.A., SANTOS NETO, E.V. & EILER, J.M. 2018. A reconnaissance study of $^{13}C–^{13}C$ clumping in ethane from natural gas. *Geochimica et Cosmochimica Acta*, **223**, 229–244.

CRAIG, H. 1953. The geochemistry of the stable carbon isotopes. *Geochimica et Cosmochimica Acta*, **3**, 53–92.

CURIALE, J.A. 1994. Correlation of oils and source rocks: a conceptual and historical perspective. *In*: MAGOON, L. & DOW, W. (eds) *The Petroleum System – From Source to Trap*. American Association of Petroleum Geologists, Memoirs, **60**, 252–260.

CURRY, W.H. 2005. Paleotopography at the top of the Tensleep Formation, Bighorn Basin, Woming. *In*: GOOLSBY, J. & MORTON, D. (eds) *Wyoming Geological Association, 36th Annual Field Conference, 1984, Guidebook*. Wyoming Geological Association, Casper, Wyoming, 199–211.

DE BRUIN, R.H. 1997. An overview of Bighorn Basin oil and gas fields, with emphasis on Badger Basin Field. *In*: CAMPEN, E.B. (ed.) *Bighorn Basin; 50 Years on the Frontier; Evolution of the Geology of the Bighorn Basin; Part II, Improved Exploration for Natural Gas; Wyoming Institute for Energy Research & Wyoming Science, Technology and Energy Authority*. Wyoming Geological Association, Casper, WY, 7–13.

EGLINTON, G. & CALVIN, M. 1967. Chemical fossils. *Scientific American*, **216**, 32–43.

EILER, J.M. 2007. 'Clumped-isotope' geochemistry: the study of naturally-occurring, multiply-substituted isotopologues. *Earth and Planetary Science Letters*, **262**, 309–327.

EILER, J.M., CLOG, M. *ET AL.* 2013. A high-resolution gas-source isotope ratio mass spectrometer. *International Journal of Mass Spectrometry*, **335**, 45–56.

EILER, J.M., CLOG, M., LAWSON, M., LLOYD, M., PIASECKI, A., PONTON, C. & XIE, H. 2017. The isotopic structures of geological organic compounds. *In*: LAWSON, M., FORMOLO, M.J. & EILER, J.M. (eds) *From Source to Seep: Geochemical Applications in Hydrocarbon Systems*. Geological Society, London, Special Publications, **468**. First published online December 14, 2017, https://doi.org/10.1144/SP468.4

ESESTIME, P., HEWITT, A. & HODGSON, N. 2016. Zohr – A newborn carbonate play in the Levantine Basin, East-Mediterranean. *First Break*, **34**, 87–93.

ESLINGER, E. & PEVEAR, D. 1988. *Clay Minerals for Petroleum Geologists and Engineers*. Society of Economic Paleontolgists and Mineralogists, Tulsa, OK.

EXXONMOBIL CORPORATE REPORT 2016. *The Outlook for Energy: A View to 2040*. ExxonMobil, Irving, TX.

FARLEY, K.A., WOLF, R.A. & SILVER, L.T. 1996. The effects of long alpha-stopping distances on (U–Th)/He ages. *Geochimica et Cosmochimica Acta*, **60**, 4223–4229.

FERRY, J.M., PASSEY, B.H., VASCONCELOS, C. & EILER, J.M. 2011. Formation of dolomite at 40–80°C in the Latemar carbonate buildup, Dolomites, Italy, from clumped isotope thermometry. *Geology*, **39**, 571–574.

GAO, L., HE, P., JIN, Y., ZHANG, Y., WANG, X., ZHANG, S. & TANG, Y. 2016. Determination of position-specific carbon isotope ratios in propane from hydrocarbon gas mixtures. *Chemical Geology*, **435**, 1–9.

GAO, Y., CASEY, J.F., BERNARDO, L.M., YANG, W. & BISSADA, K.K. 2017. Vanadium isotope composition of crude oil: effects of source, maturation and biodegradation. *In*: LAWSON, M., FORMOLO, M.J. & EILER, J.M. (eds) *From Source to Seep: Geochemical Applications in Hydrocarbon Systems*. Geological Society, London, Special Publications, **468**. First published online December 14, 2017, https://doi.org/10.1144/SP468.2

GEHRELS, G.E., VALENCIA, V.A. & PULLEN, A. 2006. Detrital zircon geochronology by laser-ablation multicollector ICPMS at the Arizona LaserChron Center. *In*: LOSZEWSKI, T. & HUFF, W. (eds) *Geochronology: Emerging Opportunities*. The Paleontological Society Papers, **12**, 67–76.

GEORGIEV, S.V., STEIN, H.J., HANNAH, J.L., GALIMBERTI, R., NALI, M., YANG, G. & ZIMMERMAN, A. 2016. Re–Os dating of maltenes and asphaltenes within single samples of crude oil. *Geochimica et Cosmochimica Acta*, **179**, 53–75.

GHOSH, P., ADKINS, J. *ET AL.* 2006. $^{13}C–^{18}O$ bonds in carbonate minerals: a new kind of paleothermometer. *Geochimica et Cosmochimica Acta*, **70**, 1439–1456.

GILBERT, A., YAMADA, K., SUDA, K., UENO, Y. & YOSHIDA, N. 2016. Measurement of position-specific 13°C isotopic composition of propane at the nanomole level. *Geochimica et Cosmochimica Acta*, **177**, 205–216.

GILFILLAN, S.M., LOLLAR, B.S. ET AL. 2009. Solubility trapping in formation water as dominant CO_2 sink in natural gas fields. *Nature*, **458**, 614–618.

GOLDSTEIN, R.H. & REYNOLDS, T.J. 1994. *Systematics of Fluid Inclusions in Diagenetic Minerals*. Society for Sedimentary Geology (SEPM), Short Course, **31**.

GRUNAU, H.R. 1984. Natural gas in major basins worldwide attributed to source type, thermal maturity and bacterial origin. *Proceedings of the 11th World Petroleum Congress*, **21**, 293–302.

HAGEN, E.S. & SURDAM, R.C. 1984. Maturation history and thermal evolution of Cretaceous source rocks of the Bighorn Basin, Wyoming and Montana. *In*: WOODWARD, J., MEISSNER, F.F. & CLAYTON, J.L. (eds) *Hydrocarbon Source Rocks of the Greater Rocky Mountain Region*. Rocky Mountain Association of Geologists, Denver, CO, 321–338.

HANTSCHEL, T. & KAUERAUF, A.I. 2009. *Fundamentals of Basin and Petroleum Systems Modeling*. Springer, Berlin.

HEDBERG, H.D. 1964. Geological aspects of origin of petroleum. *AAPG Bulletin*, **48**, 1755–1803.

HOLBA, A.G., TEGELAAR, E.W., HUIZINGA, B.J., MOLDOWAN, J.M., SINGLETARY, M.S., McCAFFREY, M.A. & DZOU, L.I.P. 1998. 24-Norcholestanes as age-sensitive molecular fossils. *Geology*, **26**, 783–786.

HOLLAND, G., LOLLAR, B.S., LI, L., LACRAMPE-COULOUME, G., SLATER, G.F. & BALLENTINE, C.J. 2013. Deep fracture fluids isolated in the crust since the Precambrian era. *Nature*, **497**, 357–360.

HUNT, A.G., DARRAH, T.H. & POREDA, R.J. 2012. Determining the source and genetic fingerprint of natural gases using noble gas geochemistry: a northern Appalachian Basin case study. *AAPG Bulletin*, **96**, 1785–1811.

HUNT, J.M. 1996. *Petroleum Geochemistry and Geology*. W.H. Freeman, New York.

HUNT, J.M., PHILP, R.P. & KVENVOLDEN, K.A. 2002. Early developments in petroleum geochemistry. *Organic Geochemistry*, **33**, 1025–1052.

HUNT, T.S. 1861. Notes on the history of petroleum or rock oil. *Canadian Naturalist and Geologist*, **6**, 241–255.

HUNTINGTON, K.W., BUDD, D.A., WERNICKE, B.P. & EILER, J.M. 2011. Use of clumped-isotope thermometry to constrain the crystallization temperature of diagenetic calcite. *Journal of Sedimentary Research*, **81**, 656–669.

KHATIB, H. 2012. IEA world energy outlook 2011 – a comment. *Energy Policy*, **48**, 737–743.

KVENVOLDEN, K.A. 2006. Organic geochemistry: a retrospective of its first 70 years. *Organic Geochemistry*, **37**, 1–11.

KVENVOLDEN, K.A. 2008. Origins of organic geochemistry. *Organic Geochemistry*, **39**, 905–909.

LAWSON, M., SHENTON, B.J. ET AL. 2017. Deciphering the diagenetic history of the El Abra Formation of eastern Mexico using reordered clumped isotope temperatures and U–Pb dating. *GSA Bulletin*, first published online October 19, 2017, https://doi.org/10.1130/B31656.1

LEWAN, M.D. 1985. Evaluation of petroleum generation by hydrous pyrolysis experimentation. *Philosophical Transactions of the Royal Society of London, Series A*, **315**, 123–134.

LEWAN, M.D. 1994. Assessing natural oil expulsion from source rocks by laboratory pyrolysis. *In*: MAGOON, L.

& DOW, W. (eds) *The Petroleum System – From Source to Trap*. American Association of Petroleum Geologists, Memoirs, **60**, 201–210.

LILLIS, P. & SELBY, D. 2013. Evaluation of the rhenium–osmium geochronometer in the Phosphoria petroleum system, Bighorn Basin of Wyoming and Montana, USA. *Geochimica et Cosmochimica Acta*, **118**, 312–330.

LINK, W.K. 1952. Significance of oil and gas seeps in world oil exploration. *AAPG Bulletin*, **36**, 1505–1540.

MACDONALD, J.M., JOHN, C.M. & GIRARD, J.P. 2017. Testing clumped isotopes as a reservoir characterization tool: a comparison with fluid inclusions in a dolomotized sedimentary carbonate reservoir buried to 2–4 km. *In*: LAWSON, M., FORMOLO, M.J. & EILER, J.M. (eds) *From Source to Seep: Geochemical Applications in Hydrocarbon Systems*. Geological Society, London, Special Publications, **468**. First published online December 14, 2017, https://doi.org/10.1144/SP468.7

MACKENZIE, A.S., MAXWELL, J.R., COLEMAN, M.L. & DEEGAN, C.E. 1984. Biological markers and isotope studies of North Sea crude oils and sediments. *In*: *Proceedings of the 11th World Petroleum Congress, Volume 2. Geology Exploration Reserves*. Wiley, Chichester, 45–56.

MAGOON, L.B. & DOW, W.G. 1994. The petroleum system. *In*: MAGOON, L.B. & DOW, W.G. (eds) *The Petroleum System – From Source to Trap*. American Association of Petroleum Geologists, Memoirs, **60**, 3–24.

MALIVA, R.G., HERRMANN, R., COULIBALY, K. & GUO, W. 2015. Advanced aquifer characteristics for optimization of managed aquifer recharge. *Environmental Earth Science*, **73**, 7759–7767.

MAUGHAN, E.K. 1984. Geological setting and some geochemistry of petroleum source rocks in the Permian Phosphoria Formation. *In*: WOODWARD, J., MEISNER, F.F. & CLAYTON, J.L. (eds) *Hydrocarbon Source Rocks of the Greater Rocky Mountain Region, 1984*. Rocky Mountain Association of Geologists, Denver, CO, 281–294.

MAY, S., KLEIST, R., KNELLER, E., JOHNSON, C. & CREANEY, S. 2010. Global petroleum systems in space and time. *In*: VINING, B.A. & PICKERING, S.C. (eds) *Petroleum Geology: From Mature Basins to New Frontiers – Proceedings of the 7th Petroleum Geology Conference*. Geological Society, London, 1–9, https://doi.org/10.1144/0070001

MAY, S.R., GRAY, G.G., SUMMA, L.L., STEWART, N.R., GEHRELS, G.E. & PECHA, M.E. 2013. Detrital zircon geochronology from the Bighorn Basin, Wyoming, USA: implications for tectonostratigraphic evolution and paleogeography. *GSA Bulletin*, **125**, 1403–1422.

MILKOV, A.V. 2010. Methanogenic biodegradation of petroleum in the West Siberian Basin (Russia): Significance for formation of giant Cenomanian gas pools. *AAPG Bulletin*, **94**, 1485–1541.

MOLDOWAN, J.M., SEIFERT, W.K. & GALLEGOS, E.J. 1985. Relationship between petroleum composition and depositional environment of petroleum source rocks. *AAPG Bulletin*, **69**, 1255–1268.

MOLDOWAN, J.M., DAHL, W., HUIZINGA, B.J., FAGO, F.J., HICKLEY, L.J., PEAKMAN, T.M. & TAYLOR, D.W. 1994. The molecular fossil record of oleanane and its relation to angiosperms. *Science*, **265**, 768–771.

MOORE, M.T., VINSON, D.S., WHYTE, C.J., EYMOLD, W.K., WALSH, T.B. & DARRAH, T.H. 2018. Differentiating between biogenic and thermogenic sources of natural gas in coalbed methane reservoirs from the Illinois Basin using noble gas and hydrocarbon geochemistry. *In*: LAWSON, M., FORMOLO, M.J. & EILER, J.M. (eds) *From Source to Seep: Geochemical Applications in Hydrocarbon Systems*. Geological Society, London, Special Publications, **468**. First published online January 18, 2018, https://doi.org/10.1144/SP468.8

NEMČOK, M. 2016. *Rifts and Passive Margins: Structural Architecture, Thermal Regimes, and Petroleum Systems*. Cambridge University Press, New York.

ODEDRA, A., BURLEY, S.D., LEWIS, A., HARDMAN, M. & HAYNES, P. 2005. The world according to gas. *In*: DORÉ, A.G. & VINING, B.A. (eds) *Petroleum Geology: North-West Europe and Global Perspective – Proceedings of the 6th Petroleum Geology Conference*. Geological Society, London, 571–586, https://doi.org/10.1144/0060571

OLDENBURG, T.B.P., SILVA, R.C., RADOVIC, J., SNOWDON, R.W., GONZALEZ-ARISMENDI, G.P., BROWN, M. & LARTER, S.R. 2015. FTICR-MS – towards reaction systems models in petroleum geochemistry. *Goldschmidt Abstracts*, **2015**, 2336.

ONO, S., WANG, D.T., GRUEN, D.S., SHERWOOD LOLLAR, B., ZAHNISER, M.S., MCMANUS, B.J. & NELSON, D.D. 2014. Measurement of a doubly substituted methane isotopologue, $^{13}CH_3D$, by tunable infrared laser direct absorption spectroscopy. *Analytical Chemistry*, **86**, 6487–6494.

PAULL, R.A. & PAULL, R.K. 1986. Depositional history of lower Triassic Dinwoody Formation, Bighorn Basin, Wyoming and Montana. *In: Montana Geological Society and Yellowstone Bighorn Research Association Joint Field Conference and Symposium Geology of the Beartooth Uplift and Adjacent Basins*. Montana Geological Society, Billings, MT, 13–25.

PEDENTCHOUK, N. & TURICH, C. 2017. Carbon and hydrogen isotopic compositions of *n*-alkanes as a tool in petroleum exploration. *In*: LAWSON, M., FORMOLO, M.J. & EILER, J.M. (eds) *From Source to Seep: Geochemical Applications in Hydrocarbon Systems*. Geological Society, London, Special Publications, **468**. First published online December 14, 2017, https://doi.org/10.1144/SP468.1

PETERS, K.E., MOLDOWAN, J.M., SCHOELL, M. & HEMPKINS, W.B. 1986. Petroleum isotopic and biomarker composition related to source rock organic matter and depositional environment. *In*: LEYTHAEUSER, D. & RULLKOTTER, J. (eds) *Advances in Organic Geochemistry*. Organic Geochemistry, **10**, 73–84.

PETERS, K.E., WALTERS, C.C. & MOLDOWAN, J.M. 2005. *The Biomarker Guide Volume 2: Biomarkers and Isotopes in Petroleum Exploration and Earth History*. Cambridge University Press, Cambridge.

PEVEAR, D.R. 1992. Illite age analysis, a new tool for basin thermal history analysis. *In*: KHARAKA, Y.K. & MAEST, A.S. (eds) *Water–Rock Interaction*. A.A. Balkema, Rotterdam, The Netherlands, 1251–1254.

PEVEAR, D.R. 1998. Illite and hydrocarbon exploration. *Proceedings of the National Academy of Sciences of the United States of America*, **96**, 3440–3446.

PIASECKI, A., SESSIONS, A., LAWSON, M., FERREIRA, A.A., NETO, E.S. & EILER, J.M. 2016. Analysis of the site-specific carbon isotope composition of propane by gas source isotope ratio mass spectrometer. *Geochimica et Cosmochimica Acta*, **188**, 58–72.

PRATT, W.E. 1943. *Oil in the Earth*. University of Kansas Press, Lawrence, KS.

RICE, D.D. 1993. Biogenic gas – controls, habits and resource potential. *In*: HOWELL, D.G. (ed.) *The Future of Energy Gases*. United States Geological Survey, Professional Papers, **1570**, 583–606.

RICE, D.D. & CLAYPOOL, G.E. 1981. Generation, accumulation, and resource potential of biogenic gas. *AAPG Bulletin*, **65**, 5–25.

ROBERTS, L.N., FINN, T.M., LEWAN, M.D. & KIRSCHBAUM, M.A. 2008. *Burial History, Thermal Maturity, and Oil and Gas Generation History of Source Rocks in the Bighorn Basin, Wyoming and Montana*. United States Geological Survey, Scientific Investigations Report, **2008–5037**.

SCHOELL, M. 1980. The hydrogen and carbon isotopic composition of methane from natural gases of various origins. *Geochimica et Cosmochimica Acta*, **44**, 649–661.

SHELDON, R.P. 1967. Long distance migration of oil in Wyoming. *The Mountain Geologist*, **4**, 53–65.

SHENTON, B.J., GROSSMAN, E.L. *ET AL.* 2015. Clumped isotope thermometry in deeply buried sedimentary carbonates: The effects of bond reordering and recrystallization. *GSA Bulletin*, **127**, 1036–1051, https://doi.org/10.1130/B31169.1

SILJESTRÖM, S., LAUSMAA, J., SJÖVALL, P., BROMAN, C., THIEL, V. & HODE, T. 2010. Analysis of hopanes and steranes in single oil-bearing fluid inclusions using time-of-flight secondary ion mass spectrometry (ToF-SIMS). *Geobiology*, **8**, 37–44.

SILJESTRÖM, S., POTTORF, R., DREYFUS, S. & THIEL, V. 2016. *Determination of Single Hydrocarbon Inclusions Source Rock Facies*. International Workshop of Organic Geochemistry, Osaka, Japan.

SIMMONS, S.P. & SCHOLLE, P.A. 1990. Late Paleozoic uplift and sedimentation, Northeast Bighorn Basin, Wyoming. *In*: SPECHT, R.W. (ed.) *Wyoming Sedimentation and Tectonics*. Wyoming Geological Association, Casper, Wyoming, 41st Annual Field Conference Guidebook, 39–55.

STOLPER, D.A., LAWSON, M. *ET AL.* 2014*a*. The formation temperature of methane in natural environments. *Science*, **344**, 1500–1503.

STOLPER, D.A., SESSIONS, A.L. *ET AL.* 2014*b*. Combined $^{13}C–$D and D–D clumping in methane: methods and preliminary results. *Geochimica et Cosmochimica Acta*, **126**, 169–191.

STOLPER, D., MARTINI, A. *ET AL.* 2015. Distinguishing and understanding thermogenic and biogenic sources of methane using multiply substituted isotopologues. *Geochimica et Cosmochimica Acta*, **161**, 219–247.

STOLPER, D.A., LAWSON, M., FORMOLO, M.J., DAVIS, C.L., DOUGLAS, P.M.J. & EILER, J.M. 2017. The utility of methane clumped isotopes to constrain the origins of methane in natural gas accumulations. *In*: LAWSON, M., FORMOLO, M.J. & EILER, J.M. (eds) *From Source to Seep: Geochemical Applications in Hydrocarbon Systems*. Geological Society, London, Special Publications, **468**. First published online December 14, 2017, https://doi.org/10.1144/SP468.3

STONE, D.S. 1967. Theory of Paleozoic oil and gas accumulation in Big Horn Basin, Wyoming. *AAPG Bulletin*, **51**, 2056–2114.

SUMMA, L. 2015. From source to seep: New challenges and emerging technologies in petroleum systems analysis. *Goldschmidt Abstracts*, **2015**, 3028.

TISSOT, B.P. & WELTE, D.H. 1984. *Petroleum Formation and Occurrence*. Springer, Berlin.

TISSOT, B.P., PELET, R. & UNGERER, P. 1987. Thermal history of sedimentary basins, maturation indices and kinetics of oil and gas generation. *AAPG Bulletin*, **12**, 1445–1466.

TREIBS, A. 1936. Chlorophyll and hemin derivatives in organic mineral substances. *Angewandte Chemie*, **49**, 682–686.

VANDENBROUCKE, M., BEHAR, F. & RUDKIEWICZ, J.L. 1999. Kinetic modelling of petroleum formation and cracking: implications from the high pressure/high temperature Elgin Field (UK, North Sea). *Organic Geochemistry*, **30**, 1105–1125.

VENTURA, G.T., GALL, L., SIEBERT, C., PRYTULAK, J., SZATMARI, P., HÜRLIMANN, M. & HALLIDAY, A.N. 2015. The stable isotope composition of vanadium, nickel, and molybdenum in crude oils. *Applied Geochemistry*, **59**, 104–117.

WANG, D.T., GRUEN, D.S. *ET AL.* 2015. Nonequilibrium clumped isotope signals in microbial methane. *Science*, **348**, 428–431.

WAPLES, D.W. 1994. Maturity modeling: Thermal indicators, hydrocarbon generation and oil cracking. *In*: MAGOON, L. & DOW, W. (eds) *The Petroleum System – From Source to Trap*. American Association of Petroleum Geologists, Memoirs, **60**, 307–322.

WENGER, L., DAVIS, C.L. & ISAKSEN, G.H. 2002. Multiple controls on petroleum biodegradation and impact on oil quality. *SPE Reservoir Evaluation and Engineering*, **5**, 375–383.

WEO 2011. *The IEA World Energy Outlook 2011*. International Energy Agency, Paris.

WHITICAR, M.J. 1994. Correlation of natural gases with their sources. *In*: MAGOON, L. & DOW, W. (eds) *The Petroleum System – From Source to Trap*. American Association of Petroleum Geologists, Memoirs, **60**, 261–283.

WHITICAR, M.J., FABER, E. & SCHOELL, M. 1986. Biogenic methane formation in marine and freshwater environments: CO_2 reduction v. acetate fermentation – isotope evidence. *Geochimica et Cosmochimica Acta*, **50**, 693–709.

WOLF, R.A., FARLEY, K.A. & SILVER, L.T. 1996. Helium diffusion and low-temperature thermochronometry of apatite. *Geochimica et Cosmochimica Acta*, **60**, 4231–4240.

WOLF, R.A., FARLEY, K.A. & SILVER, L.T. 1997. Assessment of (U–Th)/He thermochronometry: The low-temperature history of the San Jacinto mountains, California. *Geology*, **25**, 65–68.

ZARTMAN, R.E., WASSERBURG, G.H. & REYNOLDS, J.H. 1961. Helium, argon and carbon in some natural gases. *Journal of Geophysical Research*, **66**, 277–306.

ZHANG, T., ELLIS, G.S., WANG, K., WALTERS, C.C., KELEMEN, S.R., GILLAIZEUA, B. & TANG, Y. 2007. Effect of hydrocarbon type of thermochemical sulfate reduction. *Organic Geochemistry*, **38**, 897–910.

The utility of methane clumped isotopes to constrain the origins of methane in natural gas accumulations

DANIEL A. STOLPER[1]*, MICHAEL LAWSON[2], MICHAEL J. FORMOLO[2], CARA L. DAVIS[2], PETER M. J. DOUGLAS[3] & JOHN M. EILER[4]

[1]*Department of Earth and Planetary Science, University of California, Berkeley, CA 94720, USA*

[2]*ExxonMobil Upstream Research Company, Spring, TX 77389, USA*

[3]*Department of Earth and Planetary Sciences, McGill University, Montreal, Quebec H3A 0E8, Canada*

[4]*Division of Geological and Planetary Sciences, California Institute of Technology, Pasadena, CA 91125, USA*

Correspondence: dstolper@berkeley.edu

Abstract: Methane clumped-isotope compositions provide a new approach to understanding the formational conditions of methane from both biogenic and thermogenic sources. Under some conditions, these compositions can be used to reconstruct the formational temperatures of the gas, and this capability can be applied to common subsets of both biogenic and thermogenic systems. Additionally, there are examples in which clumped-isotope compositions do not reflect gas-formation temperatures but instead mixing effects and kinetic phenomena; such kinetic effects also occur in common and recognizable subtypes of biogenic and thermogenic gases. Here we review the use of methane clumped-isotope measurements for understanding the origin of methane in the subsurface. We review methane clumped-isotope measurements from numerous biogenic and thermogenic natural gas reservoirs. We then place these measurements in the context of common frameworks for identifying the formational conditions of methane including the use of methane $\delta^{13}C$ and δD values and C_1/C_{2-3} ratios. Finally, we propose a framework for how methane clumped isotopes can be used to identify the origin of methane accumulations.

Methane is an important greenhouse gas, reactant and product of microbial metabolisms, and energy resource. In general, it is the most abundant alkane in any natural gas accumulation (e.g. Mango *et al.* 1994; Hunt 1996). Methane in natural gas accumulations generally has one of two origins. First, methane can be generated by the thermally induced breakdown (also termed pyrolysis, or 'cracking') of larger hydrocarbon molecules from solid (e.g. kerogen), liquid or gaseous hydrocarbons (e.g. Tissot & Welte 1978; Hunt 1996; Seewald 2003). This methane is classified here as 'thermogenic.' Second, methane can be produced by microorganisms known as methanogens. Methanogens make methane via either the net reaction of CO_2 with H_2 (hydrogenotrophic methanogenesis) or the cleavage and reduction of a methyl group from larger organic molecules like acetate or methanol (e.g. Claypool & Kaplan 1974; Rice & Claypool 1981; Thauer 1998). Methane produced by these organisms is termed 'biogenic' or microbial methane. Additionally, methane can be formed 'abiotically' through, for example, the hydrogenation of CO_2 in the absence of life. Such abiotic methane is not known to contribute substantial quantities of gas to economically significant hydrocarbon accumulations (e.g. Etiope & Sherwood Lollar 2013).

Establishing the origin of methane (i.e. biogenic v. thermogenic), whether for economic, biogeochemical, or environmental reasons, is generally one of the first steps in the study of hydrocarbon systems (Bernard *et al.* 1976; Schoell 1980, 1983; Whiticar *et al.* 1986; Whiticar 1999; Vinson *et al.* 2017). For example, in the study of greenhouse gas emissions, the origin of methane to the atmosphere from a given environment or setting is needed to develop emission reduction strategies (e.g. Miller *et al.* 2013). Alternatively, during petroleum exploration, the origin of methane can provide information on potential source-rock locations (if thermogenic) and the possibility of other accumulations in the area (e.g. Magoon & Dow 1994). A variety of approaches are used to identify the origin and history of

From: LAWSON, M., FORMOLO, M. J. & EILER, J. M. (eds) 2018. *From Source to Seep: Geochemical Applications in Hydrocarbon Systems*. Geological Society, London, Special Publications, **468**, 23–52.
First published online December 14, 2017, https://doi.org/10.1144/SP468.3
© 2018 ExxonMobil Upstream Research Company. Published by The Geological Society of London. All rights reserved.
For permissions: http://www.geolsoc.org.uk/permissions. Publishing disclaimer: www.geolsoc.org.uk/pub_ethics

hydrocarbon gases including molecular and isotopic measurements. The complexity of these approaches ranges from qualitative fingerprinting (e.g. Bernard *et al.* 1976; Schoell 1980, 1983; Whiticar *et al.* 1986) to sophisticated, quantum-mechanically grounded models of gas generation (e.g. Tang *et al.* 2000; Ni *et al.* 2011). This information is integrated with thermal histories provided by basin modelling, the stratigraphic and sedimentological evidence for the environment of deposition, and the structural analysis of trap and seal formation to understand the history of the petroleum system. This integrated approach is required to predict the potential distribution of hydrocarbons both spatially and temporally.

Here, we review the utility of measuring methane isotopologues with multiple heavy isotopes (e.g. both ^{13}C and deuterium (D)), which are colloquially referred to as 'clumped' isotopologues (Eiler 2007), to study the origin of methane. We review the conditions under which methane clumped-isotope measurements yield quantitative and usefully precise measures of a gas's formation temperature. We additionally address examples where the clumped-isotopic composition does not yield a meaningful formation temperature but instead relates to formational mechanisms. We specifically highlight the constraints methane clumped isotopes can provide on the origin of both biogenic and thermogenic natural gas deposits. This review places methane clumped isotopes into the broader framework of tools used to study hydrocarbon systems including standard molecular and isotopic techniques. Douglas *et al.* (2017) provides a broader review for methane clumped-isotope measurements beyond hydrocarbon systems.

Conventional molecular and isotopic techniques used to study the origins of hydrocarbon gases

Here we review two frameworks commonly used to identify the origin of methane in a given environment. Later in the review we place clumped-isotope measurements of methane into the context of these frameworks. We do not attempt to review all classification schemes here (e.g. Schoell 1983; Chung *et al.* 1988; Ballentine & O'Nions 1994; Prinzhofer & Huc 1995; Lorant *et al.* 1998).

One composition space used to map out the origin of methane in natural gas deposits is the 'Whiticar' plot (Fig. 1a; Whiticar *et al.* 1986; Whiticar 1999). It is empirical in nature and involves the comparison of an unknown sample with a two-dimensional isotopic map that delineates gas origins (biogenic, thermogenic or a mixture of the two) using the δD and δ^{13}C values of methane. This map is based on isotopic measurements of methane with independently known origins. The composition space defined by the Whiticar plot separates biogenic from thermogenic methane as follows: biogenic methane always has, at a given δD value, a lower δ^{13}C value than thermogenic methane. Depending on the δD value, the δ^{13}C cut-off for defining biogenic v. thermogenic gas can vary from −45 to −60‰. Additionally, although thermogenic and biogenic methane overlap extensively in δD values (overlapping from *c.* −150 to −275‰), biogenic methane can have significantly lower δD values (below −400‰) while thermogenic gases can have δD values as high as −100‰. Finally, the category of 'mixed/overlap' in the Whiticar plot indicates gases that are either mixtures of biogenic

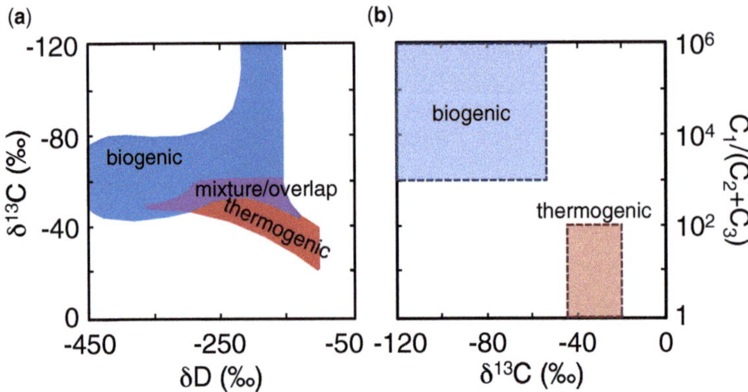

Fig. 1. Common composition spaces used to identify the origin of methane from environmental samples. (**a**) The 'Whiticar' plot (Whiticar *et al.* 1986; Whiticar 1999) uses the δD and δ^{13}C values of methane to distinguish biogenic and thermogenic gases from one another. (**b**) The 'Bernard' plot (Bernard *et al.* 1976) uses the C_1/C_{2-3} ratio v. the δ^{13}C value of methane to distinguish thermogenic from biogenic gas. Many versions of the Bernard plot exist, with different boundaries (e.g. Vinson *et al.* 2017). The boundaries given here are those given in Martini *et al.* (1996). Additionally, pathways of diffusion, mixing and methane oxidation are sometimes plotted in these spaces (e.g. Whiticar 1999; Etiope & Sherwood Lollar 2013; Vinson *et al.* 2017).

and thermogenic gas, or a composition where thermogenic and biogenic gases cannot be distinguished.

Another composition space used to establish the origin of a sample of methane is the 'Bernard' plot (Fig. 1b; Bernard *et al.* 1976). The Bernard plot differentiates thermogenic from biogenic gases using methane $\delta^{13}C$ values and C_1/C_{2-3} ratios. C_1/C_{2-3} ratios quantify the relative abundance of methane (C_1) to the sum of ethane (C_2) and propane (C_3). C_1/C_{2-3} ratios are used to distinguish biogenic from thermogenic gases because, in biogenic systems, methane is generally the main hydrocarbon generated (generally >99%). Thermogenic gases, in contrast can contain significant quantities (up to tens of per cent) of alkanes larger than methane (C_2+ alkanes). In the Bernard plot, biogenic gases are defined to have C_1/C_{2-3} values greater than 1000 while thermogenic gases have C_1/C_{2-3} values less than 100. Higher C_1/C_{2-3} values for thermogenic gases may be achieved via compositional fractionations induced during migration (Bernard *et al.* 1977).

In addition to these composition spaces, it is generally assumed that both the $\delta^{13}C$ and δD values of methane and C_1/C_{2-3} ratios increase over the course of petroleum generation. The extent of hydrocarbon generation generally increases with increasing burial temperature (and time spent at a given temperature) of the source rock. Consequently, the position of a thermogenic gas in the Bernard and Whiticar plots within the nominally thermogenic field is indicative of the 'thermal maturity' of the gas's source rock at the time of the gas's generation (Stahl & Carey 1975; Schoell 1980, 1983; Chung *et al.* 1988; Rooney *et al.* 1995; Hunt 1996; Tang *et al.* 2000; Ni *et al.* 2011).

The Whiticar and Bernard plots are useful for identifying the origin of methane in the environment. However, the criteria used on these plots to delineate biogenic from thermogenic methane are not always definitive. Specifically, in these plots, a gas with a C_1/C_{2-3} ratio greater than 1000 and $\delta^{13}C$ value less than −60‰ will always be classified as biogenic and a gas with a C_1/C_{2-3} ratio less than 100 and $\delta^{13}C$ greater than −50‰ will always be classified as thermogenic. However, some suspected biogenic methane occurrences have methane $\delta^{13}C$ values as high as −45‰ (Martini *et al.* 1996). Additionally, C_1/C_{2-3} values in modern marine sediments can be as low as 2 under conditions where ethane and propane are also biogenic in origin (Hinrichs *et al.* 2006). Moreover, the model of Tang *et al.* (2000) predicts that thermogenic methane generated at temperatures less than 180°C in nature can have $\delta^{13}C$ values between −60 to −70‰. Finally, thermogenic gases trapped in 'unconventional' shale-gas deposits ('shale gases') can have C_1/C_{2-3} values greater than 1000 (e.g. Stolper *et al.* 2014a). These counter-examples illustrate that, while empirical classification schemes

such as the Whiticar and Bernard plots are useful for interpreting the origins of natural gas deposits, they are not always clear-cut. Such counter-examples have led to alternative delineations of methane sources in δD v. $\delta^{13}C$ and C_1/C_{2-3} v. $\delta^{13}C$ plots (e.g. Etiope & Sherwood Lollar 2013; Vinson *et al.* 2017). This demonstrates that such spaces serve as useful starting points for the identification of methane sources, but they are not always definitive.

One physical variable often associated with the origin of a gas is its formation temperature. Biogenic gases are generally thought to form in nature from *c.* 0°C to 80°C (Wilhelms *et al.* 2001; Valentine 2011), though pure cultures of methanogens can grow in the laboratory at temperatures at least as high as 122°C (Takai *et al.* 2008). In contrast, thermogenic gases are thought to form at temperatures above *c.* 60°C (Tissot & Welte 1978; Quigley & Mackenzie 1988; Hunt 1996; Seewald *et al.* 1998; Seewald 2003). Furthermore, some models predict that most thermogenic gases form above 150°C (Quigley & Mackenzie 1988). Thus, measurements of methane formation temperatures could provide an additional parameter for understanding the genesis of hydrocarbon gases. Additionally, such a tool would allow for insights into the geological history of sedimentary basins by providing constraints on the minimum temperatures a given source rock (that generated methane) reached during burial. We now discuss how clumped-isotope measurements of methane, can, under some circumstances, be used to measure methane formational temperatures in a variety of hydrocarbon systems. First we briefly review the theory and measurement of methane clumped isotopes.

Theory and nomenclature of methane clumped-isotope measurements

Clumped isotopologues are any molecules with two or more rare (generally heavy) isotopes (Eiler & Schauble 2004; Wang *et al.* 2004; Eiler 2007). An example of a methane clumped isotopologue is $^{13}CH_3D$. Such isotopologues are of geological and geochemical interest because, for a population of molecules in isotopic equilibrium with one another, their abundance is controlled solely by the system's average or 'bulk' isotopic composition (constrained by its δD and $\delta^{13}C$ values) and the system's temperature (e.g. Wang *et al.* 2004).

Take the following isotope-exchange reaction between various methane isotopologues:

$$^{13}CH_3D + {}^{12}CH_4 \leftrightarrow {}^{13}CH_4 + {}^{12}CH_3D. \quad (1)$$

At equilibrium, the relative abundances of the isotopologues in equation (1) are controlled by the reaction's equilibrium constant. This equilibrium constant is a monotonic function of temperature (Ma *et al.* 2008; Cao & Liu 2012; Ono *et al.* 2014; Stolper *et al.* 2014*b*; Webb & Miller III 2014; Liu & Liu 2016). Additionally, for an isotopically equilibrated system the left side of the equation, which contains the clumped isotopologue ($^{13}CH_3D$), is always favoured relative to the right side at finite temperatures. This leads to a unique excess of $^{13}CH_3D$ at a given temperature (with larger excesses at lower temperatures) compared to that expected for a random distribution of isotopes amongst all isotopologues. This random distribution is constrained by the average isotopic composition of the methane (i.e. the δD and $\delta^{13}C$ values, which are measureable themselves). Consequently, if the abundance of $^{13}CH_3D$ or other clumped isotopologues can be constrained along with the δD and $\delta^{13}C$ values, an 'apparent' clumped-isotope-based methane formation temperature can be calculated. This apparent temperature will only reflect a true formation temperature if the methane formed in isotopic equilibrium and maintained that composition up until analysis. This is a key requirement that, as we discuss below, is apparently met in some but not all cases. We note that the actual reactions that allow isotopic equilibrium to be achieved between various methane isotopologues are not restricted to reactions involving only methane (i.e. reactions like those in equation 1), but instead are probably achieved via isotope-exchange reactions with other molecules including H_2O, H_2 or CO_2. Rather, the sorts of isotope-exchange reactions as given in equation (1) provide a framework for calculating equilibrium constants for systems in isotopic equilibrium regardless of the reactions that allow that equilibrium to be achieved (Urey 1947). More thorough reviews of the theory of clumped isotopes can been found in Wang *et al.* (2004), Eiler (2007), Affek (2012) and Eiler (2013).

The abundance of clumped isotopologues is reported relative to a calculated abundance that assumes all isotopes are randomly distributed amongst all isotopologues. This distribution is referred to as random or 'stochastic'. The clumped-isotopic composition of a sample is reported using Δ notation (Wang *et al.* 2004) such that, for example,

$$\Delta_{^{13}CH_3D} = \left(\frac{^{13CH_3D}R}{^{13CH_3D}R^*} - 1 \right) \times 1000. \qquad (2)$$

Here, $^{13CH_3D}R = [^{13}CH_3D]/[^{12}CH_4]$ and the * indicates the ratio that would be observed if all isotopes are randomly distributed amongst all isotopologues. As discussed in Wang *et al.* (2004) and specifically for methane in Stolper *et al.* (2014*b*), Δ values are related to the equilibrium constants of

isotope-exchange reactions. Specifically, $\Delta_{^{13}CH_3D}$ is related the equilibrium constant for reaction (1), $K_{^{13}CH_3D}$, as follows:

$$\Delta_{^{13}CH_3D} \cong -1000 \times \ln{(K_{^{13}CH_3D})}. \qquad (3)$$

The approximate sign is included due to the potential second-order effects of non-random proportions of $^{13}CH_4$ and $^{12}CH_3D$. This is unimportant for samples with δD and $\delta^{13}C$ values typical of those found in natural materials (see discussion in Stolper *et al.* (2014*b*)). Thus, a $\Delta_{^{13}CH_3D}$ value is directly related to the equilibrium constant for the clumped-isotope reaction in equation (1) and can be used to calculate a gas-formation temperature.

For reasons related to how the majority of methane clumped-isotope measurements presented here were made (discussed in the next section), this review primarily uses a subtly different Δ notation. Specifically, the majority of methane clumped-isotope abundances reported in this review combine the abundances of both of methane's mass-18 clumped isotopologues, $^{13}CH_3D$ and $^{12}CH_2D_2$. We report their combined abundances relative to $^{12}CH_4$ v. the random isotopic distribution using Δ_{18} notation such that:

$$\Delta_{18} = \left(\frac{^{18}R}{^{18}R^*} - 1 \right) \times 1000. \qquad (4)$$

Here, $^{18}R = ([^{13}CH_3D] + [^{12}CH_2D_2])/[^{12}CH_4]$.

As 98% of mass-18 methane is $^{13}CH_3D$, inclusion of $^{12}CH_2D_2$ is unimportant for most applications within the error of the measurement (though the difference could be important in certain cases where significant kinetic isotope effects are expressed) – this is discussed in detail in Stolper *et al.* (2014*b*). We provide a calculation of the dependence of Δ_{18} on temperature from 0 to 300°C in Figure 2. This range delineates the field of values where biogenic and thermogenic gases would likely be found if the methane formed in isotopic equilibrium.

We note that before the first clumped-isotope measurements of methane were made it was not clear whether Δ_{18} values would reflect equilibrium-based formation temperatures or kinetic processes reflecting methane generation, migration or extraction. Formation temperatures can only be meaningfully calculated from clumped-isotope abundances if the methane formed in internal isotopic equilibrium. Most previous interpretations of $\delta^{13}C$ and δD values of both thermogenic and biogenic methane invoked kinetic isotope effects to describe the observed isotopic compositions (Whiticar *et al.* 1986; Espitalie *et al.* 1988; Clayton 1991; Hunt 1996; Whiticar 1999; Tang *et al.* 2000; Xiao 2001;

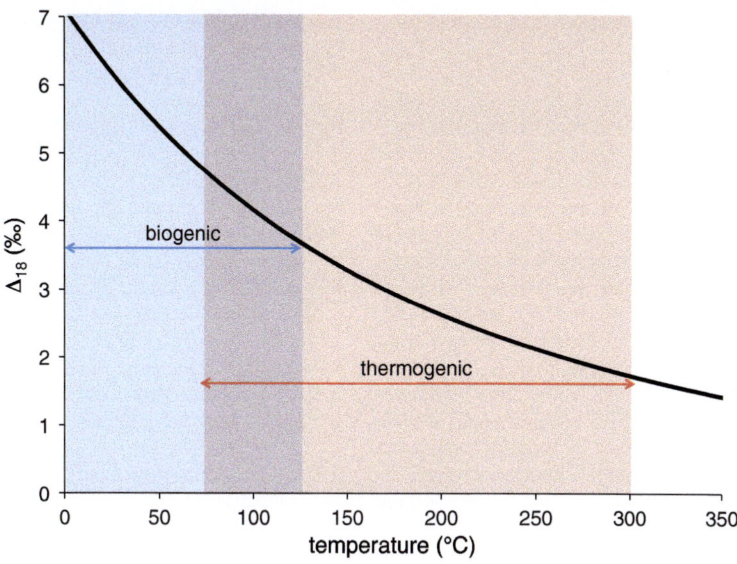

Fig. 2. Equilibrium dependence of Δ_{18} on temperature. Modified from Stolper *et al.* (2014*b*). Expected ranges are given for biogenic and thermogenic gases formed in clumped-isotopic equilibrium. The overall thermogenic range is derived from the start of the oil window of *c.* 60°C (Hunt 1996) through *c.* 300°C; 300°C is the approximate modelled maximum temperature of methane generation (e.g. Behar *et al.* 1992; Tsuzuki *et al.* 1999; Vandenbroucke *et al.* 1999; Tang *et al.* 2000; Dominé *et al.* 2002; Burruss & Laughrey 2010), though the maximum temperature of thermogenic gas generation is poorly constrained (Seewald 2003). The maximum temperature of biogenic methane generation (122°C) is taken from the experiments of Takai *et al.* (2008).

Seewald 2003; Valentine *et al.* 2004; Ni *et al.* 2011). Therefore the capacity to use clumped-isotope abundances to measure meaningful methane formation temperatures was not obvious prior to the study of natural gases and those generated in the lab to mimic gas-formational conditions.

Measurements of methane clumped isotopologues

Multiple techniques currently exist for accurate and precise (sub per mille) variations in mass-18 methane isotopologue abundances. The first such technique was described by Stolper *et al.* (2014*b*), who used a prototype 'high-resolution' gas-source isotope-ratio mass spectrometer (the Thermo Scientific MAT 253 Ultra) detailed in Eiler *et al.* (2013). The critical enabling feature of this mass spectrometer is its capacity to cleanly separate $H_2^{16}O$ (18.011 amu), a ubiquitous contaminant in all mass spectrometers, from $^{13}CH_3D$ (18.041 amu). The external precision (1 standard deviation; σ) for Δ_{18} measurements made on *c.* 50 micromoles of pure methane (*c.* 1 ml at STP) is ±0.25 ‰. The majority (83%) of data discussed here come from measurements made on this instrument.

Following this mass spectrometric technique, Ono *et al.* (2014) described a technique for measuring abundances of $^{13}CH_3D$ (constraining $\Delta_{^{13}CH_3D}$) using infrared spectroscopy accurately and precisely. Although precision depends on the amount of gas analysed, internal precisions of ±0.1‰ (1σ) can be achieved using *c.* 450 micromoles (10 ml at STP) of pure methane. Most of the data discussed here were derived using these two techniques (96%) with the remainder from the recently commissioned Nu Instruments Panorama (see below).

More advanced mass spectrometric techniques for measuring both $^{13}CH_3D$ and $^{12}CH_2D_2$ abundances separately are now ready. Production-line versions of the Mat 253 Ultra (Clog *et al.* 2015; Ellam *et al.* 2015) as well as another prototype high-resolution gas-source isotope-ratio mass spectrometer (the Nu Panorama; Young *et al.* 2016) have both been shown to be capable of resolving $^{13}CH_3D$ and $^{12}CH_2D_2$ from each other and various ion adducts. Additionally, advances in infrared spectroscopy may also permit analysis of both of these clumped isotopologues of methane. Thus, methane clumped-isotope measurements involving separate measurements of $^{13}CH_3D$ and $^{12}CH_2D_2$ abundances will be one of the next forefronts of methane clumped-isotope geochemistry.

Methane clumped-isotope measurements of thermogenic gases

Here, we review the dataset of current methane clumped-isotope measurements from both experimental simulations of hydrocarbon generation and environmental settings. We classify thermogenic gases using two criteria related to the physical and chemical conditions present in a hydrocarbon deposit. The first criterion is whether or not the gas was formed in place in the reservoir, or migrated from its formational location. Natural gas deposits in which the stratigraphic location where the gas forms and remains trapped are identical are termed 'unconventional' deposits (Curtis 2002). In most unconventional deposits, the reservoir (and source) rock is shale and it must be hydraulically fractured in order to create permeability for gas or oil extraction. In contrast, 'conventional' reservoirs are those in which hydrocarbons formed elsewhere (the 'source rock') and became trapped during migration from their source. These are termed conventional because they have been, historically, the typical accumulations of hydrocarbon liquids and gases.

These designations are significant for our purposes. Conventional reservoirs can accumulate gases from a variety of different sources generated over a range of conditions and temperatures. The geological and geochemical history of a conventional reservoir may have no direct relationship to the formational conditions of the trapped hydrocarbons. In contrast, gases in an unconventional reservoir are retained in the source and so experience the same thermal and burial history as the source rock after gas formation.

In both conventional and unconventional reservoirs, not all generated and/or trapped hydrocarbons are retained. For example, in thermally mature unconventional systems, the majority of generated hydrocarbons (including both oils and gases) are thought to be expelled over the course of oil and gas generation (e.g. Jarvie *et al.* 2007; Xia 2014). Similarly, in conventional systems if there is not a sufficient sealing lithology and trapping geometry then hydrocarbons will not be trapped. Thus, the gases trapped in both unconventional and conventional reservoirs may only represent a snapshot of the system's gas generation and accumulation history.

The second criterion we use to distinguish between various types of natural gas deposits is whether the gas was 'associated' (i.e. found) with or without liquid hydrocarbons in the reservoir. Associated or 'oil-associated' gases are either dissolved in oil ('solution gas') or present as a gaseous phase overlying a liquid phase in the reservoir (a 'gas cap'). 'Non-associated' gases are present in the gaseous phase in the reservoir and are not in contact with any liquid hydrocarbons in the subsurface. We apply a single-phase gas/oil ratio (GOR) of 6000 standard cubic feet/stock tank barrel (scf/stb) to differentiate non-associated gases in a subsurface accumulation from oil-associated gases. If there are two phases present in the subsurface (e.g. an oil leg with a gas cap), a 6000 scf/stb gas could be associated. While we recognize this here, we do not consider it further in this review.

There are additional classification schemes that use similar terms (oil-associated and non-associated) to indicate whether the gas was originally generated with oil or not (e.g. Schoell 1983). Other terms used to categorize and describe natural gas reservoirs include 'primary' gases that form directly from the breakdown of kerogen and 'secondary' gases that form via the breakdown of oil and hydrocarbon gases. These frameworks require *a priori* knowledge of the processes by and conditions under which a gas formed. Problematically, such information is generally not available and cannot always be inferred. Our usage of the terms conventional v. unconventional and associated v. non-associated are based solely on known conditions in the reservoir at the time of drilling and hydrocarbon extraction. We note that in some cases a gas has been examined that is known to be thermogenic, but is not currently in a reservoir (e.g. a gas seep). Such gases are included in the larger comparison of thermogenic and biogenic gases, but not in the more specific comparisons of different types of thermogenic gases. The locations of the various accumulations are given in Figure 3.

The first thermogenic methane clumped-isotope results

The first evidence that methane clumped-isotope measurements may yield meaningful gas-formation temperatures came from Stolper *et al.* (2014*b*). Specifically, isotopic (including clumped-isotope) measurements were made on methane sampled from a commercial high-purity gas cylinder. Based on its bulk isotopic composition ($\delta D = -175.5‰$ and $\delta^{13}C = -42.9‰$), the sample was assumed to be thermogenic in origin (Fig. 1a). Methane from the cylinder yielded a clumped-isotope temperature of 170°C. This temperature is within the range expected for thermogenic gas-formation temperatures (*c.* 60–300°C; e.g. Behar *et al.* 1992; Hunt 1996; Tsuzuki *et al.* 1999; Vandenbroucke *et al.* 1999; Tang *et al.* 2000; Dominé *et al.* 2002; Burruss & Laughrey 2010) and is thus a geologically reasonable gas-formation temperature. Following this, additional measurements of gases from cylinders or laboratory gas lines with assumed thermogenic origins (again based on bulk isotopic compositions) using the spectroscopic technique yielded clumped-isotope-based

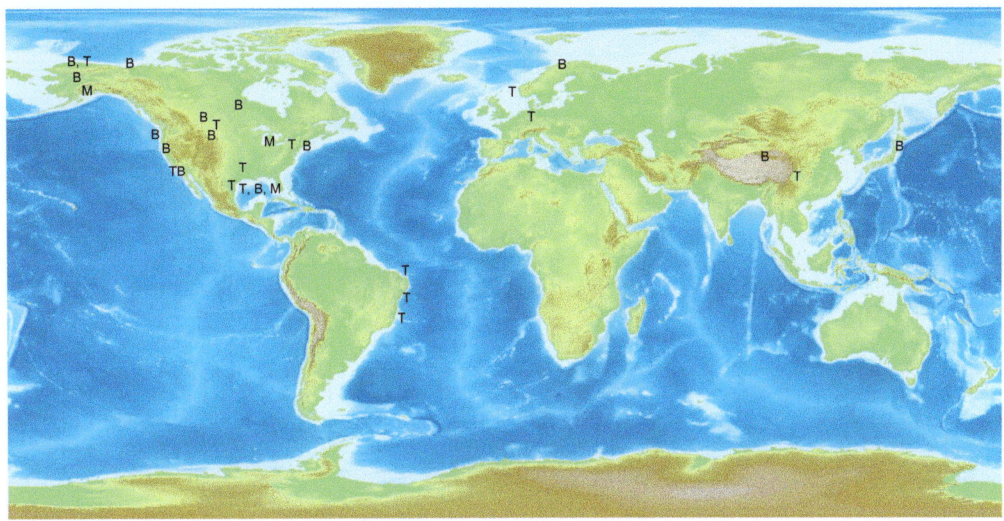

T: Thermogenic
B: Biogenic (Microbial)
M: Mixed Thermogenic and Biogenic

Fig. 3. Locations of sampled gases discussed here. Data derived from Stolper *et al.* (2014*a*, 2015), Wang *et al.* (2015), Inagaki *et al.* (2015), Douglas *et al.* (2016, 2017) and Young *et al.* (2017). Specific accumulations and locations are discussed in these references.

temperatures from 151 to 212°C (Ono *et al.* 2014). These results are consistent with methane clumped-isotope temperatures reflecting thermogenic gas-formation temperatures, but, due to lack of knowledge of the samples' origins, they could not be evaluated further.

Stolper *et al.* (2014*a*) presented the first methane clumped-isotope measurements of thermogenic gases from experimental simulations of thermogenic gas generation and from samples taken from unconventional non-associated and conventional oil-associated gas deposits. For reasons discussed above, unconventional deposits were targeted because the thermal history of the gas (once formed) and the source/reservoir are identical.

Non-associated gases extracted from unconventional deposits hosted in the Haynesville Shale (Texas; Hammes *et al.* 2011) and Marcellus Shale (Pennsylvania; Lash & Engelder 2011) were studied. The Haynesville Shale deposits currently remain near their maximum burial depths and temperatures (Stolper *et al.* 2014*a*). The measured clumped-isotope temperatures ranged from 169 to 207°C and are within 2σ of the measurement precision of current reservoir temperatures (163–190°C), modelled maximum burial temperatures (175–207°C), and independently modelled gas-formation temperatures (168–173°C; Fig. 4; Stolper *et al.* 2014*a*). It is important to note here that the methane clumped-isotope temperatures of gases sampled from a

subsurface hydrocarbon accumulation reflect the bulk weighted average temperature of all methane that was generated and stored (i.e. it is a cumulative measure).

In contrast to the Hayneville Shale, the section of the studied Marcellus Shale reached maximum burial temperatures, which are modelled to have been between 183 and 219°C, and then was uplifted and cooled to current temperatures between 60 and 70°C (Stolper *et al.* 2014*a*). The clumped-isotope temperatures for these Marcellus Shale samples were found to range from 179 to 207°C (Stolper *et al.* 2014*a*). This range overlaps the Haynesville Shale clumped-isotope temperature range and all measured temperatures are within 2σ of model-based average gas-formation temperatures (171–173°C).

Although these clumped-isotope temperatures are consistent with expected gas-formation temperatures in the studied Haynesville and Marcellus Shale samples, an alternative interpretation is that the temperatures reflect partially or completely the effects of isotope-exchange reactions that occurred after gas formation. In other words, if methane, after formation, continued to isotopically re-equilibrate, the clumped-isotopic composition would reflect the gas's thermal history post formation. Re-equilibration would require methane to exchange hydrogen isotopes with methane molecules or other compounds including, for example, H_2O, H_2 or other hydrogen-bearing compounds. The rate of this exchange will

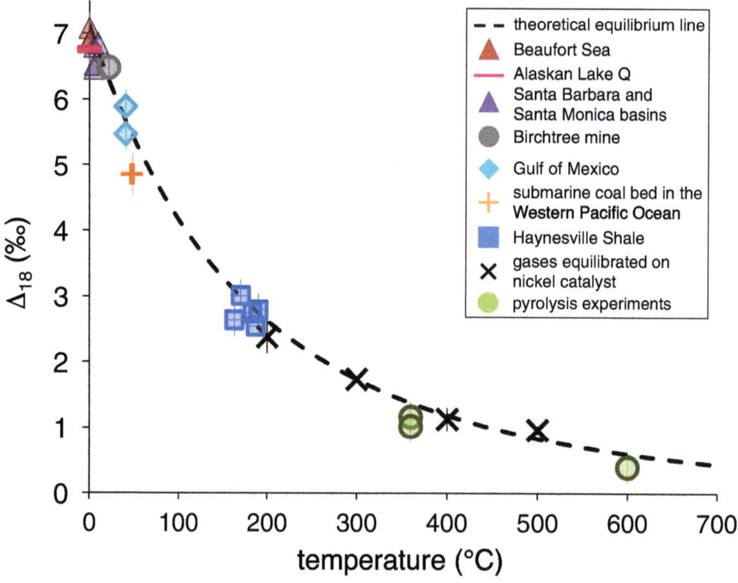

Fig. 4. Samples of biogenic and thermogenic gases from environmental systems or experimental products with independently constrained formation temperatures (given by the x-axis) compared to measured Δ_{18} values. Data from Stolper *et al.* (2014*a*, *b*, 2015), Inagaki *et al.* (2015), Douglas *et al.* (2016) and Young *et al.* (2017). Data from Inagaki *et al.* (2015) and Young *et al.* (2017) were converted to the Δ_{18} reference frame based on $\Delta_{^{13}CH_3D}$ temperatures. Error bars are 1σ.

scale with temperature, with faster rates at higher temperatures. The temperature below which such reactions cease to occur at a significant rate is termed an isotopic 'blocking temperature.' Such blocking temperatures have been studied previously for carbonate clumped isotopes (e.g. Ghosh *et al.* 2006; Dennis & Schrag 2010; Passey & Henkes 2012; Stolper & Eiler 2015). The clumped-isotope results from the Marcellus and Haynesville Shale samples indicate that the methane clumped-isotope blocking temperature in these unconventional deposits is above the shales' maximum burial temperatures (>200°C). Specifically, if the methane clumped-isotope blocking temperature was less than 200°C for geologically relevant timescales, the Marcellus Shale samples would presumably have yielded clumped-isotope-based temperatures consistently lower than the Haynesville Shale samples due to uplift and cooling of the examined Marcellus Shale gases after gas generation. This was not observed. We note though, that the blocking temperatures for methane, whatever they are, may and likely will vary based on the mineralogy of the reservoir rock (e.g. presence of transition metals like nickel) and chemical conditions (e.g. presence or absence of H_2O or oil) in the reservoir (Seewald 2003). Such different conditions will affect the reaction pathways available to promote C—H exchange.

Taken together, the clumped-isotope temperatures from the Haynesville and Marcellus Shale samples reported by Stolper *et al.* (2014*a*) represent geologically reasonable temperatures for thermogenic gas formation. They appear unaffected by subsequent cooling of the systems and agree with independent metrics and models for thermogenic gas-formation temperatures (e.g. Tissot & Welte 1978; Quigley & Mackenzie 1988; Hunt 1996; Seewald *et al.* 1998; Seewald 2003). Thus, the measured clumped-isotope temperatures are, to first order, conceivable average gas-formation temperatures. This was explored further in Stolper *et al.* (2014*a*) with measurements of oil-associated thermogenic gases from a conventional reservoir in the Potiguar Basin, Brazil. Methane from Potiguar Basin samples yielded clumped-isotope temperatures from 157 to 221°C. These temperatures are also in the expected range of thermogenic gas-formation temperatures. Additionally, the clumped-isotope temperatures correlated positively with the $\delta^{13}C$ values of methane – such a correlation was expected based on conventional interpretations of $\delta^{13}C$ as a maturity indicator (discussed in the sections entitled 'Conventional molecular and isotopic techniques used to study the origins of hydrocarbon gases' briefly above and 'Values of $\delta^{13}C$ and clumped-isotope temperatures of thermogenic gases' more extensively below).

It is noteworthy that some methane clumped-isotope temperatures from the Potiguar Basin samples, all of which are oil-associated, fall outside of the commonly assumed temperature range for oil generation (the oil window) of *c.* 60 to 160°C (Hunt 1996) and even more recent maximum estimates of *c.* 200°C; (e.g. Vandenbroucke *et al.* 1999; Lewan & Ruble 2002). Stolper *et al.* (2014*a*) proposed that the 'hot' methane did not form with oil but instead, during migration, ended up in the same reservoir as the oil. Indeed, source rocks of high maturity (past oil-window maturity) are found in the Potiguar Basin, making such an idea plausible (Stolper *et al.* 2014*a*). As will be discussed in the next section, thermogenic gases from different hydrocarbon systems frequently yield methane clumped-isotope temperatures from 60 to 160°C indicating methane clumped-isotope temperatures consistent with co-formation of gas and oil are common.

Finally, clumped isotopes were measured on methane generated experimentally by closed-system hydrous pyrolysis of organic-rich shale at 360°C or closed-system anhydrous pyrolysis of pure propane at 600°C. The temperatures were found to be within 2σ measurement precisions of the known formation temperatures. Additionally, these temperatures were 5σ away from measured clumped-isotope temperatures from environmental thermogenic methane samples (Fig. 4). These experiments thus independently support the hypothesis that methane clumped-isotope temperatures can reflect formation temperatures. In all, these results led Stolper *et al.* (2014*a*) to hypothesize that clumped-isotope-based temperatures of thermogenic gases could be used to infer, at least for the systems examined, average gas-formation temperatures.

Further studies of thermogenic gases

Additional clumped-isotope studies of thermogenic gases from a variety of different hydrocarbon systems have now been made (e.g. Stolper *et al.* 2015; Wang *et al.* 2015; Douglas *et al.* 2016, 2017; Young *et al.* 2017). In this section we review the overall distribution of clumped-isotope-based temperatures derived from thermogenic gases (study locations given in Fig. 3). All measured thermogenic conventional oil-associated and non-associated gases and unconventional non-associated gases are included. Unconventional oil-associated gases appear to yield non-equilibrium clumped-isotope distributions and are not included in the present discussion. They are treated separately below. The geological and production histories can be found in the original studies from which the measurements are taken (see Douglas *et al.* 2017).

The observed range of clumped-isotope temperatures for thermogenic methane is from 72 to 298°C (Fig. 5c). The temperatures are normally distributed with a peak value of 175°C ± 47°C (1σ; Fig. 5c). This distribution can be compared to model-based expectations for both the expected range of methane formation temperatures and the amount of methane produced as a function of temperature. In models of methane generation, the amount of methane produced at different temperatures varies because it is assumed that rates of methane generation increase with increasing temperature but decrease as methane precursors are converted to either methane or oxidized forms of carbon such as graphite (Seewald 2003).

We use the model of Hunt (1996), which is reproduced in Figure 5d, to make the comparison between measured v. expected gas-formation temperatures and the expected distribution of those temperatures. The model of Hunt (1996) is a general model of methane generation kinetics. We note, though, that there are other models of gas generation distributions as a function of temperature and they do not always agree (e.g. Seewald 2003). Regardless, the overall range and distribution of measured v. modelled temperatures are similar indicating that the two are in general agreement. We note that direct comparisons of all measured thermogenic gas clumped-isotope-based temperatures with modelled formation temperatures and thermal burial histories of the source rocks, as was done for unconventional unassociated gases in Stolper *et al.* (2014*a*), is not generally possible. This is because most examined samples do not have independently constrained thermal and generation histories (e.g. because they have migrated from their source rock to the reservoir rock and/or lack independent constraints to calibrate a model of their thermal histories), thus preventing such a comparison.

The measured clumped-isotope temperatures extend to higher temperatures (*c.* 300°C) than the *c.* 250°C cut-off indicated by the model of Hunt (1996). However, recent models, the discovery of high-temperature oil reservoirs, and experiments all indicate that oils can be stable up to *c.* 200°C (e.g. Tsuzuki *et al.* 1999; Vandenbroucke *et al.* 1999; Lewan & Ruble 2002) and that methane generation occurs at temperatures above 250°C in nature (e.g. Behar *et al.* 1992; Tsuzuki *et al.* 1999; Vandenbroucke *et al.* 1999; Tang *et al.* 2000; Dominé *et al.* 2002; Burruss & Laughrey 2010). Although the maximum temperature of thermogenic methane generation remains an open question, the clumped-isotope temperatures are in line with these recent indications that it may extend up to at least *c.* 300°C. Alternatively, these higher temperatures could result from subtle kinetic isotope effects expressed during gas generation or migration.

Specifically, recent pyrolysis experiments on coal (Shuai *et al.* in press) indicate that disequilibrium

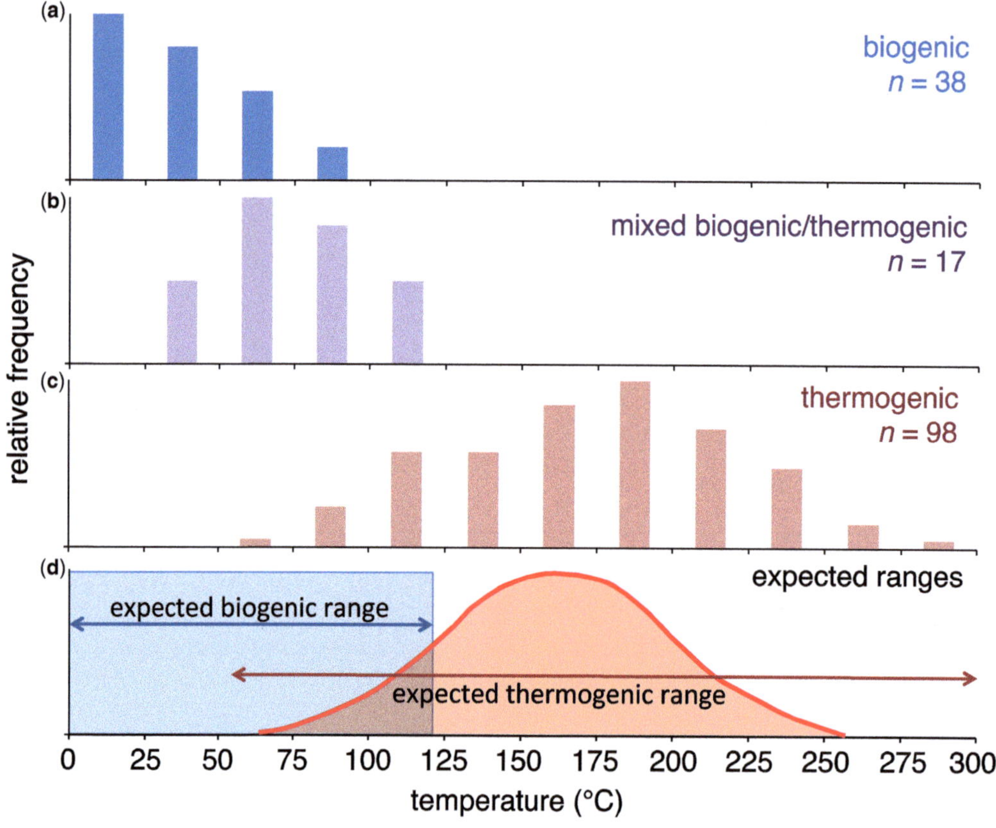

Fig. 5. Distribution of measured clumped-isotope-based temperatures from biogenic (**a**), mixed thermogenic and biogenic (**b**) and thermogenic (**c**) gases compared to expected formation ranges of biogenic and thermogenic gases (**d**); the thermogenic distribution is derived from Hunt (1996). Data from Stolper *et al.* (2014a, 2015), Wang *et al.* (2015), Inagaki *et al.* (2015), Douglas *et al.* (2016, 2017) and Young *et al.* (2017). Number of samples measured is given by *n*. All histograms are normalized such that the maximum box has a height of 1.

effects can be expressed during methane generation. In these experiments coal was heated in sealed gold tubes at different rates (i.e. increase in temperature per unit time) to a final temperature between 400 and 620°C. Samples quenched between 400 and 520°C yield Δ_{18} values consistent with the generation of methane at equilibrium. Above 520°C, δD values increase rapidly and Δ_{18} values first decrease (becoming negative) and then increase at 600°C to Δ_{18} values similar to those expected for clumped-isotope equilibrium for the final experimental temperature. These Δ_{18} values indicate that non-equilibrium processes during methane generation can occur at elevated temperatures in experimental samples. Interestingly, the transition to non-equilibrium Δ_{18} values coincides with the onset of ethane cracking. The precise mechanisms controlling the non-equilibrium Δ_{18} values of the evolved methane are currently unknown, nor is it clear whether they are relevant to natural systems.

However, they show that laboratory pyrolysis experiments can create both equilibrium and non-equilibrium methane clumped-isotope compositions depending on the details of the experiment.

The following sections focus on how the clumped-isotope temperatures reflect both the formational conditions and trapping histories of oil-associated and non-associated gases in conventional and unconventional reservoirs. We do not discuss cases where the origin of the thermogenic gases is not well constrained (e.g. thermogenic gas seeps).

Conventional oil-associated gases

Oil-associated methane trapped in conventional reservoirs (from six distinct hydrocarbon systems) yield clumped-isotope temperatures that range from 103 to 266°C, with an average value of 167°C (±40°C, 1σ; Fig. 6a). This range spans nearly the entire expected range of thermogenic gas generation temperatures

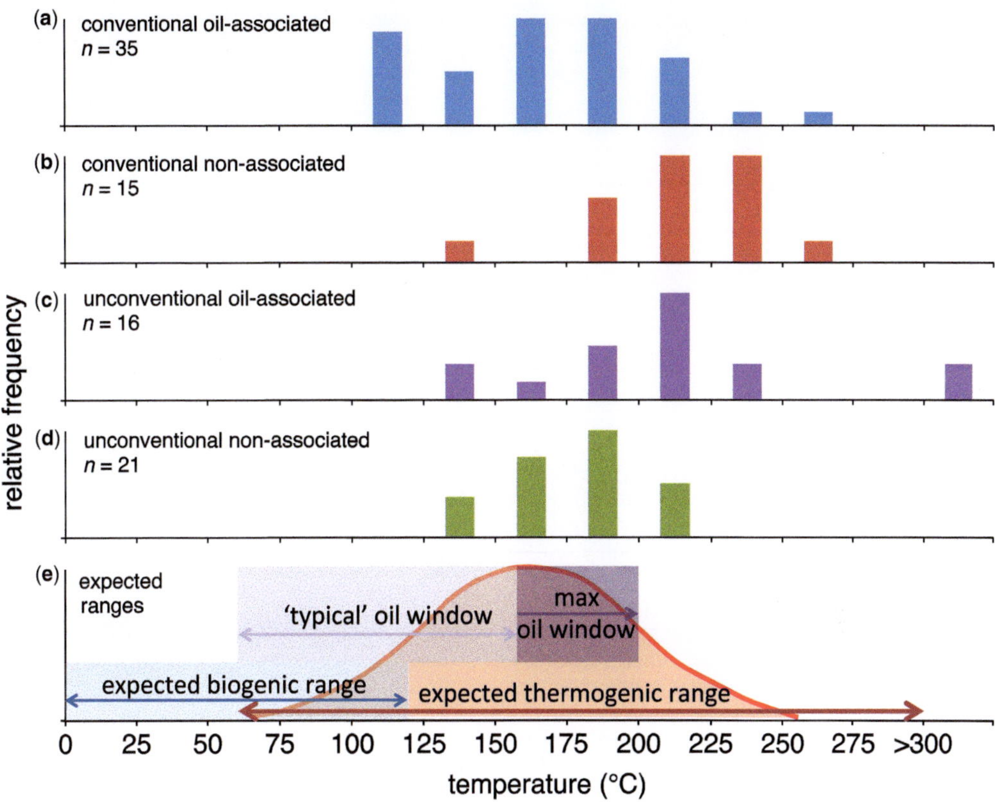

Fig. 6. Distribution of measured clumped-isotope-based temperature from thermogenic deposits including: (**a**) conventional oil-associated deposits; (**b**) conventional non-associated deposits; (**c**) unconventional oil-associated deposits; and (**d**) unconventional non-associated deposits. These are compared to expected formation ranges of biogenic and thermogenic gases in (**e**). Data from Stolper *et al.* (2014*a*, 2015), Wang *et al.* (2015), Young *et al.* 2017 and Douglas *et al.* (2017). Number of samples measured is given by *n*. All histograms are normalized such that the maximum box has a height of 1. The typical range of oil generation is often given from *c.* 60 to 160°C (Hunt 1996), but some models indicate that oils can be generated and are stable up to *c.* 200°C (Tsuzuki *et al.* 1999; Lewan & Ruble 2002). Other ranges are described in Figure 5.

(Fig. 6a v. Fig. 6e). Clumped-isotope temperatures of oil-associated gases from conventional reservoirs are equally common from 100 to 150°C as 150 to 225°C, with a sharp drop-off in frequency for temperatures above 225°C (Fig. 6a). Thus, the examined oil-associated gases, to first order, yield temperatures consistent with generation in the oil window (< *c.* 160°C) and at temperatures above oil stability. The overall range of clumped-isotope temperatures is probably the result of the capacity for conventional reservoirs to trap hydrocarbons generated over the full spectrum of potential gas generation temperatures. For example, if the gases in the reservoir were co-generated with the oils and remained associated during migration, the methane clumped-isotope temperatures might be expected to range from about 60 to 160°C. However, if the reservoirs also capture gases generated at higher temperatures (e.g. from the

breakdown of oil and gas or kerogen above the oil window), then the measured methane clumped-isotope temperatures could be higher. Indeed, as some models predict that most natural gas forms above oil generation temperatures (>150°C; Quigley & Mackenzie 1988), it may not be surprising that conventional reservoirs often contain methane with clumped-isotope temperatures above 150°C.

An alternative interpretation for the observed temperature ranges measured in conventional oil-associated gases is that they contain a component of methane formed in equilibrium and a component formed with Δ_{18} values lower than expected for isotopic equilibrium, as occurs in some coal-pyrolysis experiments (Shuai *et al.* in press). We do not favour this interpretation for the data summarized in Figure 6a for two reasons: (1) methane in conventional oil-associated deposits has not yielded clumped-isotope temperatures that exceed plausible gas generation temperatures (unlike the non-equilibrium temperature intervals of the pyrolysis experiments discussed above or some of the more extreme findings for unconventional oil-associated gases described below); and (2) correlations between $\delta^{13}C$ and methane temperature in the conventional oil-associated gases (presented below) generally resemble those expected for models of isotopic evolution during petroleum generation.

The observed range in clumped-isotope temperatures in conventional oil-associated deposits suggests one use of clumped-isotope measurements of methane is to establish whether the gases present in the deposit were co-generated with the oil or formed later (e.g. from the cracking of residual oil in the source rock, from oil in deeper reservoir intervals, or from a different source interval entirely) and mixed into the oils during migration. As will be discussed below (section entitled 'A synthesis of clumped-isotope studies and their utility in the study of economic accumulations of hydrocarbons'), this information can be used in an exploratory framework to detect the generation of hydrocarbons at greater depths and higher temperatures than oil generation, even in systems where the deeper-formed gas has dissolved into oils at shallower depths. Such information could indicate the potential for gas reservoirs at greater depths.

Conventional non-associated gases

Methane clumped-isotope measurements have been made on conventional non-associated gases from two systems. The first is the Rotliegend system in Germany (McCann 1998), where gases are coal-derived (a 'type III' kerogen source): see, for example, Killops & Killops (2013) or Peters *et al.* (2005) for further explanation on kerogen types. The second is from the North Sea (Sleipner Vest; mixture of algal [type II] and coal [type III] sources; Ranaweera 1990). Together, these systems yield an average clumped-isotope temperature of 213°C (±30, 1σ; Fig. 6b) with a range from 144 to 267°C. More specifically, the Rotliegend samples have an average temperature of 217°C (±34, 1σ) and the North Sea samples have an average temperature of 205°C (±21, 1σ).

These average clumped-isotope temperatures are higher than the average for conventional oil-associated gases (167°C). Indeed, only one sample (from Rotliegend) yields a temperature (144°C) consistent with generation below the oil window (<160°C). A simple interpretation of this result is that these hydrocarbon reservoirs captured gases dominantly generated at temperatures above oil generation. If broadly applicable, this would indicate that the majority of non-associated gases form at temperatures above oil generation. This hypothesis will require the examination of a wider range of conventional non-associated gases.

Alternatively, the difference between the clumped-isotope temperatures of conventional oil-associated v. non-associated gases may be due to different source organic types. The oil-associated gases were sourced dominantly from lacustrine and marine, organic carbon (type I and II kerogens). In contrast, the non-associated gases are sourced in part from dominantly gas-prone coals (type III kerogens). Some experiments indicate that gas generation kinetics for coals v. marine and lacustrine sources can differ (e.g. Burnham 1989; Pepper & Dodd 1995; Behar *et al.* 1997) and that coals yield methane at higher temperatures than other kerogen types (e.g. Pepper & Dodd 1995; Behar *et al.* 1997). Thus, the higher average temperatures in the studied non-associated conventional gases relative to associated gases may result from different kerogen sources.

Unconventional non-associated gases

Methane clumped-isotope temperatures from unconventional non-associated gases (sampled from five different basins) have an average value of 179°C (±23°C, 1σ; Fig. 5d) and range from 144°C to 207°C. All temperatures are at the top or above the nominal oil window (*c.* 160°C). Because gases in unconventional systems are thought to have formed *in situ* (Curtis 2002), the clumped-isotope temperatures, if interpreted as average gas-formation temperatures, indicate that these systems retain gases dominantly formed after oil generation either from oil, gas, or residual kerogen cracking.

The higher average temperature and tighter distribution (smaller standard deviation) of unconventional non-associated gases compared to oil-associated gases are consistent with the trapping histories of these different reservoir types. Specifically, unconventional systems retain the gases formed within the reservoir. Additionally, these systems are thought, in many cases, to expel oil and gas formed during oil generation at lower temperatures (e.g. Jarvie *et al.* 2007; Xia 2014), which would release methane formed at lower temperatures during oil generation (<160°C). After any such expulsion event(s), the system would then preferentially retain gases formed either from oil and gas cracking or the breakdown of residual kerogen at temperatures above 160°C. This does not indicate that all non-associated unconventional systems only retain higher-maturity gases, but that those that have been examined, which are economically productive, appear to preferentially sample gases with higher (>150°C) clumped-isotope temperatures. The clumped-isotope temperatures of non-associated gases from unconventional reservoirs could be used to place quantitative constraints on the timing and mass of oil and gas expelled from the unconventional reservoir. This requires incorporating methane clumped-isotope temperatures into current quantitative models of oil and gas generation.

We note that clumped-isotope temperatures of methane from non-associated unconventional reservoirs do not yield the elevated temperatures (>220°C) we observe in some conventional accumulations. It is possible that this is simply the result of the specific unconventional systems that have been studied to date, which, where known, do not exceed modelled maximum burial temperatures greater than 220°C (Stolper *et al.* 2014*a*). Future studies could focus on comparing conventional gas accumulations sourced from shales that are currently unconventional to the gases still retained in the shale.

An observation that we discuss below is that the samples with the most enriched $\delta^{13}C$ values for methane (i.e. greater than −32‰), which are generally taken to indicate generation at elevated maturities and thus higher temperatures, do not have the highest clumped-isotope-based temperatures. Instead, these samples yield clumped-isotope temperatures that are commonly *c.* 150°C. This is unexpected if clumped-isotope-based temperatures are interpreted as average formation temperatures. We return to this intriguing discrepancy below (section entitled 'Values of $\delta^{13}C$ and clumped-isotope temperatures of thermogenic gases') when plots of $\delta^{13}C$ v. clumped-isotope temperatures are discussed in detail.

Unconventional associated gases

Two unconventional reservoirs with oil-associated gases have been studied: the Eagle Ford Shale (Texas; Mullen 2010), and the Bakken Shale (North Dakota; Meissner 1978). The presence of oil in these systems constrains the maximum formation temperatures of the methane to temperatures below those at which oil is stable on geological timescales. Although this temperature is not agreed upon, it is probably no higher than *c.* 200°C under geological conditions (Quigley & Mackenzie 1988; Hunt 1996; Vandenbroucke *et al.* 1999; Lewan & Ruble 2002). In contrast, the clumped-isotope temperatures of gases from these reservoirs range from 140 to 380°C, with an average value of 215°C (±59, 1σ). Thus, many of the measured temperatures exceed what would be expected for oil stability. Furthermore, the temperatures are higher, on average, than unconventional non-associated gases, which are derived from systems modelled in many cases to have reached burial temperatures above 200°C. This strongly indicates that the clumped-isotope compositions of the unconventional oil-associated deposits studied do not reflect equilibrium conditions during gas generation. Instead they probably reflect kinetic isotope effects expressed during methane generation, storage in the reservoir (including leakage or phase changes), or extraction of hydrocarbons.

Kinetic isotope effects expressed during gas generation could have created the observed non-equilibrium clumped-isotope compositions. Evidence in favour of this possibility comes from pyrolysis experiments on coal discussed above which result in lower Δ_{18} values (and thus hotter apparent temperatures) during some experimental conditions. If this process occurs in natural environments it could explain the high apparent methane clumped-isotope temperatures from the Eagle Ford and Bakken formations. This scenario requires that the kinetics and reaction mechanisms for gas generation in these unconventional systems differ from other systems which do not yield such high apparent clumped-isotope temperatures, but instead temperatures generally consistent with generation in the oil window. This is because a significant proportion (35%) of all thermogenic gases yield clumped-isotope temperatures less than 160°C, i.e. below the typically assumed maximum range for the oil window (Fig. 6a).

Finally, we note that although the coal-pyrolysis experiments only yield non-equilibrium Δ_{18} values at the onset of ethane cracking, ethane is considered the second most stable alkane (with methane the most stable) with respect to thermally activated breakdown reactions (Behar *et al.* 1992). For geologically relevant thermal histories, ethane breaks down only after oil has cracked to hydrocarbon smaller gases (Behar *et al.* 1992). Thus, the relevance of the experiments to oil-associated, unconventional gases is not clear.

Alternatively, the non-equilibrium clumped-isotope compositions could result from kinetic isotope effects expressed during gas extraction. The extraction of gas and oil from conventional systems v. unconventional systems can differ. Specifically, unconventional systems are usually hydraulically fractured in order to release hydrocarbons whereas most conventional reservoirs are not. During oil and gas recovery from unconventional reservoirs, the pressure in the reservoir declines rapidly over the first year, lowering the total yield of oil and gas that can be extracted over time (e.g. Lee *et al.* 2011). Larger molecules such as liquid hydrocarbons require larger pressure gradients for recovery compared to gases. Thus, it is possible that oils that contain some dissolved methane are left behind in the reservoir during production of hydrocarbons or that a two-phase hydrocarbon system develops within the subsurface if the pressure in the reservoir drops below the bubble point. Nonetheless, kinetic isotope effects associated with the dissolution or degassing of methane into or out of this oil could result in the observed non-equilibrium clumped-isotopic compositions of the methane. The apparently meaningful gas-formation temperatures from unconventional non-associated systems in which methane is the dominant constituent and no oil is presented supports this idea. This hypothesis could be tested by taking samples from a well at different times in its production history as the pressure in the unconventional reservoir declines.

Additionally, kinetic isotope effects may be expressed during the migration of gases from the shale into the fractures created during hydraulic fracturing of the shales. For example both diffusion and adsorption/desorption of hydrocarbons are thought to occur during migration in shales and could result in the expression of kinetic isotope effects (Xia & Tang 2012). If such kinetic isotope effects are preferentially expressed when oil is present in a system (e.g. because a liquid hydrocarbon phase changes the isotope effects associated with methane's diffusivity or sorption behavior), they could influence the methane clumped-isotope compositions. Regardless of this, future work on oil-associated unconventional deposits should focus on understanding why these gases show different clumped-isotope systematics as compared to higher-maturity unconventional deposits and conventional reservoirs.

Thermogenic methane in clumped-isotopic equilibrium?

The bulk isotopic composition (i.e. δD and $\delta^{13}C$) of thermogenic methane is generally considered to be set by kinetically controlled processes (Sackett 1978; Chung *et al.* 1988; Clayton 1991; Tang *et al.* 2000; Ni *et al.* 2011). Only at the highest gas maturities (temperatures $>200°C$) have equilibrium exchange processes between methane and water been suggested to influence the δD values of thermogenic methane (Burruss & Laughrey 2010). However, year-long incubations of methane at $323°C$ in deuterium-labelled water do not result in the definitive occurrence of exchange of hydrogen isotopes between water and methane (Reeves *et al.* 2012). Nonetheless, it has generally been assumed that under most conditions, thermogenic methane's isotopic composition is controlled by kinetic isotope effects. Thus, the interpretation of some thermogenic methane clumped-isotope temperatures as gas-formation temperatures from 70 to 300°C requires the presence of unanticipated isotope-exchange processes during methane generation.

To explain this, we focus on mechanisms that could allow C—H bonds in methyl precursors to exchange hydrogen. It is important to note that achievement of C—H isotope equilibrium within methane or between methane and another phase does not require methane to be in carbon isotopic equilibrium with any other molecules during formation. It only requires reversible hydrogen isotope-exchange reactions to take place (perhaps only at a local molecular scale, such that evolved methane is still out of hydrogen isotope-exchange equilibrium with the bulk residue). Even so, it is possible for methane to be in internal isotopic equilibrium and still express kinetic carbon-isotope effects. We note that the framework of Helgeson *et al.* (2009), where hydrocarbons are hypothesized to be in metastable chemical equilibrium with each other and with CO_2, may imply that carbon bonds could be breaking and forming between organic molecules. In such a case, the organic molecules could achieve carbon isotopic equilibrium both within and between organic species. Carbon-isotope equilibrium between different molecules in thermogenic deposits could be tested by comparing methane clumped-isotope temperatures to known carbon-isotope equilibrium compositions between methane and CO_2 or other hydrocarbons like ethane and propane (Singh & Wolfsberg 1975; Horita 2001). This has not been done but could serve as a test of the ideas of Helgeson *et al.* (2009).

One mechanism that could generate methane in internal isotopic equilibrium is via the isotopic preequilibration of methyl groups before methane generation via free radical migration in the backbone of the hydrocarbon precursors. As free radicals move from carbon atom to carbon atom, hydrogen atoms are added to and removed from the organic molecule. This process is how deuterium and hydrogen from water are thought to be incorporated into organic matter (Hoering 1984; Lewan 1997). If these hydrogen-exchange reactions proceed before C—C bond cleavage, the methyl groups could be in or near internal isotopic equilibrium before methane formation. Methyl groups and methane have been modelled to have similar clumped-isotopic compositions at internal isotopic equilibrium (Wang *et al.* 2015). Thus methane clumped-isotope compositions could partly reflect ^{13}C—D and D—D clumping in a methyl group precursor at the temperature of cleavage.

This hypothesis requires that hydrogen addition to a released methyl group allows this internal equilibrium to be maintained. Alternatively, hydrogen isotope-exchange reactions available to the methyl free radical could promote internal isotopic equilibrium. This could occur through catalytically induced exchange reactions of methane or methyl groups during formation on other phases including clays, organic surfaces or transition metals. For example, clays have been experimentally demonstrated to catalyse hydrogen exchange in larger organic molecules (Alexander *et al.* 1982, 1984) while transition metals (Mango *et al.* 1994; Mango & Hightower 1997) have been implicated in the catalysis of methane generation. In laboratory experiments, transition metals like nickel and platinum can equilibrate the hydrogen isotopes of methane and H_2 (Horibe & Craig 1995) and allow methane to achieve internal isotopic equilibrium at temperatures above 150°C (Ono *et al.* 2014; Stolper *et al.* 2014*b*).

Regardless of mechanism, in order for thermogenic methane clumped-isotope compositions to reflect gas-formation temperatures, methane must form in internal isotopic equilibrium and retain that signature after formation. Thus, the processes that allow methane to form in clumped-isotopic equilibrium cannot promote the continued isotopic equilibration after formation. If correct, this indicates that hydrogen-isotope exchange reactions for methyl precursors or during methane generation exist, but cease to be available to methane after it is formed.

Biogenic gases

Biogenic gases in clumped-isotopic equilibrium

Stolper *et al.* (2014*a*) presented the first methane clumped-isotope measurements from biogenic sources. Biogenic methane from kilometre-deep reservoirs in the Gulf of Mexico undergoing biodegradation and methane generation yielded methane clumped-isotope temperatures from 40 to 48°C, which are within 1σ measurement precision of the known reservoir temperatures of 42 to 48°C. Additionally, methane dominantly biogenic in origin (although with some contribution of thermogenic methane) from the Antrim Shale yielded a clumped-isotope temperature of 40°C. Based on these results (Fig. 4), Stolper *et al.* (2014*a*) hypothesized that clumped-isotope temperatures of biogenic methane can reflect gas-formation temperatures in some settings.

The correspondence between biogenic gas formation and clumped-isotope-based temperatures has been subsequently corroborated in other samples with independently known methane formation temperatures (Fig. 4). For example, measurements of biogenic methane seeps from the Santa Barbara and Santa Monica basins yield clumped-isotope temperatures between 6 and 16°C, within 2σ in all cases of the seep temperatures (5–9°C). Two measurements of biogenic methane from submarine coal beds yielded clumped-isotope temperatures of 70°C (with precisions of ±5 and 12°C, 1σ) in environments with temperatures from 45 to 50°C (Inagaki *et al.* 2015). Samples emitted from underwater seeps on the Alaskan shelf in the Beaufort Sea have clumped-isotope temperatures from 0 to 5°C, all within 1σ error of local environmental

temperatures (−1.5 to 0°C; Fig. 3; Douglas *et al.* 2016). Finally, a methane sample from the Birchtree nickel mine (Manitoba, Canada), which is thought to be biogenic in origin, yielded both Δ_{13CH_3D}- and $\Delta_{12CH_2D_2}$-based temperatures of 16 ± 5 and 12 ± 2°C, similar to environmental temperatures of 20 to 23°C (Young *et al.* 2017).

Biogenic gases from systems without independently constrained gas-formation temperatures also yield clumped-isotope temperatures consistent with a biogenic origin. For example, biogenic methane from Northern Cascadia Margin methane clathrates yielded clumped-isotope-based temperatures between 12 and 42°C (Wang *et al.* 2015); coal-bed methane from the Powder River Basin has clumped-isotope temperatures from 35 to 52°C (Wang *et al.* 2015); and low, 9–30°C, clumped-isotope temperatures are observed in some gases collected from Alaskan lakes (Douglas *et al.* 2016).

Additional support for the hypothesis that biogenic gases can yield clumped-isotope temperatures that reflect generation temperatures comes from systems that are mixtures of biogenic and thermogenic gases or lack independent constraints on gas-formation temperatures. For example, in the Antrim Shale, gases are known to be a mixture of biogenic and thermogenic gases. In this system, gases independently interpreted to be dominantly biogenic (higher $C_1/C_{2–3}$) yielded lower clumped-isotope temperatures than gases enriched in ethane and propane (Fig. 7; Stolper *et al.* 2014*a*, 2015).

Taken together, along with measurements from other biogenic reservoirs (Douglas *et al.* 2016,

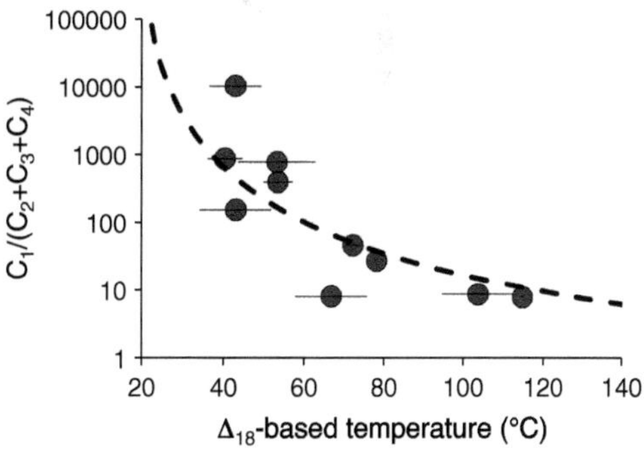

Fig. 7. Clumped-isotope measurements from Antrim Shale deposits that contain a mixture of biogenic and thermogenic gases. As would be expected, samples with higher $C_1/(C_{2–4})$ ratios, which are more similar to the biogenic end member yield lower clumped-isotope-based temperatures and vice versa. Data are fitted to a model (dotted line) that includes mixing between a biogenic and thermogenic source with biological consumption of C_2+ alkanes. Data and the model are described fully in Stolper *et al.* (2015). Error bars are 1σ.

2017), some biogenic gases – specifically those that sample subsurface biogenic systems – show a range from −1 to 95°C (Fig. 5a). This range is within that generally expected for biogenic gases (<80°C; Wilhelms *et al.* 2001; Valentine 2011) and well within maximum observed temperatures from laboratory pure cultures of up to 122°C (Fig. 5d; Takai *et al.* 2008). It further supports the hypothesis that biogenic methane can reflect gas-formation temperatures in many environmental settings.

Biogenic gases out of isotopic equilibrium

Although some biogenic gases yield clumped-isotope-based temperatures consistent with their known formation temperatures, other samples both from the environment and from laboratory experiments do not. Specifically, methane generated by hydrogenotrophic methanogens in the laboratory (Stolper *et al.* 2014a, b; Wang *et al.* 2015; Young *et al.* 2017) and methylotrophic methanogens (Douglas *et al.* 2016; Young *et al.* 2017) universally yield non-equilibrium clumped-isotope-based temperatures. The calculated temperatures are either too hot compared to the laboratory growth temperatures or have negative Δ_{18} or Δ_{13CH_3D} values (Fig. 8). Negative Δ_{18} or Δ_{13CH_3D} values are impossible for a system in internal isotopic equilibrium. In environmental systems, biogenic methane from ponds (Stolper *et al.* 2015), mid-latitude lakes, cow rumens, ophiolites that generate sufficient H_2 for methanogenesis to occur (Wang *et al.* 2015), Arctic lakes (Douglas *et al.* 2016) and various mines (Young *et al.* 2017) yield clumped-isotope temperatures that are unreasonably high or Δ_{18} and Δ_{13CH_3D} values that are negative.

What controls the clumped-isotopic compositions of biogenic methane?

Stolper *et al.* (2014a) proposed that the relative reversibility of the methanogenic enzymes (e.g. methyl co-enzyme M reductase) involved in biogenic methane generation could control the ultimate clumped-isotope composition of the gas. For example, high enzymatic reversibility would promote exchange of H and D in methane C—H bonds and allow biogenic methane to form in internal isotopic equilibrium. Stolper *et al.* (2015) and Wang *et al.* (2015) created quantitative models that could

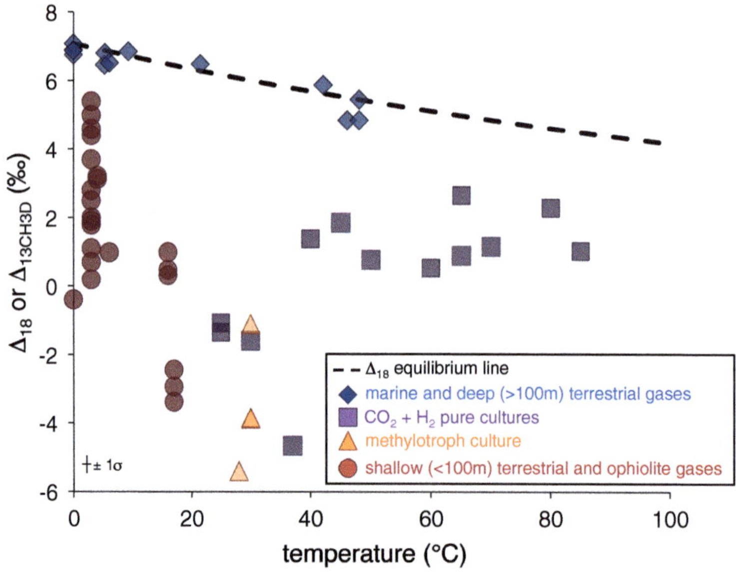

Fig. 8. Comparison of measured Δ_{18} or Δ_{13CH_3D} values of biogenic gases to their known formational or environmental sampling temperatures. Marine and deep (>100 m) terrestrial gases yield Δ_{18} values consistent with their formational temperatures. In contrast pure-culture hydrogenotrophic methanogens (that use H_2 and CO_2) and methylotrophic methanogens (that cleave methyl groups from larger organics), and biogenic methane from shallow terrestrial systems and ophiolites yield methane out of clumped-isotope equilibrium. Samples come from Stolper *et al.* (2014a, 2015), Wang *et al.* (2015), Inagaki *et al.* (2015), Douglas *et al.* (2016) and Young *et al.* 2017. Data from Inagaki *et al.* (2015) and the Birchtree mine datum of Young *et al.* (2017) were converted to the Δ_{18} reference frame using the Δ_{13CH_3D}-based temperature to compare them to the equilibrium line (which is for Δ_{18}) as these samples were interpreted to have formed in clumped-isotope equilibrium. The typical 1σ error bar for Δ measurements is given in the figure.

evaluate this hypothesis and found that the relative reversibility of enzymes could explain the biogenic clumped-isotope data. These models predict that non-equilibrium methane clumped-isotope values occur when enzymes are less reversible. The validity of these models is strengthened by their ability to relate differences in the hydrogen-isotopic composition of water and methane with the degree of disequilibrium for the clumped-isotope values.

The key insight provided by these models and environmental observations for the purposes of this review is that systems where growth rates are slow appear to show more reversibility in their enzymes (Valentine *et al.* 2004; Wing & Halevy 2014; Stolper *et al.* 2015; Wang *et al.* 2015) and thus create methane in clumped-isotopic equilibrium. Such systems (i.e. where growth rates are slow) would include marine sediments and deeply buried (e.g. >100 m deep) terrestrial systems – such systems tend to contain organic matter that is less reactive than organic matter found near the sediment–water interface. This less reactive organic carbon results in slower growth rates for the fermentative organisms that provide H_2, acetate and other methylated molecules to methanogens, lowering methanogen growth rates in the process. As many of these slow-growth systems occur at great depths (hundreds of metres to kilometres), where competent seals have the potential to develop, they could lead to the generation of biogenic gas accumulations in reservoir rocks above or within the locus of methanogenesis.

In contrast, systems in which methane generation rates (per cell) are high, show more irreversibility in their enzymes (Valentine *et al.* 2004; Wing & Halevy 2014; Stolper *et al.* 2015; Wang *et al.* 2015), and yield methane out of clumped-isotopic equilibrium. As discussed, this disequilibrium is manifested by lower Δ_{18} values (i.e. higher clumped-isotope-based temperatures) than expected for isotopic equilibrium. Such systems tend to be found in shallow terrestrial settings, including lakes, wetlands, and ponds as well as in cow rumens, where large amounts of fresh, labile organic carbon are available for conversion by fermentative microorganisms to H_2, acetate and other methylated organic molecules used by methanogens (Stolper *et al.* 2015; Wang *et al.* 2015; Douglas *et al.* 2016). Such shallow systems are unlikely to yield economic reservoirs of methane given the short transit distance to the atmosphere or sediment/water interface and the fact that sealing rocks are yet to develop the necessary competence to hold a significant column of free gas within a potential reservoir interval at such depths – this is discussed further below. Additionally, deeper terrestrial settings with large amounts of H_2 and CO_2 (or acetate), found for example in ophiolites, can also generate methane out of clumped-isotopic equilibrium (Wang *et al.* 2015).

Mixed biogenic and thermogenic deposits

Thermogenic gases generally yield clumped-isotope-based temperatures from 70 to 300°C. Biogenic gases from deeply buried marine or continental systems, on the other hand, yield clumped-isotope-based temperatures less than 100°C. Finally, shallowly sourced terrestrial systems yield variable temperatures, all of which are too hot for their given setting. Given these end-member compositions, a question is: can clumped-isotope-based temperatures be used to distinguish and quantify mixtures of biogenic gases and thermogenic gases?

Several economic hydrocarbon accumulations with contributions of both biogenic and thermogenic gases have been studied (Fig. 5b) and they yield temperatures that fall between the biogenic and thermogenic fields. Specifically, the clumped-isotope-based temperatures using data from economic deposits with mixtures of biogenic and thermogenic gases range from 40 to 118°C with an average of 73°C (± 25°C, 1σ). Importantly, in these examples, the biogenic gases appear to have formed at low metabolic rates such that the clumped-isotope-based temperatures and formation temperatures are similar for the biogenic end member. This probably indicates that in most economic hydrocarbon accumulations, the biogenic methane was formed close to isotopic equilibrium. For this comparison, we only included samples where the δD and δ^{13}C values of the end members are sufficiently close such that mixing of gases results in a pseudo-linear dependence of clumped-isotope temperature on mixing ratio (Stolper *et al.* 2014a, b, 2015; Wang *et al.* 2015; Douglas *et al.* 2016). This generally requires that the end members do not differ by more than a few tens of per mille in their isotopic composition (although this generally has to be established on a case-by-case basis). Non-linear mixing of Δ_{18} values occurs in the terrestrial Arctic as discussed below.

In these 'mixed' systems, clumped-isotope-based temperatures often correlate with other measured parameters that can differ between thermogenic and biogenic gases. For example, in the Antrim Shale, which contains mixtures of biogenic and thermogenic gases (Martini *et al.* 1996, 1998, 2003; Stolper *et al.* 2015), an inverse relationship between the clumped-isotope-based temperature and the C_1/C_{2-3} ratio is observed (Fig. 7). As discussed above, biogenic gases generally have higher C_1/C_{2-3} ratios than thermogenic gases. In the Antrim Shale, the clumped-isotope temperatures were used to calculate the relative amounts of biogenic and thermogenic gases in samples taken from different wells (Stolper *et al.* 2015).

A distinctive scenario occurs when thermogenic and biogenic gases with significantly different δD and δ^{13}C values mix. This scenario can occur in terrestrial settings in which both the δD and δ^{13}C of the biogenic gases can be substantially lower (up to *c.* 200‰ for δD and *c.* 50‰ for δ^{13}C) compared to thermogenic gases (e.g. Whiticar *et al.* 1986). A mixture of two gases where one has significantly lower δD (hundreds of per mille) and δ^{13}C (many tens of per mille) values than the other results in obvious non-linear mixing of Δ_{18} values (Stolper *et al.* 2014*a*, *b*, 2015; Wang *et al.* 2015; Douglas *et al.* 2016). For such a mixture (i.e. where one end member has both lower δD and δ^{13}C values than the other) the Δ_{18} values of the mixture will be elevated (i.e. yield colder clumped-isotope temperatures) compared to a linear mixture of Δ_{18} values. As a result, such mixtures can create non-physical, sub-freezing clumped-isotope-based temperatures. Examples of this sort of mixing scenario (including sub-freezing clumped-isotope temperatures) are given in Douglas *et al.* (2016) for terrestrial Alaskan samples. In the case of Douglas *et al.* (2016), these mixing relationships were corroborated using

radiocarbon measurements where the thermogenic gases were radiocarbon-dead while the biogenic end member had measurable ^{14}C. These sorts of mixing scenarios, especially when sub-freezing clumped-isotope-based temperatures are measured, identify the likely presence of thermogenic gases mixing into a shallow biogenic system.

Clumped-isotope temperatures in the context of Whiticar and Bernard plots

We now place samples with measured clumped-isotope compositions into the interpretive frameworks of the Whiticar and Bernard plots introduced earlier (Figs 9 & 10). To do this, we categorize samples as follows: biogenic gases are characterized as either 'non-equilibrium' or 'equilibrium' biogenic gases. 'Non-equilibrium' indicates that the clumped-isotope temperatures do not yield meaningful formation temperatures while 'equilibrium' indicates

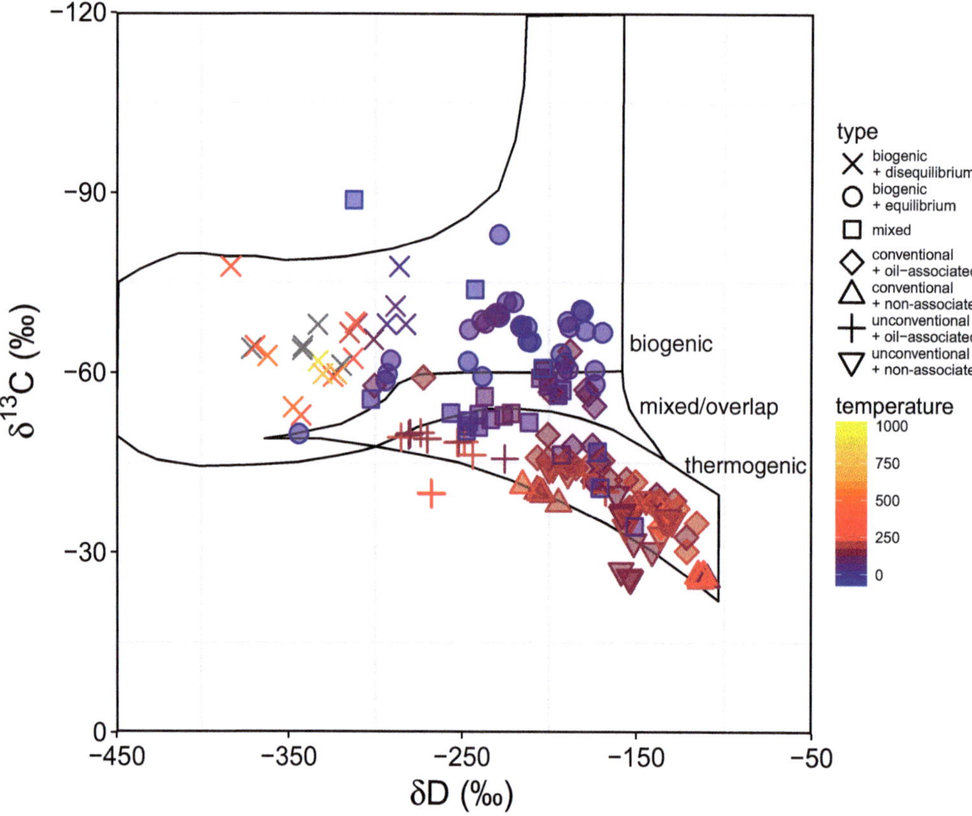

Fig. 9. Comparison of measured clumped-isotope temperatures from samples of known origin to their position in the Whiticar plot (Whiticar *et al.* 1986; Whiticar 1999). Measured temperatures are indicated by the colour of the data point. Grey colours indicate either negative Δ_{18} or Δ_{13CH_3D} values or sub-freezing temperatures. Data from Stolper *et al.* (2014*a*, 2015), Wang *et al.* (2015), Inagaki *et al.* (2015), Douglas *et al.* (2016, 2017) and Young *et al.* (2017).

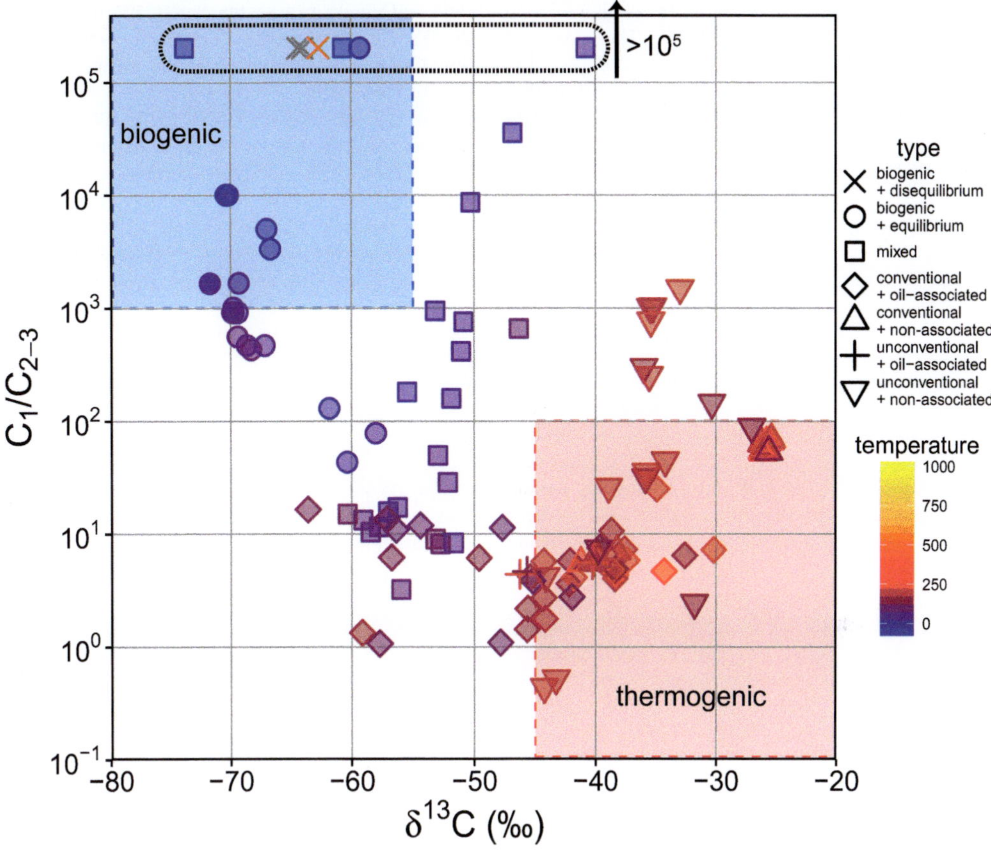

Fig. 10. Comparison of measured clumped-isotope temperatures from samples of known origin to their position in the Bernard plot (Bernard *et al.* 1976). Measured temperatures are indicated by the colour of the data point. Grey colours indicate either negative Δ_{18} or Δ_{13CH_3D} values or sub-freezing temperatures. Circled samples have C_1/C_{2-3} values greater than 10^5. Data from Stolper *et al.* (2014a, 2015) and Douglas *et al.* (2016, 2017).

the clumped-isotope temperatures reflect meaningful formation temperatures. Thermogenic gases are separated as either conventional or unconventional and oil-associated or non-associated. Finally, we include gases that are a mixture of thermogenic and biogenic sources. Symbol fill colours represent the clumped-isotope based temperature. Blue colours indicate a lower clumped-isotope temperature while red colours indicate higher clumped-isotope temperatures. Gases with negative Δ_{18} or Δ_{13CH_3D} values or sub-freezing temperatures are coloured grey.

Clumped-isotope temperatures and the Whiticar plot

Figure 9 displays samples with measured clumped-isotope compositions in the Whiticar plot. Gases of known origins (thermogenic, biogenic and mixed origins – given by the symbols) fall within or close to the respective fields outlined by this space. Setting aside samples with non-equilibrium clumped-isotope values, one visible pattern is that there is a general clustering of samples with colder clumped-isotope-based temperatures (blue fill colours) with lower δD and $\delta^{13}C$ values in the biogenic field to moderate clumped-isotope-based temperatures (purple fill colours) in the mixed field and then the warmest clumped-isotope-based temperatures (red fill colours) in the thermogenic field with the highest δD and $\delta^{13}C$ values. Consequently, the clumped-isotope temperatures fit with the general expectations of gas origins predicted by the Whiticar plot.

There are, however, a few noticeable exceptions. For example conventional oil-associated gases can be found within the biogenic field with low δD values (*c.* 300‰) and/or low $\delta^{13}C$ values (<−60‰).

Significantly, these designations are not based on the clumped-isotope temperatures. The clumped-isotope temperatures (all >100°C) support these thermogenic designations. This is an example that shows thermogenic gases can exist outside the thermogenic regions delineated by the Whiticar plot and that clumped-isotope temperatures can provide the necessary fidelity to interpret these gases as thermogenic.

Figure 9 shows a clear separation of equilibrium v. non-equilibrium biogenic gases based on δD values, but not based on $\delta^{13}C$ values – note that these figures only show measurements of environmentally derived samples and our discussion here is confined to such samples. Specifically, most biogenic gases with δD values less than −300‰ have non-equilibrium clumped-isotope temperatures. In contrast, biogenic gases with δD values greater than −280‰ yield meaningful clumped-isotope formation temperatures.

This delineation probably occurs because the same biochemical reactions control both the δD and the clumped-isotope values of biogenic methane. Specifically, when methanogens grow slowly, they appear to form methane in both clumped-isotopic equilibrium and hydrogen-isotopic equilibrium with water (Stolper *et al.* 2015; Wang *et al.* 2015). Consequently, both the δD value of the environmental waters as well as the methane–water hydrogen-isotope equilibrium fractionation factor will influence the δD value of the resultant biogenic methane. Specifically, at room temperature (*c.* 25°C), methane δD values are *c.* 180‰ lower than in water at isotopic equilibrium (Stolper *et al.* 2015). Most waters, except those at latitudes higher than 60°, have δD values greater than −100‰ (Bowen & Revenaugh 2003). Consequently, δD values of methane formed in isotopic equilibrium (including clumped) are typically restricted in the environment to values greater than *c.*−300‰ (Fig. 9). In contrast, the formation of biogenic methane out of clumped-isotopic equilibrium is caused by kinetic isotope effects that result in methane δD values that are lower than values for isotopic equilibrium between methane and water (Stolper *et al.* 2015; Wang *et al.* 2015; Douglas *et al.* 2016). As a result, non-equilibrium clumped-isotope compositions commonly have δD values less −300‰. Thus, the biogenic field of the Whiticar plot when combined with clumped-isotope temperatures reveals the thermodynamic conditions (equilibrium v. non-equilibrium) of microbial methanogenesis. We note that the only exception to the *c.* −300‰ methane δD delineation for samples in and out of clumped-isotopic equilibrium is a sample from the Birchtree mine (Young *et al.* 2017), which appears to have formed in clumped-isotopic equilibrium, but has a δD of −343‰. However, this sample is from a mine in Canada with highly depleted fluid δD values of −122‰; Bottomley *et al.* 1994). Such fluids probably result in the exceptionally low methane δD values for the reasons given above.

Finally, samples that are a mixture of biogenic and thermogenic gas span a large range of δD and $\delta^{13}C$ values. The clumped-isotope temperatures help differentiate these samples. For example, gases with mixed compositions and δD and $\delta^{13}C$ values that fall in the thermogenic field yield colder temperatures (bluer colours) than the pure thermogenic gases with similar δD and $\delta^{13}C$ values (Fig. 9). Thus, clumped-isotope temperatures appear to be a useful tool for identifying gases with mixed origins in combination with the Whiticar plot.

Clumped isotopes and the Bernard plot

Figure 10 places clumped-isotope temperatures into the framework of the Bernard plot using the same symbols and colour scheme as in Figure 9. Fewer data are available for this comparison because not all gases with measured clumped-isotope temperatures have measured C_1/C_{2-3} ratios. The overall location of samples with known origin (thermogenic, biogenic or mixed) is generally consistent with those predicted by the Bernard plot.

As discussed previously, it has been suggested that thermogenic and biogenic gases can sit outside of the boundaries given by the Bernard plot (e.g. Martini *et al.* 1996; Vinson *et al.* 2017) – and this is supported by data with measured clumped-isotope values. For example, C_1/C_{2-3} values of non-associated unconventional gases can be greater than 100. Additionally, some conventional thermogenic gases lie outside of the thermogenic field. Specifically, these gases contain elevated C_2^+ alkane contents ($C_1/C_{2-3} < c.$ 10) but also have $\delta^{13}C$ values (−50 to −64‰), which are lower than 'typical' thermogenic gases. These gases have clumped-isotope temperatures that are consistent with their thermogenic origin (100–135°C; purple-red colours in Fig. 10). This insight adds uncertainty to the common assumption that $\delta^{13}C$ values less than −60‰ positively identify biogenic methane (Bernard *et al.* 1976; Whiticar *et al.* 1986; Whiticar 1999), but supports models that predict methane formed at temperatures below *c.* 180°C can have low (<−60‰) $\delta^{13}C$ values (Tang *et al.* 2000).

Finally biogenic gases generally have C_1/C_{2-3} values >100, as predicted by the Bernard plot. However, there are biogenic samples with C_1/C_{2-3} values as low as 43. These low C_1/C_{2-3} values may result from the presence of small amounts of thermogenic gas in these specific samples. However, the biogenic nature of these samples is supported by low (<50°C) clumped-isotope temperatures. Additionally, biogenic gases from sediments with far lower C_1/C_{2-3} values than 43 (as low as 2) have been described in which the ethane and propane are also thought to be biogenic in origin (Hinrichs *et al.* 2006). Thus the ethane in the examined samples is potentially biogenic. Consequently, the clumped-isotope temperatures may aid in interpreting whether a gas is an early formed thermogenic gas (clumped temperature >60°C), or a biogenic gas with a lower temperature regardless of the C_1/C_{2-3} ratio.

Values of $\delta^{13}C$ and clumped-isotope temperatures of thermogenic gases

Here, we compare thermogenic methane $\delta^{13}C$ values to clumped-isotope temperatures. We focus on methane $\delta^{13}C$ values over δD values as the relationship between methane $\delta^{13}C$ values to gas generation kinetics is better understood than for δD values, both from an observational and theoretical standpoint. The $\delta^{13}C$ value of thermogenic methane is thought to be controlled by both the $\delta^{13}C$ value of the source organic carbon and the thermal maturity at which the methane was generated. The relationship between $\delta^{13}C$ and source maturity is generally explained by two mechanisms. First, it is assumed that elevated source-rock thermal maturities are achieved via exposure to higher burial temperatures or longer time spent at a given temperature (e.g. Burnham & Sweeney 1989; Sweeney & Burnham 1990). The kinetic isotope effects that describe the relative differences in the rate of methyl cleavage for ^{12}C or ^{13}C methyl groups have been modelled to decrease in magnitude with increasing temperature (Cramer *et al.* 1998; Tang *et al.* 2000; Xiao 2001). As methane is lower in $\delta^{13}C$ than the source organic carbon (exhibiting a 'normal' isotope effect), higher temperatures result in methane with higher $\delta^{13}C$ values relative to lower temperatures for a given methane source.

Second, the generation of methane, which has lower $\delta^{13}C$ values than the source organic matter, due to mass balance considerations, will cause the residual organic carbon to increase in its $\delta^{13}C$ value. This, in turn, causes newly generated methane to increase in $\delta^{13}C$ values with increased hydrocarbon generation. Models describe this process using Rayleigh distillation frameworks of varying complexity (Clayton 1991; Berner *et al.* 1992, 1995; Cramer *et al.* 1998; Tang *et al.* 2000).

Ultimately, given the proposed dependence of methane's $\delta^{13}C$ value on its source's thermal maturity, a positive relationship between methane $\delta^{13}C$ values v. the clumped-isotope temperatures could be expected. We examine this idea using specific examples from Brazilian reservoirs and the full suite of data.

Specific examples from Brazilian reservoirs

A correlation between $\delta^{13}C$ values and clumped-isotope temperatures of methane was first observed in Potiguar Basin samples (Brazil; conventional, oil-associated gases; Fig. 11a; Stolper *et al.* 2014*a*). This correlation was explained by the expected relationship between thermal maturity of the source rock and the $\delta^{13}C$ value of the methane discussed above (Stolper *et al.* 2014*a*).

We have observed correlations between methane $\delta^{13}C$ values and clumped-isotope temperatures in two other conventional oil-associated systems in Brazil (referred to as the Northeastern Onshore Basin and Southeastern Offshore Basin; Douglas *et al.* 2017). The dependence of $\delta^{13}C$ on formation temperature for these three systems ranges from 5.3 to 7.3°C/‰, but all are within 1σ error of each other. Theoretical estimates for the change in $\delta^{13}C$ of methane v. temperature solely due to temperature-dependent isotope effects for ^{12}C v. ^{13}C bond cleavage yield slopes from 8.8 to 9.4°C Ma^{-1} (Tang *et al.* 2000; Ni *et al.* 2011; Stolper *et al.* 2014*a*). The lower slope measured for environmental samples compared to theory may result from the importance of distillation effects discussed above. Specifically, distillation increases the $\delta^{13}C$ of methane as a function of thermal maturity in excess of that simply expected based on temperature-dependent kinetic isotope effects (which decreases the observed slope).

An interesting aspect of the clumped-isotope temperature v. $\delta^{13}C$ relationships is that, although the slopes are all similar, the relationships are offset from each other. For example, the Potiguar Basin samples (Fig. 11a) v. the Northeastern Onshore Basin samples (Fig. 11b), at a given clumped-isotope temperature, differ by *c.* 6‰ in $\delta^{13}C$, with the Potiguar Basin sample being lower. This difference probably reflects a difference in the $\delta^{13}C$ of the source organic carbon, which can vary among typically encountered source-rock compositions by *c.* 10‰ (e.g. Schoell 1984; Chung *et al.* 1992). As the clumped-isotope-based temperatures are not a function of the $\delta^{13}C$ or δD of the source organic carbon, methane clumped isotopes can provide constraints on gas-formation temperatures (and source-rock thermal maturity) that do not require knowledge of the $\delta^{13}C$ value of the source organic carbon.

General $\delta^{13}C$ v. clumped-isotope trends

Thermogenic methane $\delta^{13}C$ values v. clumped-isotope-based temperatures are compared in Figure 12 for all samples measured (as opposed to the Brazilian samples shown in Fig. 11). One general trend is that the $\delta^{13}C$ and clumped-isotope temperatures of conventional oil-associated gases positively

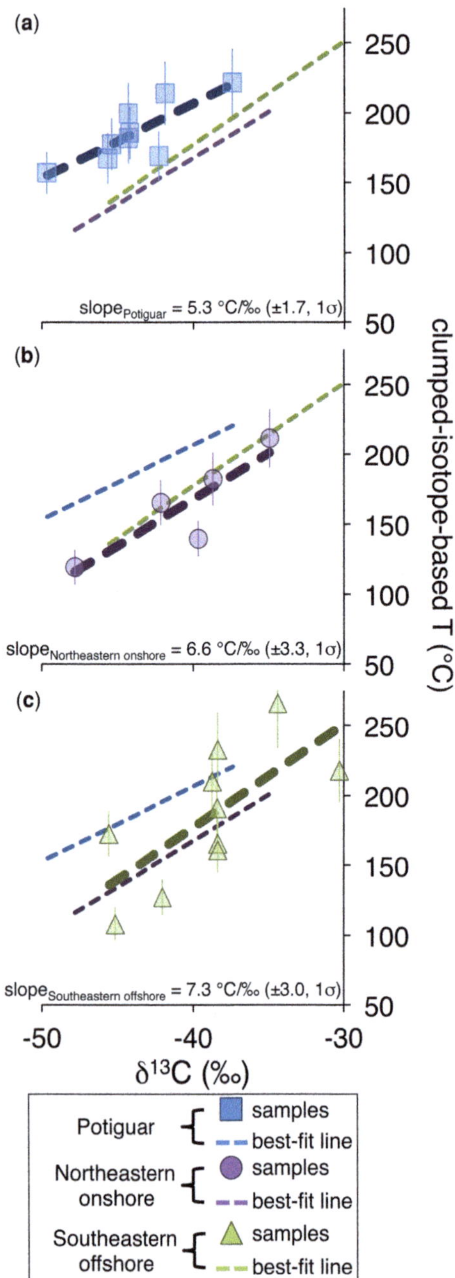

Fig. 11. $\delta^{13}C$ values v. clumped-isotope-based temperatures from various Brazilian basins including (**a**) the Potiguar Basin, (**b**) the Northeastern Onshore Basin and (**c**) the Southeastern Offshore Basin. Dashed lines are best-fit lines. For clarity, regression lines are shown for other datasets when comparisons between data are made. In all cases a positive relationship between $\delta^{13}C$ and the clumped-isotope-based temperature is observed. Data from Stolper *et al.* (2014*a*) and Douglas *et al.* (2017). Error bars are 1σ.

co-vary. There is significant scatter in this relationship that is not apparent when looking at specific gas accumulations (e.g. Fig. 11). This increased scatter is probably due to differences in the $\delta^{13}C$ of the original source organic carbon of the different oil-associated reservoirs. For example, kerogen and oil $\delta^{13}C$ values commonly vary from −32 to −22‰ (e.g. Schoell 1984; Chung *et al.* 1992).

The conventional non-associated gases have generally higher $\delta^{13}C$ values and clumped-isotope temperatures than conventional oil-associated gases. For $\delta^{13}C$ values greater than −30‰, there is significant scatter of clumped-isotope temperatures. This comparison, though, is complicated by the fact that all of the conventional non-associated gases are derived fully or partially from coal sources. Coal-derived methane is known to have a different (and weaker) $\delta^{13}C$ v. maturity relationship compared to other source-rock types (e.g. Type I and II kerogens; Schoell 1980).

The unconventional non-associated samples generally have elevated $\delta^{13}C$ values and clumped-isotope temperatures compared to the conventional gases. Interestingly, these samples exhibit a different trend between $\delta^{13}C$ and clumped-isotope-based temperature as compared to the oil-associated gases. For $\delta^{13}C$ values less than −32‰, the conventional oil-associated gases and unconventional non-associated gases occupy a similar space (Fig. 12). However, all unconventional non-associated gases with $\delta^{13}C$ values above −32‰ yield clumped-isotope-based temperatures below 165°C. This is the opposite of what would be expected based on the increasing trend defined by the conventional oil-associated gases. Furthermore, such elevated $\delta^{13}C$ values are generally thought to be indicative of gas generation at temperatures above 200°C (Tang *et al.* 2000). We suggest that this apparent decline in clumped-isotope-based temperature as $\delta^{13}C$ values increase above *c.* −32‰ for unconventional non-associated gases is a result of mixing of gases produced at different maturities within the unconventional reservoir.

As discussed above, mixing of gases lower in δD and $\delta^{13}C$ values with gases higher in δD and $\delta^{13}C$ but identical Δ_{18} values results in a mixture with a higher Δ_{18} value (and thus lower clumped-isotope temperature) than either of the end members. Take, for example, the following mixing scenario between two end-member gases. Let the first end-member gas be a typical unconventional non-associated gas with a $\delta^{13}C$ of −38‰, δD of −160‰ and clumped-isotope temperature of 200°C (Fig. 12.). Let the second end member be a high-maturity gas generated at 250°C with a $\delta^{13}C$ value of −10‰ and δD value of −60‰ (the approximate maximum values of thermogenic gases predicted by the models of Tang *et al.* (2000) and Ni *et al.* (2011)). Mixing of this high-maturity gas into a larger reservoir of the

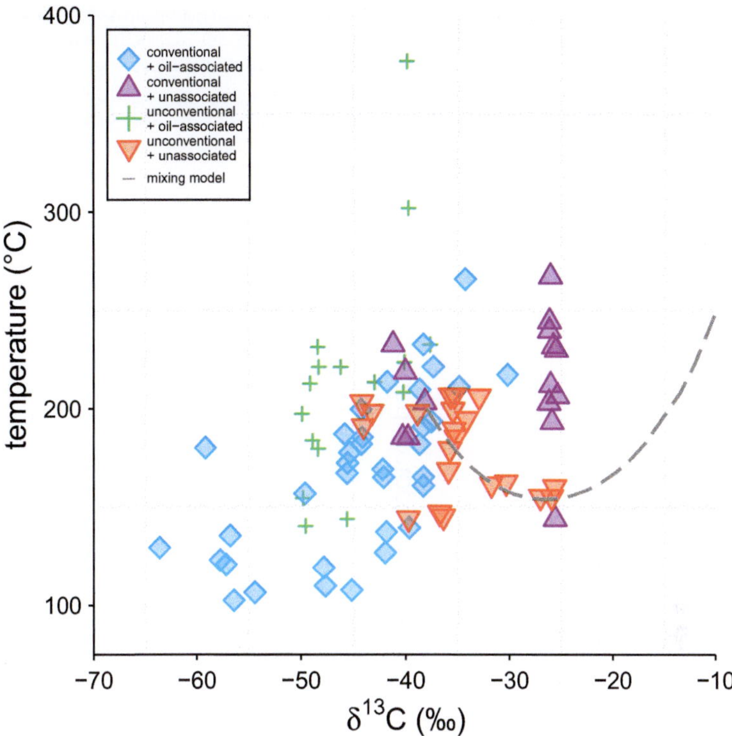

Fig. 12. Comparison of $\delta^{13}C$ values v. clumped-isotope temperatures from thermogenic gases. Data from Stolper *et al.* (2014*a*, *b*, 2015), Wang *et al.* (2015), Young *et al.* (2017) and Douglas *et al.* (2017). The modelled line shows the clumped-isotope-based temperature of a mixture of two end members. The first end-member gas has a $\delta^{13}C$ of −38‰, δD and −160‰ and clumped-isotope temperature of 200°C, similar to the composition of numerous measured unconventional, non-associated gases. The second end member, a hypothetical high-maturity thermogenic gas, has a $\delta^{13}C$ value of −10‰ and δD value of −60‰ (the approximate maximum values of thermogenic gases predicted by the models of Tang *et al.* (2000) and Ni *et al.* (2011)), and clumped-isotope formation temperature of 250°C. See text for details and discussion of the mixing line.

typical nonconventional gas, due to the non-linear dependence of Δ_{18} on the $\delta^{13}C$ and δD values of mixtures, causes the clumped-isotope temperatures to initially decline. The trajectory of this mixing relationship is given in Figure 12. Only when the mixture is made up of 50% of this high-maturity gas do temperatures start increasing again.

Because modern models predict that methane formed at the end of hydrocarbon generation (i.e. the final *c.* 5% of total methane generated) can be significantly elevated in δD (>−60‰) and $\delta^{13}C$ (>−10‰; Tang *et al.* 2000; Ni *et al.* 2011), we consider this conceptual model plausible to first order in explaining the lower clumped-isotope temperatures of the unconventional non-associated gases that have the highest $\delta^{13}C$ values. Thus, production (and mixing) of a small amount of high maturity, high $\delta^{13}C$ and δD gases into an unconventional gas reservoirs could conceivably lower the clumped-isotope based temperatures and explain the

relationships observed between the clumped-isotope temperatures and methane $\delta^{13}C$ values for non-associated, unconventional gases at the highest thermal maturities.

A synthesis of clumped-isotope studies and their utility in the study of economic accumulations of hydrocarbons

Here we examine a few simple scenarios that describe how methane clumped-isotope temperatures could be used to determine the origin of methane in naturally occurring hydrocarbon accumulations. Proper identification of the origin of methane accumulations could be used to constrain a model of the hydrocarbon system of interest by providing insight into gas generation and migration timing. Byrne *et al.* (2017) provide a more general review of how geochemical techniques are used to study hydrocarbon systems.

Scenario 1: shallow microbial source. Many microbial sources of methane are shallow in origin, generating methane within the first few hundreds of metres below the sediment–water interface. Because both a trap and seal are required to trap gas and because seals (other than methane hydrates) do not tend to develop sufficient capacity to hold hydrocarbons until sediments reach depths of *c.* 500 m (Rice & Claypool 1981), the generation of shallow microbial gas is unlikely to result in an economic gas accumulation. This situation is depicted as 'Scenario 1' in Figure 13 and Table 1. A gas of shallow microbial origin can be identified using its clumped-isotope composition in two ways. First, if the gas is generated slowly and is in clumped-isotope equilibrium, it will yield a low methane clumped-isotope temperature indicative of the shallow temperatures in the sediments (e.g. <25°C). Additionally, such a gas would be expected to be low in $\delta^{13}C$ (e.g. <50‰), elevated in δD (> *c.* −300‰), and have a high C_1/C_{2-3} ratio (e.g. >100). Second, the gas could be

generated out of clumped-isotope equilibrium. In such a case the clumped-isotope composition could be negative. However, non-equilibrium biogenic methane has been measured with apparent temperatures within the thermogenic gas range (e.g. *c.* 100–200°C; Wang *et al.* 2015; Douglas *et al.* 2016) and the origin of the gas could be incorrectly identified as thermogenic. Importantly, the $\delta^{13}C$ of these gases is likely to be less than −50‰, the δD less than −300‰, and the C_1/C_{2-3} ratio greater than 100. A thermogenic gas with these characteristics is unlikely (Bernard *et al.* 1976; Whiticar 1999).

Scenario 2: deep microbial source. Alternatively, microbial gases could be generated at sufficient depths (>500 m) to be trapped in the subsurface – such a situation is given by Scenario 2 in Figure 13 and Table 1. For this scenario, the clumped-isotope temperatures are expected to be between *c.* 30 to 80°C, with the upper limit derived from observations from nature (Wilhelms *et al.* 2001; Valentine 2011). These temperatures would correspond to maximum biogenic sources between 1500 and 3000 m (Valentine 2011). Such a gas may be distinguished from

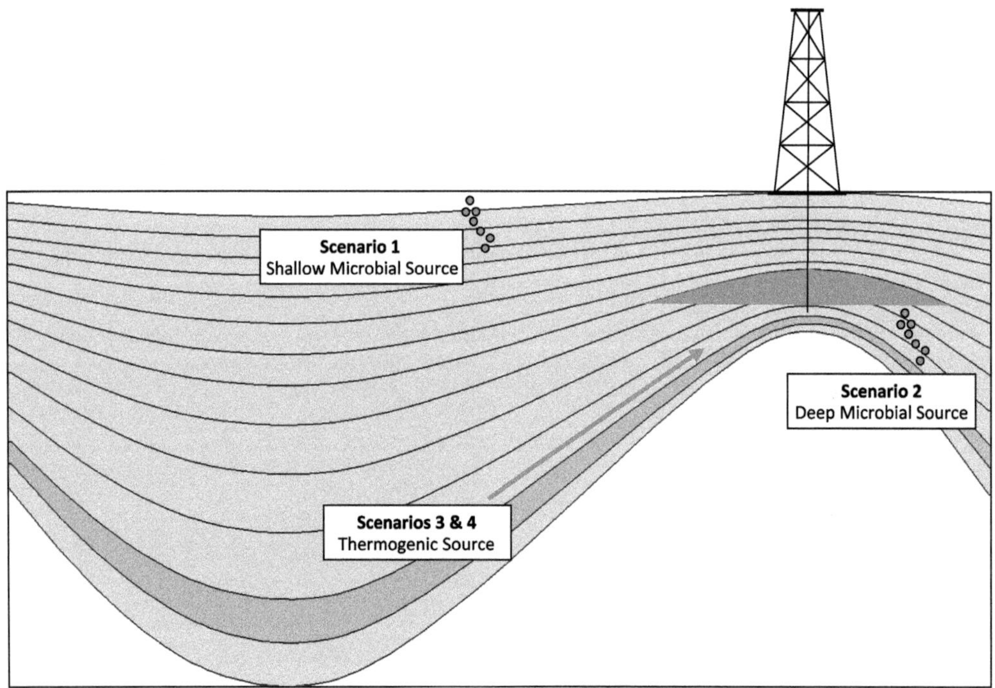

Fig. 13. Conceptual figure of how methane clumped-isotope temperatures in combination with δD and $\delta^{13}C$ values of methane and C_1/C_{2-3} values can be used to identify the source of natural gas in a sedimentary system as discussed in the section 'A synthesis of clumped-isotope studies and their utility in the study of economic accumulations of hydrocarbons'. Scenarios are given in both that section and Table 1. The arrow from Scenarios 3 and 4 represents migration of hydrocarbons to a reservoir rock (dark shaded area penetrated by the well).

Table 1. *Origin, isotopic and compositional characteristics of gases*

Scenario	Source	In clumped-isotope equilibrium?	$\delta^{13}C$ (‰)	δD (‰)	C_1/C_{2-3} ratio	Clumped-isotope temperature (°C)
1	Shallow biological source	Yes	<−50	>−300	>100	<25
		No	<−50	<−300	>100	>100 and/or negative Δ_{18}
2	Deep biological source	Yes	<−50	>−300	>100	c. 30–80
3	Thermogenic (oil-associated gas)	Yes (formed in oil window)	>−60	–	<10	60–160
		Yes (formed above oil window)	>−60	–	<10	>160
4	Thermogenic (non-associated gas)	Yes (formed in oil window)	>−60	–	>10	60–160
		Yes (formed above oil window)	>−60	–	>10	>160

These gases are discussed in the section 'A synthesis of clumped-isotope studies and their utility in the study of economic accumulations of hydrocarbons' and presented in Figure 13.

low maturity thermogenic gases generated at the onset of oil generation by having high (>100) C_1/C_{2-3} ratios.

Scenarios 3 and 4: thermogenic gas. The third and fourth scenarios describe gas with a thermogenic origin (Fig. 13 and Table 1). Scenario 3 is for an oil-associated gas, and scenario 4 is for a non-associated gas. Methane clumped-isotope temperatures for either scenario would be expected to be greater than *c.* 60°C, and probably above 100°C. If the gas were oil-associated, it would probably be distinguished by having low C_1/C_{2-3} ratios (<10). This gas may also be associated with liquid hydrocarbon components (i.e. oil would be recovered with the gas). A non-associated gas would be expected to have elevated C_1/C_{2-3} ratios (e.g. >10). In both scenarios (3 and 4) the methane would probably have $\delta^{13}C$ values >−60‰. In either scenario, a key question is whether the gas formed with or without oil and whether after migration it became associated or non-associated. For example, an originally non-associated gas could become associated it if migrates to a reservoir that contains oil generated from a different source rock or at a different time than the gas. Alternatively, an originally oil-associated gas could become non-associated during migration during a phase separation. If the methane were formed with oil, the clumped-isotope temperatures should range from 60 to 160°C. If the methane formed at higher temperatures than oil generation, it would have higher temperatures (>160°C). This difference (formation with or without oil) dictates the generation and accumulation history of hydrocarbons in the basin. Such information could be used to constrain models of the basin's thermal history and the timing of hydrocarbon generation, migration and

accumulation, and provides information on the original formation conditions of the gas.

In the scenarios given above, the sources were considered pure (i.e. the methane had one source). In the real world, traps can commonly contain mixtures of gases, both biogenic and thermogenic. Clumped isotopes are well suited to test for such mixtures and even enable the calculation of the contribution of the end members to the mixture. We explore one such example here. As discussed above, mixing of gases with the same clumped-isotope temperature but different δD and $\delta^{13}C$ values such that one end member is lower in both, leads to a lowering of the clumped-isotope temperature as compared to linear mixing of Δ_{18} values. The most extreme examples of such mixtures occur in systems in which a thermogenic gas with elevated δD and $\delta^{13}C$ values mixes with a biogenic gas formed in or out of clumped-isotopic equilibrium, but with a low $\delta^{13}C$ (<−60‰) and low δD value (<−300‰). In such cases, in the Arctic for example, clumped-isotope temperatures become exceptionally negative (as low as −60°C; Douglas *et al.* 2016). Such negative temperatures immediately imply that thermogenic gases exist in the system and are mixing with biogenic gases generated at shallow depths. Measurements of multiple gas samples in the area with different mixing ratios of these gases can be used to calculate the bulk and clumped-isotopic composition of the thermogenic end member and thus infer the temperature (and if a geotherm is known, the depth) at which the thermogenic component formed. An example of such a deconvolution is given in Douglas *et al.* (2016). Nonetheless, negative clumped-isotope temperatures, below the freezing point of water, are an immediate indication of the mixture of thermogenic and biogenic gases.

Questions moving forward

It is apparent that the clumped-isotope measurements of methane can provide new constraints on the conditions of methane generation. For biogenic gases formed slowly, thermogenic conventional oil-associated and non-associated gases, and unconventional non-associated gases, clumped-isotope temperatures reflect the gas-formation temperatures. This allows clumped-isotope temperatures to be used to reconstruct the formational environments of the gases. Microbial gases generated rapidly are formed out of clumped-isotope equilibrium. This non-equilibrium signal can be used as a tracer of the rate of methanogenesis in the environment. We have outlined how these measurements can be placed in more typical gas interpretation frameworks that use the isotopic composition ($\delta^{13}C$ and δD) of methane and the molecular composition of the gas (C_1/C_{2-3} ratios) to constrain a sample's origins. The clumped-isotope compositions of methane often provide complementary and distinct information not contained in such frameworks. Finally, we have outlined a conceptual framework for how to interpret the meaning of clumped-isotope compositions in the context of gas exploration, highlighting how the clumped-isotope measurements provide unique constraints on the potential occurrence of hydrocarbon accumulations.

These examples indicate that methane clumped-isotope measurements are a promising tool for studying the generation and accumulation history of hydrocarbon deposits in nature. But many questions and future work exist and we outline a few here.

(1) Why do oil-associated gases from unconventional reservoirs sometimes yield distinctly disequilibrium clumped-isotope compositions? Above we proposed that this may be the result of kinetic isotope effects associated with early stages of oil or bitumen cracking, or the means by which unconventional reservoirs are produced (via hydraulic fracking). This latter hypothesis could be tested by sampling gas from an oil-associated conventional reservoir over the course of its production history and examining if there are changes in both the bulk and clumped-isotopic composition. Nonetheless, more of these systems must be studied to understand why these systems are distinct from other thermogenic hydrocarbon accumulations.

(2) Sophisticated models grounded in chemical kinetics exist to describe the formation of thermogenic oils and gases as a function of source organic type and thermal history. These models are routinely used to predict, based on an understanding of the geology of a system, where and when gas accumulations could have formed (Tissot & Welte 1978; Quigley & Mackenzie 1988; Burnham 1989; Behar *et al.* 1992, 1997; Pepper & Corvi 1995; Vandenbroucke *et al.* 1999; Tang *et al.* 2000; Lewan & Ruble 2002). Such models have been extended to include the evolution of $\delta^{13}C$ and δD values of thermogenic gases (Tang *et al.* 2000; Ni *et al.* 2011). The addition of methane clumped-isotope constraints to such models is needed, but will require the effects of mixing to be considered. For example, extensive Rayleigh distillation of the source organic matter (which will result in methane with high $\delta^{13}C$ and δD values) could create significant non-linear mixing effects for Δ_{18} values.

(3) How will independent $^{12}CH_2D_2$ measurements contribute to the already measured Δ_{18} and $\Delta_{^{13}CH_3D}$ measurements discussed above? One aspect they will be able to test immediately is whether the interpretations put forward that many thermogenic and biogenic gases formed in isotopic equilibrium are correct. If the $^{13}CH_3D$ and $^{12}CH_2D_2$ values yield identical clumped-isotope temperatures, then this will provide strong support for the hypothesis that methane clumped-isotope temperatures can reflect thermogenic and biogenic gas-formation temperatures. Initial results of gases from pure-tank cylinders with δD and $\delta^{13}C$ values consistent with a thermogenic origin as well as two natural gas samples, one from a Marcellus Shale deposit and one from a Utica Shale deposit, yield indistinguishable temperatures based on $^{13}CH_3D$ and $^{12}CH_2D_2$ values (Young *et al.* 2016) and thus provide initial support to this hypothesis. Additionally, biogenic gas from the Birchtee mine yields methane in clumped-isotope equilibrium for both $^{13}CH_3D$ and $^{12}CH_2D_2$ with temperatures (12–16°C) similar to the environmental temperatures (20–23°C), supporting the hypothesis that biogenic methane can form in clumped-isotopic equilibrium. Values of $^{12}CH_2D_2$ will also play a useful role in distinguishing two-component mixing problems, as the non-linearity of mixing for $^{12}CH_2D_2$ values depends solely on the δD value of the end member and not the $\delta^{13}C$ values, which is not the case for $^{13}CH_3D$. The utility and power of these measurements will be realized in the near future as more measurements are made (Young *et al.* 2017). The frameworks presented above will need to incorporate these results.

(4) Finally, other clumped and 'position-specific' isotopic measurements of hydrocarbons are

beginning to be explored including $^{13}C-^{13}C$ ethane clumped-isotope measurements (Clog *et al.* 2014) and the propensity of ^{13}C to be found on the exterior or interior carbon positions of propane (Gilbert *et al.* 2016; Piasecki *et al.* 2016). These measurements are reviewed in Eiler *et al.* (2017). As the information encoded by these measurements is understood new frameworks that incorporate the growing number of new isotopic measurements of hydrocarbons will need to be created.

This work was supported by the NSF, ExxonMobil, Petrobras, Statoil and Caltech. We thank ExxonMobil for permission to publish. This work was the outgrowth of a session at the 2015 Goldschmidt Conference. We wish to thank the other participants in the session for stimulating discussions of the work.

References

AFFEK, H.P. 2012. Clumped isotope paleothermometry: principles, applications, and challenges. *The Paleontological Society Papers*, **18**, 101–114.

ALEXANDER, R., KAGI, R.I. & LARCHER, A.V. 1982. Clay catalysis of aromatic hydrogen-exchange reactions. *Geochimica et Cosmochimica Acta*, **46**, 219–222.

ALEXANDER, R., KAGI, R.I. & LARCHER, A.V. 1984. Clay catalysis of alkyl hydrogen exchange reactions – reaction mechanisms. *Organic Geochemistry*, **6**, 755–760.

BALLENTINE, C.J. & O'NIONS, R.K. 1994. The use of natural He, Ne and Ar isotopes to study hydrocarbon-related fluid provenance, migration and mass balance in sedimentary basins. *In*: PARNELL, J. (ed.) *Geofluids: Origin, Migration and Evolution of Fluids in Sedimentary Basins*. Geological Society, London, Special Publications, **78**, 347–361, https://doi.org/10.1144/GSL.SP.1994.078.01.23

BEHAR, F., KRESSMANN, S., RUDKIEWICZ, J. & VANDENBROUCKE, M. 1992. Experimental simulation in a confined system and kinetic modelling of kerogen and oil cracking. *Organic Geochemistry*, **19**, 173–189.

BEHAR, F., VANDENBROUCKE, M., TANG, Y., MARQUIS, F. & ESPITALIE, J. 1997. Thermal cracking of kerogen in open and closed systems: determination of kinetic parameters and stoichiometric coefficients for oil and gas generation. *Organic Geochemistry*, **26**, 321–339.

BERNARD, B.B., BROOKS, J.M. & SACKETT, W.M. 1976. Natural gas seepage in the Gulf of Mexico. *Earth and Planetary Science Letters*, **31**, 48–54.

BERNARD, B.B., BROOKS, J.M. & SACKETT, W.M. 1977. A geochemical model for characterization of hydrocarbon gas sources in marine sediments. *9th Offshore Technology Conference Proceedings*, Houston, TX, **3**, 435–438.

BERNER, U., FABER, E. & STAHL, W. 1992. Mathematical simulation of the carbon isotopic fractionation between huminitic coals and related methane. *Chemical Geology*, **94**, 315–319.

BERNER, U., FABER, E., SCHEEDER, G. & PANTEN, D. 1995. Primary cracking of algal and landplant kerogens: kinetic models of isotope variations in methane, ethane and propane. *Chemical Geology*, **126**, 233–245.

BOTTOMLEY, D.J., GREGOIRE, D.C. & RAVEN, K.G. 1994. Saline ground waters and brines in the Canadian Shield: geochemical and isotopic evidence for a residual evaporite brine component. *Geochimica et Cosmochimica Acta*, **58**, 1483–1498.

BOWEN, G.J. & REVENAUGH, J. 2003. Interpolating the isotopic composition of modern meteoric precipitation. *Water Resources Research*, **39**, 9-1–9-13.

BURNHAM, A. 1989. A Simple Kinetic Model of Petroleum Formation and Cracking. Lawrence Livermore National Lab, Report UCID **21665**.

BURNHAM, A.K. & SWEENEY, J.J. 1989. A chemical kinetic model of vitrinite maturation and reflectance. *Geochimica et Cosmochimica Acta*, **53**, 2649–2657.

BURRUSS, R. & LAUGHREY, C. 2010. Carbon and hydrogen isotopic reversals in deep basin gas: evidence for limits to the stability of hydrocarbons. *Organic Geochemistry*, **41**, 1285–1296.

BYRNE, D.J., BARRY, P.H., LAWSON, M. & BALLENTINE, C.J. 2017. Noble gases in conventional and unconventional petroleum systems. *In*: LAWSON, M., FORMOLO, M.J. & EILER, J.M. (eds) *From Source to Seep: Geochemical Applications in Hydrocarbon Systems*. Geological Society, London, Special Publications, **468**. First published online December 14, 2017, https://doi.org/10.1144/SP468.5

CAO, X. & LIU, Y. 2012. Theoretical estimation of the equilibrium distribution of clumped isotopes in nature. *Geochimica et Cosmochimica Acta*, **77**, 292–303.

CHUNG, H., GORMLY, J. & SQUIRES, R. 1988. Origin of gaseous hydrocarbons in subsurface environments: theoretical considerations of carbon isotope distribution. *Chemical Geology*, **71**, 97–104.

CHUNG, H., ROONEY, M., TOON, M. & CLAYPOOL, G.E. 1992. Carbon isotope composition of marine crude oils (1). *AAPG Bulletin*, **76**, 1000–1007.

CLAYPOOL, G.E. & KAPLAN, I. 1974. The origin and distribution of methane in marine sediments. *In*: KAPLAN, I.R. (ed.) *Natural Gases in Marine Sediments*. Marine Science, **3**, Springer, Boston, MA, 99–139.

CLAYTON, C. 1991. Carbon isotope fractionation during natural gas generation from kerogen. *Marine and Petroleum Geology*, **8**, 232–240.

CLOG, M., MARTINI, A., LAWSON, M. & EILER, J. 2014. Doubly ^{13}C-substituted ethane in shale gases. Paper presented at the Goldschmidt Conference, Sacramento, CA, 435.

CLOG, M., ELLAM, R., HILKERT, A., SCHWIETERS, J. & HAMILTON, D. 2015. New developments in high-resolution gas source isotope ratio mass spectrometers. Paper presented at the AGU Fall Meeting, San Francisco.

CRAMER, B., KROOSS, B.M. & LITTKE, R. 1998. Modelling isotope fractionation during primary cracking of natural gas: a reaction kinetic approach. *Chemical Geology*, **149**, 235–250.

CURTIS, J.B. 2002. Fractured shale-gas systems. *AAPG Bulletin*, **86**, 1921–1938.

DENNIS, K.J. & SCHRAG, D.P. 2010. Clumped isotope thermometry of carbonatites as an indicator of diagenetic alteration. *Geochimica et Cosmochimica Acta*, **74**, 4110–4122.

DOMINÉ, F., BOUNACEUR, R., SCACCHI, G., MARQUAIRE, P.-M., DESSORT, D., PRADIER, B. & BREVART, O. 2002. Up to what temperature is petroleum stable? New insights from a 5200 free radical reactions model. *Organic Geochemistry*, **33**, 1487–1499.

DOUGLAS, P., STOLPER, D. *ET AL*. 2016. Diverse origins of Arctic and Subarctic methane point source emissions identified with multiply-substituted isotopologues. *Geochimica et Cosmochimica Acta*, **188**, 163–188.

DOUGLAS, P., STOLPER, D. *ET AL*. 2017. Methane clumped isotopes: progress and potential for a new isotopic tracer. *Organic Geochemistry*, **113**, 262–282, https://doi.org/10.1016/j.orggeochem.2017.07.016

EILER, J.M. 2007. 'Clumped-isotope' geochemistry – The study of naturally-occurring, multiply-substituted isotopologues. *Earth and Planetary Science Letters*, **262**, 309–327.

EILER, J.M. 2013. The isotopic anatomies of molecules and minerals. *Annual Review of Earth and Planetary Sciences*, **41**, 411–441.

EILER, J.M. & SCHAUBLE, E. 2004. $^{18}O^{13}C^{16}O$ in Earth's atmosphere. *Geochimica et Cosmochimica Acta*, **68**, 4767–4777.

EILER, J.M., CLOG, M. *ET AL*. 2013. A high-resolution gas-source isotope ratio mass spectrometer. *International Journal of Mass Spectrometry*, **335**, 45–56.

EILER, J.M., CLOG, M., LAWSON, M., LLOYD, M., PIASECKI, A., PONTON, C. & XIE, H. 2017. The isotopic structures of geological organic compounds. *In*: LAWSON, M., FORMOLO, M.J. & EILER, J.M. (eds) *From Source to Seep: Geochemical Applications in Hydrocarbon Systems*. Geological Society, London, Special Publications, **468**. First published online December 14, 2017, https://doi.org/10.1144/SP468.4

ELLAM, R., NEWTON, J., HILKERT, A., SCHWIETERS, J. & DEERBERG, M. 2015. Initial results from the SUERC 253 ultra: a new high resolution isotope Ratio Mass Spectrometer for Isotopologue Analysis. *Goldschmidt Conference Abstracts*, Prague, CZ, 822.

ESPITALIE, J., UNGERER, P., IRWIN, I. & MARQUIS, F. 1988. Primary cracking of kerogens. Experimenting and modeling C_1, C_2–C_5, C_6–C_{15+} classes of hydrocarbons formed. *Organic Geochemistry*, **13**, 893–899.

ETIOPE, G. & SHERWOOD LOLLAR, B. 2013. Abiotic methane on Earth. *Reviews of Geophysics*, **51**, 276–299.

GHOSH, P., ADKINS, J. *ET AL*. 2006. ^{13}C–^{18}O bonds in carbonate minerals: a new kind of paleothermometer. *Geochimica et Cosmochimica Acta*, **70**, 1439–1456.

GILBERT, A., YAMADA, K., SUDA, K., UENO, Y. & YOSHIDA, N. 2016. Measurement of position-specific 13 C isotopic composition of propane at the nanomole level. *Geochimica et Cosmochimica Acta*, **177**, 205–216.

HAMMES, U., HAMLIN, H.S. & EWING, T.E. 2011. Geologic analysis of the Upper Jurassic Haynesville Shale in east Texas and west Louisiana. *AAPG Bulletin*, **95**, 1643–1666.

HELGESON, H.C., RICHARD, L., MCKENZIE, W.F., NORTON, D.L. & SCHMITT, A. 2009. A chemical and thermodynamic model of oil generation in hydrocarbon source rocks. *Geochimica et Cosmochimica Acta*, **73**, 594–695.

HINRICHS, K.-U., HAYES, J.M. *ET AL*. 2006. Biological formation of ethane and propane in the deep marine subsurface. *Proceedings of the National Academy of Sciences*, **103**, 14 684–14 689.

HOERING, T. 1984. Thermal reactions of kerogen with added water, heavy water and pure organic substances. *Organic Geochemistry*, **5**, 267–278.

HORIBE, Y. & CRAIG, H. 1995. D/H fractionation in the system methane-hydrogen-water. *Geochimica et Cosmochimica Acta*, **59**, 5209–5217.

HORITA, J. 2001. Carbon isotope exchange in the system CO_2-CH_4 at elevated temperatures. *Geochimica et Cosmochimica Acta*, **65**, 1907–1919.

HUNT, J.M. 1996. *Petroleum Geochemistry and Geology*. W.H. Freeman and Company, New York.

INAGAKI, F., HINRICHS, K.-U. *ET AL*. 2015. Exploring deep microbial life in coal-bearing sediment down to *c.* 2.5 km below the ocean floor. *Science*, **349**, 420–424.

JARVIE, D.M., HILL, R.J., RUBLE, T.E. & POLLASTRO, R.M. 2007. Unconventional shale-gas systems: the Mississippian Barnett Shale of north-central Texas as one model for thermogenic shale-gas assessment. *AAPG Bulletin*, **91**, 475–499.

KILLOPS, S.D. & KILLOPS, V.J. 2013. *Introduction to Organic Geochemistry*. Blackwell, Oxford.

LASH, G.G. & ENGELDER, T. 2011. Thickness trends and sequence stratigraphy of the Middle Devonian Marcellus Formation, Appalachian Basin: implications for Acadian foreland basin evolution. *AAPG Bulletin*, **95**, 61–103.

LEE, D.S., HERMAN, J.D., ELSWORTH, D., KIM, H.T. & LEE, H.S. 2011. A critical evaluation of unconventional gas recovery from the Marcellus Shale, northeastern United States. *KSCE Journal of Civil Engineering*, **15**, 679–687.

LEWAN, M. 1997. Experiments on the role of water in petroleum formation. *Geochimica et Cosmochimica Acta*, **61**, 3691–3723.

LEWAN, M. & RUBLE, T. 2002. Comparison of petroleum generation kinetics by isothermal hydrous and nonisothermal open-system pyrolysis. *Organic Geochemistry*, **33**, 1457–1475.

LIU, Q. & LIU, Y. 2016. Clumped-isotope signatures at equilibrium of CH_4, NH_3, H_2O, H_2S and SO_2. *Geochimica et Cosmochimica Acta*, **175**, 252–270.

LORANT, F., PRINZHOFER, A., BEHAR, F. & HUC, A.-Y. 1998. Carbon isotopic and molecular constraints on the formation and the expulsion of thermogenic hydrocarbon gases. *Chemical Geology*, **147**, 249–264.

MA, Q., WU, S. & TANG, Y. 2008. Formation and abundance of doubly-substituted methane isotopologues ($^{13}CH_3D$) in natural gas systems. *Geochimica et Cosmochimica Acta*, **72**, 5446–5456.

MAGOON, L.B. & DOW, W.G. 1994. *The Petroleum System: From Source to Trap*. American Association of Petroleum Geologists.

MANGO, F.D. & HIGHTOWER, J. 1997. The catalytic decomposition of petroleum into natural gas. *Geochimica et Cosmochimica Acta*, **61**, 5347–5350.

MANGO, F.D., HIGHTOWER, J. & JAMES, A.T. 1994. Role of transition-metal catalysis in the formation of natural gas. *Nature*, **368**, 536–538.

MARTINI, A.M., BUDAI, J.M., WALTER, L.M. & SCHOELL, M. 1996. Microbial generation of economic accumulations of methane within a shallow organic-rich shale. *Nature*, **383**, 155–158.

MARTINI, A.M., WALTER, L.M., BUDAI, J.M., KU, T.C.W., KAISER, C.J. & SCHOELL, M. 1998. Genetic and temporal relations between formation waters and biogenic methane: upper Devonian Antrim Shale, Michigan Basin,

USA. *Geochimica et Cosmochimica Acta*, **62**, 1699–1720.

MARTINI, A.M., WALTER, L.M., KU, T.C., BUDAI, J.M., MCINTOSH, J.C. & SCHOELL, M. 2003. Microbial production and modification of gases in sedimentary basins: a geochemical case study from a Devonian shale gas play, Michigan basin. *AAPG Bulletin*, **87**, 1355–1375.

MCCANN, T. 1998. The Rotliegend of the NE German Basin: background and prospectivity. *Petroleum Geoscience*, **4**, 17–27, https://doi.org/10.1144/petgeo.4.1.17

MEISSNER, F.F. 1978. Petroleum geology of the Bakken formation Williston Basin, North Dakota and Montana. *Economic Geology of the Williston Basin Symposium*. Montana Geological Society, Billings, Montana, 207–227.

MILLER, S.M., WOFSY, S.C. *ET AL.* 2013. Anthropogenic emissions of methane in the United States. *Proceedings of the National Academy of Sciences*, **110**, 20 018–20 022.

MULLEN, J. 2010. Petrophysical characterization of the Eagle Ford Shale in south Texas. Paper presented at the Canadian Unconventional Resources and International Petroleum Conference, Society of Petroleum Engineers.

NI, Y., MA, Q., ELLIS, G.S., DAI, J., KATZ, B., ZHANG, S. & TANG, Y. 2011. Fundamental studies on kinetic isotope effect (KIE) of hydrogen isotope fractionation in natural gas systems. *Geochimica et Cosmochimica Acta*, **75**, 2696–2707.

ONO, S., WANG, D.T., GRUEN, D.S., SHERWOOD LOLLAR, B., ZAHNISER, M.S., MCMANUS, B.J. & NELSON, D.D. 2014. Measurement of a doubly substituted methane isotopologue, $^{13}CH_3D$, by tunable infrared laser direct absorption spectroscopy. *Analytical Chemistry*, **86**, 6487–6494.

PASSEY, B. & HENKES, G. 2012. Carbonate clumped isotope bond reordering and geospeedometry. *Earth and Planetary Science Letters*, **351–352**, 223–236.

PEPPER, A.S. & CORVI, P.J. 1995. Simple kinetic models of petroleum formation. Part I: oil and gas generation from kerogen. *Marine and Petroleum Geology*, **12**, 291–319.

PEPPER, A.S. & DODD, T.A. 1995. Simple kinetic models of petroleum formation. Part II: oil-gas cracking. *Marine and Petroleum Geology*, **12**, 321–340.

PETERS, K.E., WALTERS, C. & MOLDOWAN, J. 2005. *The Biomarker Guide: Volume II. Biomarkers and Isotopes in Petroleum Systems and Earth History*. Cambridge University Press, Cambridge, UK.

PIASECKI, A., SESSIONS, A., LAWSON, M., FERREIRA, A., SANTOS NETO, E.V. & EILER, J.M. 2016. Analysis of the site-specific carbon isotope composition of propane by gas source isotope ratio mass spectrometer. *Geochimica et Cosmochimica Acta*, **188**, 58–72.

PRINZHOFER, A.A. & HUC, A.Y. 1995. Genetic and post-genetic molecular and isotopic fractionations in natural gases. *Chemical Geology*, **126**, 281–290.

QUIGLEY, T. & MACKENZIE, A. 1988. The temperatures of oil and gas formation in the sub-surface. *Nature*, **333**, 549–552.

RANAWEERA, H. 1990. *Sleipner Vest*. Graham & Trotman, London.

REEVES, E.P., SEEWALD, J.S. & SYLVA, S.P. 2012. Hydrogen isotope exchange between n-alkanes and water under hydrothermal conditions. *Geochimica et Cosmochimica Acta*, **77**, 582–599.

RICE, D.D. & CLAYPOOL, G.E. 1981. Generation, accumulation, and resource potential of biogenic gas. *AAPG Bulletin*, **65**, 5–25.

ROONEY, M.A., CLAYPOOL, G.E. & MOSES CHUNG, H. 1995. Modeling thermogenic gas generation using carbon isotope ratios of natural gas hydrocarbons. *Chemical Geology*, **126**, 219–232.

SACKETT, W.M. 1978. Carbon and hydrogen isotope effects during the thermocatalytic production of hydrocarbons in laboratory simulation experiments. *Geochimica et Cosmochimica Acta*, **42**, 571–580.

SCHOELL, M. 1980. The hydrogen and carbon isotopic composition of methane from natural gases of various origins. *Geochimica et Cosmochimica Acta*, **44**, 649–661.

SCHOELL, M. 1983. Genetic characterization of natural gases. *AAPG Bulletin*, **67**, 2225–2238.

SCHOELL, M. 1984. Recent advances in petroleum isotope geochemistry. *Organic Geochemistry*, **6**, 645–663.

SEEWALD, J.S. 2003. Organic–inorganic interactions in petroleum-producing sedimentary basins. *Nature*, **426**, 327–333.

SEEWALD, J.S., BENITEZ-NELSON, B.C. & WHELAN, J.K. 1998. Laboratory and theoretical constraints on the generation and composition of natural gas. *Geochimica et Cosmochimica Acta*, **62**, 1599–1617.

SHUAI, Y., DOUGLAS, P.M.J. *ET AL.* in press. Equilibrium and non-equilibrium controls on the abundances of clumped isotopologues of methane during thermogenic formation in laboratory experiments: implications for the chemistry of pyrolysis and the origins of natural gases. *Geochimica et Cosmochimica Acta*.

SINGH, G. & WOLFSBERG, M. 1975. The calculation of isotopic partition function ratios by a perturbation theory technique. *The Journal of Chemical Physics*, **62**, 4165–4180.

STAHL, W.J. & CAREY, B.D. 1975. Source-rock identification by isotope analyses of natural gases from fields in the Val Verde and Delaware basins, west Texas. *Chemical Geology*, **16**, 257–267.

STOLPER, D.A. & EILER, J.M. 2015. The kinetics of solid-state isotope-exchange reactions for clumped isotopes: a study of inorganic calcites and apatites from natural and experimental samples. *American Journal of Science*, **315**, 363–411.

STOLPER, D.A., LAWSON, M. *ET AL.* 2014a. Formation temperatures of thermogenic and biogenic methane. *Science*, **344**, 1500–1503.

STOLPER, D.A., SESSIONS, A.L. *ET AL.* 2014b. Combined ^{13}C-D and D-D clumping in methane: methods and preliminary results. *Geochimica et Cosmochimica Acta*, **126**, 169–191.

STOLPER, D.A., MARTINI, A.M. *ET AL.* 2015. Distinguishing and understanding thermogenic and biogenic sources of methane using multiply substituted isotopologues. *Geochimica et Cosmochimica Acta*, **161**, 219–247.

SWEENEY, J.J. & BURNHAM, A.K. 1990. Evaluation of a simple model of vitrinite reflectance based on chemical kinetics. *AAPG Bulletin*, **74**, 1559–1570.

TAKAI, K., NAKAMURA, K. *ET AL.* 2008. Cell proliferation at 122°C and isotopically heavy CH_4 production by a hyperthermophilic methanogen under high-pressure cultivation. *Proceedings of the National Academy of Sciences*, **105**, 10 949–10 954.

TANG, Y., PERRY, J., JENDEN, P. & SCHOELL, M. 2000. Mathematical modeling of stable carbon isotope ratios in natural gases. *Geochimica et Cosmochimica Acta*, **64**, 2673–2687.

THAUER, R.K. 1998. Biochemistry of methanogenesis: a tribute to Marjory Stephenson. *Microbiology*, **144**, 2377–2406.

TISSOT, B.P. & WELTE, D.H. 1978. *Petroleum Formation and Occurrence: A New Approach to Oil and Gas Exploration*. Springer-Verlag, Berlin.

TSUZUKI, N., TAKEDA, N., SUZUKI, M. & YOKOI, K. 1999. The kinetic modeling of oil cracking by hydrothermal pyrolysis experiments. *International Journal of Coal Geology*, **39**, 227–250.

UREY, H.C. 1947. The thermodynamic properties of isotopic substances. *Journal of the Chemical Society*, 562–581.

VALENTINE, D.L. 2011. Emerging topics in marine methane biogeochemistry. *Annual Review of Marine Science*, **3**, 147–171.

VALENTINE, D.L., CHIDTHAISONG, A., RICE, A., REEBURGH, W.S. & TYLER, S.C. 2004. Carbon and hydrogen isotope fractionation by moderately thermophilic methanogens. *Geochimica et Cosmochimica Acta*, **68**, 1571–1590.

VANDENBROUCKE, M., BEHAR, F. & RUDKIEWICZ, J. 1999. Kinetic modelling of petroleum formation and cracking: implications from the high pressure/high temperature Elgin Field (UK, North Sea). *Organic Geochemistry*, **30**, 1105–1125.

VINSON, D.S., BLAIR, N.E., MARTINI, A.M., LARTER, S., OREM, W.H. & MCINTOSH, J.C. 2017. Microbial methane from in situ biodegradation of coal and shale: a review and reevaluation of hydrogen and carbon isotope signatures. *Chemical Geology*, **453**, 128–145.

WANG, D.T., GRUEN, D.S. *ET AL.* 2015. Nonequilibrium clumped isotope signals in microbial methane. *Science*, **348**, 428–431.

WANG, Z., SCHAUBLE, E.A. & EILER, J.M. 2004. Equilibrium thermodynamics of multiply substituted isotopologues of molecular gases. *Geochimica et Cosmochimica Acta*, **68**, 4779–4797.

WEBB, M.A. & MILLER III T.F., 2014. Position-specific and clumped stable isotope studies: comparison of the Urey and path-integral approaches for carbon dioxide, nitrous oxide, methane, and propane. *The Journal of Physical Chemistry A*, **118**, 467–474.

WHITICAR, M.J. 1999. Carbon and hydrogen isotope systematics of bacterial formation and oxidation of methane. *Chemical Geology*, **161**, 291–314.

WHITICAR, M.J., FABER, E. & SCHOELL, M. 1986. Biogenic methane formation in marine and freshwater environments: CO_2 reduction vs acetate fermentation – isotope evidence. *Geochimica et Cosmochimica Acta*, **50**, 693–709.

WILHELMS, A., LARTER, S., HEAD, I., FARRIMOND, P., DI-PRIMIO, R. & ZWACH, C. 2001. Biodegradation of oil in uplifted basins prevented by deep-burial sterilization. *Nature*, **411**, 1034–1037.

WING, B.A. & HALEVY, I. 2014. Intracellular metabolite levels shape sulfur isotope fractionation during microbial sulfate respiration. *Proceedings of the National Academy of Sciences*, **111**, 18 116–18 125, https://doi.org/10.1073/pnas.1407502111

XIA, X. 2014. Kinetics of gaseous hydrocarbon generation with constraints of natural gas composition from the Barnett Shale. *Organic Geochemistry*, **74**, 143–149.

XIA, X. & TANG, Y. 2012. Isotope fractionation of methane during natural gas flow with coupled diffusion and adsorption/desorption. *Geochimica et Cosmochimica Acta*, **77**, 489–503.

XIAO, Y. 2001. Modeling the kinetics and mechanisms of petroleum and natural gas generation: a first principles approach. *Reviews in Mineralogy and Geochemistry*, **42**, 383–436.

YOUNG, E.D., RUMBLE, D., FREEDMAN, P. & MILLS, M. 2016. A large-radius high-mass-resolution multiple-collector isotope ratio mass spectrometer for analysis of rare isotopologues of O_2, N_2, CH_4 and other gases. *International Journal of Mass Spectrometry*, **401**, 1–10.

YOUNG, E.D., KOHL, I.E. *ET AL.* 2017. The relative abundances of resolved $^{12}CH_2D_2$ and $^{13}CH_3D$ and mechanisms controlling isotopic bond ordering in abiotic and biotic methane gases. *Geochimica et Cosmochimica Acta*, **203**, 235–264.

The isotopic structures of geological organic compounds

JOHN M. EILER[1]*, MATTHIEU CLOG[2], MICHAEL LAWSON[3], MAX LLOYD[1],
ALISON PIASECKI[4], CAMILO PONTON[1] & HAO XIE[1]

[1]*Division of Geological and Planetary Sciences, California
Institute of Technology, Pasadena, CA 91125*

[2]*University of Glasgow, Glasgow G12 8QQ, Scotland, UK*

[3]*ExxonMobil Upstream Research Company, Spring, TX 77389*

[4]*Department of Earth Science, University of Bergen, 5020 Bergen, Norway*

**Correspondence: eiler@gps.caltech.edu*

Abstract: Organic compounds are ubiquitous in the Earth's surface, sediments and many rocks, and preserve records of geological, geochemical and biological history; they are also critical natural resources and major environmental pollutants. The naturally occurring stable isotopes of volatile elements (D, ^{13}C, ^{15}N, ^{17,18}O, ^{33,34,36}S) provide one way of studying the origin, evolution and migration of geological organic compounds. The study of bulk stable isotope compositions (i.e. averaged across all possible molecular isotopic forms) is well established and widely practised, but frequently results in non-unique interpretations. Increasingly, researchers are reading the organic isotopic record with greater depth and specificity by characterizing stable isotope 'structures' – the proportions of site-specific and multiply substituted isotopologues that contribute to the total rare-isotope inventory of each compound. Most of the technologies for measuring stable isotope structures of organic molecules have been only recently developed and to date have been applied only in an exploratory way. Nevertheless, recent advances have demonstrated that molecular isotopic structures provide distinctive records of biosynthetic origins, conditions and mechanisms of chemical transformation during burial, and forensic fingerprints of exceptional specificity. This paper provides a review of this young field, which is organized to follow the evolution of molecular isotopic structure from biosynthesis, through diagenesis, catagenesis and metamorphism.

The stable isotope compositions of natural organic compounds can potentially be used toaddress a variety of significant questions, such as identifying the sources of hydrocarbon pollutants in near-surface waters or air (Elsner *et al.* 2012), characterizing the precursor biomolecules parental to geological hydrocarbons (Ferreira *et al.* 2012), recognizing and quantifying conditions and mechanisms of hydrocarbon generation (Whiticar *et al.* 1986; Chung *et al.* 1988; Tang *et al.* 2000), or of physical or biological destruction (Martini *et al.* 2003). There is an extensive scientific literature addressing such questions using measurements of the isotopic compositions – generally D/H, ^{13}C/^{12}C, ^{15}N/^{14}N or ^{34}S/^{32}S ratios – of bulk oil or gas fractions or, more usefully, individual molecular constituents (i.e. compound-specific analysis). Virtually all such studies have focused on the average isotopic contents of these materials, summed across all of the many isotopologues of the compounds of interest. For instance, the ^{13}C/^{12}C ratio of an organic compound, averaged across isotopologues that may differ in their site of ^{13}C substitution and/or numbers of ^{13}C substitutions.

Such measurements fail to observe the information potentially recorded by differences in proportions of the many site-specific and multiply substituted species. Note that small gaseous molecules (e.g. CO_2, CH_4, N_2O) commonly have around ten isotopologues, small metabolites or hydrocarbons (e.g. alanine or *iso*-octane) typically have thousands of isotopologues, and larger biomolecules (sugars, fatty acids) have millions, to millions of millions (e.g. hopanoids, hormones, proteins and other macromolecular materials). Nearly all such isotopologues can be thought of as unique chemical species, differing in their physical, thermodynamic and chemical-kinetic properties.

In this paper, we review the small but rapidly growing body of work that examines the stable isotope distributions of hydrocarbons at intramolecular scales – i.e. 'site-specific' differences in isotopic composition between structurally non-equivalent sites in the same molecule, and 'clumped-isotope' compositions, or probabilities that two or more rare isotopes are present in the same molecule. We refer to this family of isotopic properties collectively as

From: LAWSON, M., FORMOLO, M. J. & EILER, J. M. (eds) 2018. *From Source to Seep: Geochemical Applications in Hydrocarbon Systems*. Geological Society, London, Special Publications, **468**, 53–81.
First published online December 14, 2017, https://doi.org/10.1144/SP468.4
© 2018 The Author(s). Published by The Geological Society of London. All rights reserved.
For permissions: http://www.geolsoc.org.uk/permissions. Publishing disclaimer: www.geolsoc.org.uk/pub_ethics

molecular 'isotopic structure'. There have been few papers that include measurements of the isotopic structures of natural hydrocarbons (and even fewer that include a significant number of samples with known origins). However, there is a larger body of work documenting isotopic structures of biomolecules that are likely precursors to petroleum and natural gas compounds, or examining fundamental fractionations controlling isotopic structures of organics; and, over the last 5 years, considerable effort has been put into advancing the technologies that permit such analyses (Appendix B). Taken together, this work suggests we are on the cusp of a new, potentially large sub-discipline of organic geochemistry.

This review aims to present a comprehensive, organized 'snap shot' of this field as it presently exists, and to predict some of the ways in which it could evolve over the next several years. We do not provide a detailed discussion of the clumped-isotope geochemistry of methane – arguably the largest and most mature part of our subject – because it is considered in its own review paper in this volume (see Stolper *et al.* 2017). The primary focus of this paper is on natural hydrocarbons containing two or more carbon atoms, or organic compounds that may be precursors to such natural hydrocarbons. Somewhat unusual for such a review, we have made an effort to include work that, so far, has been presented only in theses, conference proceedings and public reports to funding agencies, because significant contributions to analytical technologies and scientific findings have not yet appeared in the peer-reviewed literature. We doubtless have missed some such sources, and some of this material is likely to evolve by the time it reaches the peer-reviewed literature.

An organizing principle: the genealogy of molecular isotopic structures from biomass to petroleum and gas compounds

The body of work discussed in this review spans a great range in techniques, chemical compounds and motivating questions – many seemingly far removed from petroleum geochemistry. We attempt to lend this material a rational overarching structure by casting it as a narrative about the molecular isotopic structures of biomolecules and their geological transformation into petroleum and gas compounds. We begin this discussion with an overview of sedimentary organic matter and the changes it undergoes during diagenesis, catagenesis and metamorphism (see Fig. 1, adapted from Tissot & Welte 1984).

Sedimentary organic matter is dominated by four major components: lipids (including fatty acids and hopanoids); proteins and amino acids; carbohydrates

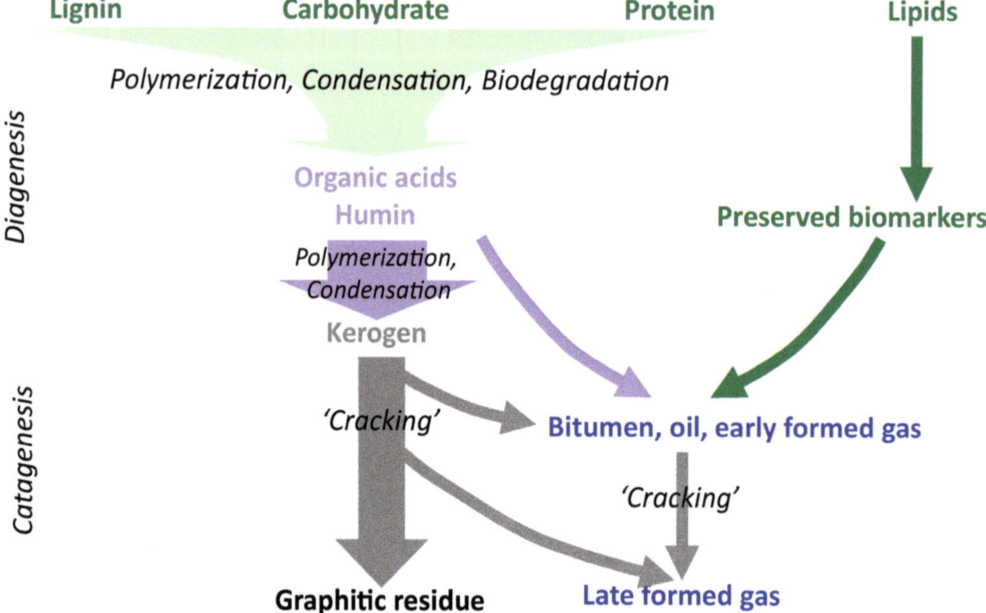

Fig. 1. Schematic representation of the evolution of sedimentary organic matter through formation, diagenesis and catagenesis, increasing in depth of burial and maximum temperature from top to bottom. Adapted from Tissot & Welte (1984).

(sugars, starches and cellulose); and lignin and other polymerized and structurally ill-defined components (Tissot & Welte 1984; Wakeham *et al.* 1997; Hedges *et al.* 2000; Freeman 2001). Cellulose and lignin are proportionately more abundant in terrestrial sedimentary organic matter, whereas proteins and lipids are proportionately more abundant in marine organic matter. The abundance ratios and chemical forms of these constituents change rapidly after formation through utilization by heterotrophs, environmental degradation, and early diagenesis during and just after burial. Soluble carbohydrate and protein decrease radically in proportion to cellulose, lignin and other polymerized components, and the remaining fraction undergoes polymerization and condensation to form humin – a refractory, structurally complex macromolecular matrix – with more reactive moieties transforming to small organic acids. Lipids and closely related compounds (e.g. hydrocarbons formed by decarboxylation of fatty acids) are relatively well preserved through these initial transformations.

Further burial and heating leads to the onset of catagenesis, marked by further polymerization, condensation and other reactions, transforming humin into kerogen and concentrating relatively labile constituents (fatty and other organic acids, alkanes) into the first generation of bitumen, oil and early-formed natural gas (which typically contains sub-equal amounts of methane and higher order volatile alkanes, along with CO_2, H_2, H_2S and other minor constituents; Tissot & Welte 1984; Marshall & Rodgers 2008). Further heating drives more extensive catagenesis, condensing the relatively open structures of immature kerogens to form progressively more graphitic networks, decomposing, or 'cracking' kerogen components to liberate oil and gas fraction compounds, and degrading existing oil and gas compounds to form smaller, simpler and more volatile species (Quigley & Mackenzie 1988; Burnham & Sweeney 1989; Lewan 1993; Mao *et al.* 2010). Overall, the proportion of gas to oil rises, methane makes up a greater fraction of all gas, and proportions of well-preserved biomarkers diminish (Price & Schoell 1995; Prinzhofer & Huc 1995). Where this process proceeds to near the onset of greenschist-facies metamorphism ($c.$ 250 + °C), oil and early-formed gas will be nearly quantitatively expelled or thermally destroyed, and kerogen begins conversion to graphite, leaving only graphitic carbon and methane-rich late-formed gas (Wada *et al.* 1994). Two significant open questions about this family of processes are: could oil-fraction compounds persist for long times at relatively high temperatures (Mango 1991), and, when water is present, does kerogen cracking continue to produce gas components at high temperatures (200 + °C; Seewald *et al.* 1998)?

We ask several questions about the evolution in molecular isotopic structure through the preceding chain of events:

(1)   What are the isotopic structures of the components of sedimentary organic matter that will ultimately transform to geological hydrocarbons?

(2)   Which of these intramolecular isotopic properties (site-specific variations; proportions of clumped-isotope species) are preserved through diagenesis and catagenesis to appear in oil and gas components?

(3)   How do the reactions of diagenesis and catagenesis subsample or modify the isotopic structures of original biomolecules?

(4)   Can we recognize new compounds formed during thermal maturation of organic matter based on their isotopic structures?

(5)   Do those compounds record the conditions (e.g. temperatures and/or pressures) of their formation?

(6)   Do exchange reactions proceed rapidly enough in geological environments to fully equilibrate the isotopic structures of biomolecules and/or derived hydrocarbons?

(7)   Do the reactions of advanced catagenesis in high-maturity systems leave some imprint on molecular isotopic structures of their residues?

(8)   Can we see the effects of microbial degradation or production of organics in the subsurface (i.e. during biodegradation of oil and gas)?

The remainder of this paper proceeds through its subject from the beginning of biosynthesis to the end of catagenesis and onset of metamorphism. In the interests of brevity, much of the relevant technical material – the nomenclature used in the study of molecular isotopic structures, and the technologies for making such measurements – is consigned to the Appendices.

In the beginning: isotopic structures of biomolecules

Nearly all organic components of petroleum and natural gas are derived directly or indirectly from biomolecules in sedimentary organic matter, and so the isotopic structures in the products of biosynthesis must be the first step in our exploration of our subject.

Equilibrium controls?

In past decades, there was a protracted debate as to whether the isotopic contents (and, when discussed, isotopic structures) of biomolecules formed in or near isotopic equilibrium. If so, they should be understood through the thermodynamic isotope effects that control many inorganic heterogeneous and homogeneous isotope exchange equilibria (Galimov 1973, 1974, 1985). If, instead, their isotopic compositions are controlled by irreversible reactions, then they can only be approached through some understanding of inheritance of isotopic compositions from precursors and the kinetic isotope effects (KIEs) associated with non-reversible chemistry (Ivlev *et al.* 1974; DeNiro & Epstein 1977; O'Leary *et al.* 1981).

This debate was long-since decided in favour of inheritance and kinetics (we will see below several good reasons why; see Hayes (2001) for a broader review of this issue). But there are at least three reasons for us to retain an interest in thermodynamically controlled isotopic distributions in biomolecules:

(1)   Some biosynthetic reactions are reversible, at least for some elements, sites and conditions, and so could achieve something close to equilibrium isotope distributions (e.g. Valentine *et al.* 2004; Gilbert *et al.* 2012*a, b*). This possibility is re-enforced by the recent discovery that biogenic methane in marine seeps and most subsurface environments forms in clumped-isotope equilibrium at its formation temperature (Stolper *et al.* 2017).

(2)   Buried organic matter may be susceptible to reversible isotopic exchange with water and other environmental constituents, meaning even some compounds that form by kinetically controlled processes may later achieve isotopic equilibrium during burial diagenesis (e.g. Schimmelmann *et al.* 2006).

(3)   Equilibrium fractionations often provide useful approximations of KIEs because both are influenced by isotopic effects on vibrational energies of molecules (for example, the C isotope fractionation associated with CO_2 reduction on the RuBisCO enzyme is similar in direction and amplitude to the equilibrium fractionation between CO_2 and graphite (Guy *et al.* 1993; Chacko *et al.* 2001)).

Few experimental studies examine equilibrium isotopic fractionations in organic molecules (for the illuminating reason that they are difficult to produce in most compounds; see Wang & Sessions (2009*a, b*) for rare examples). But there is a much larger number of theoretical studies that predict site-specific and clumped-isotope structures of organics (Galimov & Shirinskii 1975; Galimov 1985; Rustad 2009; Webb & Miller 2014; Kubicki *et al.* 2016; Piasecki *et al.* 2016*a*; Webb *et al.* 2017). Several trends observed or predicted by these studies are as follows:

(1)   The distribution of ^{13}C within and between organic molecules is predicted to follow a pattern where the $\delta^{13}C$ of carbon sites that participate in C—O bonds (carboxyl and carbonyl groups) is greater than carbon that is exclusively bonded to other carbon and hydrogen (e.g. methyl groups), both of which are greater than the $\delta^{13}C$ of carbons bonded to sulphur (e.g. H_3C—S—..., such as in methionine). The amplitudes of these effects are predicted to be tens of per mille at earth-surface conditions (Galimov & Shirinskii 1975; Galimov 1985; Rustad 2009). Where multiple C—H bonding environments are present (such as the alkanes), ^{13}C is predicted to be preferentially partitioned into —C— and —CH— v. —CH_2—, and all these are enriched compared to – CH_3 sites (Webb & Miller 2014; Kubicki *et al.* 2016; Piasecki *et al.* 2016*a*), with amplitudes on the order of 10‰.

(2)   The distribution of deuterium in hydrocarbons is generally predicted to follow a pattern where —CH_2— are *c.* 70‰ higher in δD than —CH_3— groups at ambient temperatures (Wang *et al.* 2009*a, b*; Webb & Miller 2014; Kubicki *et al.* 2016; Piasecki *et al.* 2016*a*). In branched alkanes, —CH— groups are predicted to be even more strongly enriched in D relative to methyl positions ($\varepsilon^D_{CH–CH3}$ *c.* 150‰ at 20°C). The D abundances in H-bearing sites adjacent to C=C double bonds (i.e. in alkenes) are somewhat lower than their structural equivalents in alkanes (by *c.* 20–40‰ at 100°C).

(3)   Adjacent multiple ^{13}C substitutions in ethane and propane are expected to have Δ_i values (enrichments relative to a stochastic distribution) of *c.* 0.5‰ at 20°C (Piasecki *et al.* 2016*a*; Webb *et al.* 2017). Two non-adjacent ^{13}C substitutions in hydrocarbons are predicted to have negligible enrichments at equilibrium (Δ_i value ≪0.1‰). Immediately adjacent clumping of ^{13}C with D or D with D in hydrocarbons are predicted to have associated Δ_i values of *c.* 5‰ and *c.* 20‰, respectively, at earth-surface temperatures. Predicted enrichments (Δ_i values) of clumped-isotope species that have multiple but non-adjacent $^{13}C + D$ or $D + D$ substitutions (e.g. $H_2D^{12}C$—$^{13}CH_2$—$^{12}CH_3$) are about an order of magnitude smaller, placing them near or below the limits of precision for any current analytical method.

For all of the generalizations made above, the magnitudes of these equilibrium fractionations are predicted to increase with decreasing temperature, without 'cross-overs' (changes in sign of fractionation) over the range of temperatures most relevant to natural hydrocarbons (*c.* 0–300°C). Only a small number of these predictions have been directly observed by experiment, and the level of theory used to calculate most of them has uncertainties that can be as much as *c.* 50%, relative (Kubicki *et al.* 2016). What is called for here is a body of experimental work that is only now becoming technically feasible and has no real precedence in the literature.

The (mostly) kinetically controlled isotopic structures of common biomolecules

Despite the reasons for interest in isotope exchange equilibria between and within organic molecules, the weight of evidence strongly indicates that the isotopic structures of biomolecules dominantly reflect inheritance from their precursors, modified by the KIEs of irreversible biosynthetic reactions. If features of these biosynthetic isotopic structures are inherited in some recognizable form by petroleum or gas compounds, then we should be able to use stable isotope data to identify specific precursors of petroleum compounds and the mechanisms of reactions that converted them to oil and gas components. This vision has been recognized for more than 30 years (e.g. Monson & Hayes 1982*a*). Nevertheless, we face this large, complex subject with only a few constraints and little detailed understanding:

Fatty acids. We understand more about the isotopic structures of fatty acids than any other major class of biomolecules (although the title for best-described group might go instead to the sugars; below). An early indirect insight into this subject came from an experiment conducted by DeNiro & Epstein (1977), who examined the carbon isotope effect associated with decarboxylation of pyruvate to form acetyl coenzyme A – a step common to the biosynthetic pathways of fatty acids (and other metabolites). This study found that the KIE for pyruvate decarboxylation leads to a *c.* 15‰ difference in $\delta^{13}C$ between methyl (higher $\delta^{13}C$) and CO–R (lower $\delta^{13}C$) in product acetyl groups. It is worth noting that this is an important instance where the expectations of equilibrium isotope distributions are clearly not observed (i.e. at equilibrium CO_n groups should be higher in $\delta^{13}C$ than CH_n groups). This finding leads to the expectation that biological fatty acids should be lower in $\delta^{13}C$ than associated carbohydrate (as is commonly observed); and, because most sites in biological fatty acids are derived from head-to-tail attachment of acetyl groups ('elongation'), fatty acids should exhibit a distinctive 'even/odd' carbon isotope structure, where higher $\delta^{13}C$ positions derived from the methyl ends of acetyl groups alternate with lower $\delta^{13}C$ positions from the CO–R ends.

In a series of papers that are noteworthy both for their analytical innovation and depth of mechanistic interpretation, Monson & Hayes (1980, 1982*a*, *b*) examined the carbon isotope structures of fatty acids by chemical attack (decarboxylation and ozonolysis,

followed by decarboxylation) to selectively liberate a subset of the carbon sites. They first studied *Escherichia coli* – a bacterium that lacks mitochondria – finding that terminal carboxyl groups of C14–18 fatty acids are generally lower in $\delta^{13}C$ than their bulk parent molecules (by anywhere from *c.* 1 to *c.* 14‰), and that the olefinic carbons in unsaturated fatty acids (i.e. the C=C double bond – the sites oxidized by ozonolysis) are also depleted in ^{13}C by several per mille relative to their bulk parent. (Note this is a second instance where carbon isotope variations within natural biomolecules are opposite those expected for equilibrium isotope distributions). Monson & Hayes (1980, 1982*a*) develop an argument based on our understanding of fatty acid biosynthesis, showing that these findings are consistent with an 'even/odd' carbon isotope ordering, where carbons from the CO–R sites carry a distinctive ^{13}C depletion (as one would expect from the experiment of Deniro & Epstein (1977)). We will see that this pattern could be responsible for one of the more distinctive findings to date regarding carbon isotope structures of alkanes.

However, as self consistent as these findings may be, the picture they paint is incomplete. In a follow-up study, Monson & Hayes (1982*b*) applied their approach to fatty acids from *Saccharomyces cerevisiae* (yeast; a eukaryote that possesses mitochondria), obtaining results suggesting an 'even/odd' isotopic ordering that is more subtle (approximately several per mille in amplitude) and opposite in sign, suggesting assembly from acetyl groups having ^{13}C-depleted methyl groups and ^{13}C-rich CO–R groups. A detailed explanation of this finding is beyond what we can explore here, but may generally be said to result from the citric acid cycle, which produces and consumes acetyl-CoA by means other than pyruvate decarboxylation.

High-sensitivity nuclear magnetic resonance (NMR) techniques have been used to study the hydrogen isotope structures of several natural fatty acids (Billault *et al.* 2001; Duan *et al.* 2002; Markai *et al.* 2002; Lesot *et al.* 2008). This work is complicated by the fact that hydrogen sites in the centres of saturated fatty acids are difficult to distinguish by NMR; nevertheless, through a combination of derivatization and focus on unsaturated and branched fatty acids and novel NMR approaches, a detailed picture of the pattern of natural variations has been reconstructed. Site-specific D/H variations generally have amplitudes of *c.* 300–400‰; one common pattern observed is strong (several hundred ‰) D depletion of hydrogen in methyl groups relative to CH_2 groups – at least broadly consistent in direction with predicted equilibrium controls (Wang & Sessions 2009*a*, *b*). However, several observations argue against such a simple interpretation. The amplitudes of natural variations are several times greater than expected for equilibrium effects; and, CH_2 groups adjacent to sites of desaturation can also be strongly D depleted. Furthermore, pro-S and pro-R hydrogen sites can differ in D/H ratio. All of these findings suggest the importance of KIEs and inheritance from isotopically distinct pools of H during biosynthesis. The oxygen and clumped-isotope structures of natural fatty acids are unknown.

Amino acids. Some features of the site-specific carbon isotope structures of amino acids were constrained by Abelson & Hoering (1961) – this being the first attempt to systematically examine isotopic structures of an important class of biomolecules. This study compared the bulk $\delta^{13}C$ values of several common essential amino acids with the $\delta^{13}C$ of CO_2 released from those amino acids by decarboxylation, thus constraining the difference in $\delta^{13}C$ between carboxyl and non-carboxyl carbons. (We are aware of current efforts to revisit this subject using NMR and high resolution mass spectrometry, but neither body of work has been published.)

The first-order result of this work is that primary producers (autotrophs) contain amino acids with carboxyl positions that are generally higher in $\delta^{13}C$ than other carbon sites, by roughly 10‰ on average. This is an instance where the predictions of equilibrium stable isotope effects (^{13}C enrichments in the oxidized carbon sites) are met, at least approximately. Rustad (2009) explores this relationship in some detail by comparing theoretical equilibrium predictions to measured natural products for several amino acids, noting that despite differences in detail there is a general similarity in direction and magnitude of difference in $\delta^{13}C$ between carboxyl and non-carboxyl carbon sites. However, Abelson & Hoering (1961) also examined amino acids from two heterotrophs and found they exhibit the opposite pattern – carboxyl carbons lower in $\delta^{13}C$ than non-carboxyl carbons (i.e. similar to the carboxyl/non-carboxyl difference in fatty acids from *E.coli*). To the best of our knowledge, there are no observations of the site-specific H, O, N or S isotope geochemistry or clumped-isotope geochemistry of amino acids.

Carbohydrates. Early studies of carbon isotope structures of biosynthetic fatty acids and amino acids (above) assumed, explicitly or implicitly, that primary photosynthate (3-phosphoroglyceric acid) and derived carbohydrates have relatively simple carbon isotope structures with minimal site-specific variations. However, chemical degradations of sugars followed by Isotope Ratio Mass Spectrometry (IRMS) (Rossmann *et al.* 1991) and recent NMR studies of sugars and starches (Gilbert *et al.* 2012*a*, *b*) show this not to be the case. Site-specific carbon isotope variations in common C6 sugars have

amplitudes up to *c.* 25‰, and differ systematically in the pattern of site-specific variation depending on photosynthetic pathway (C3 v. C4 v. crassulacean acid metabolism (CAM) photosynthesis). The patterns of carbon isotope variation in glucosyl and fructosyl groups are complex and diverse, but generally exhibit depletions in $\delta^{13}C$ in the C1 and (particularly) C6 positions relative to C2–5 positions. The largest systematic elaboration on this pattern is that the C2 positions of glucosyl groups are generally lower in $\delta^{13}C$ than C2 positions in fructosyl groups from the same source. These variations can be understood as consequences of primary carbon fixation (suggested to set the $\delta^{13}C$ of C6 carbons in sugars), steps in the pentose phosphate pathway, photorespiration, and reactions involving assembly of sucrose and larger carbohydrates, and breakdown of those compounds for export and utilization.

A long-standing and large body of work examines the D/H ratio of 'non-exchangeable hydrogen' in cellulose – that is, hydrogen that is bound to carbon and does not undergo isotopic exchange with water on laboratory timescales, as opposed to the highly exchangeable hydroxyl hydrogen also present in cellulose (Epstein *et al.* 1976, and many subsequent papers). While such data is not precisely a site-specific isotopic measurement, it does sample a subset of molecular sites having common properties and so has some bearing on our subject. The first-order findings of this work are that these non-exchangeable hydrogens are lower in δD than coexisting environmental waters, typically by *c.* 200‰ but with secondary dependence on aridity and temperature. This phenomenon is well explored as a palaeoclimate archive.

Detailed (i.e. truly site-specific) hydrogen isotope structures of carbohydrates are less well known in natural materials, but have been examined by NMR methods in soluble sugars, starch and cellulose from several sources (Zhang *et al.* 2002; Betson *et al.* 2006; Augusti 2007; Augusti *et al.* 2008). The most general finding is that site-specific D/H variations have high amplitudes (hundreds of per mille); that patterns of D/H variation differ by photosynthetic pathway (i.e. C3 v. C4 v. CAM plants); and that differences in these isotopic structures between sugars and starches from the same plants suggest a strong influence of reactions occurring after initial synthesis of C6 sugars. The deconvolution of primary biosynthetic signals from subsequent fractionation and/ or exchange is best understood for wood cellulose, based on experiments involving introducing labelled water to plant tissues followed by NMR analysis of hydrogen isotope structures of extracted carbohydrates (Augusti *et al.* 2008). This work suggests site-specific data could be used to deconvolve climatic records from variable metabolic effects (although this strategy has not been widely employed as yet).

A second significant finding is that the difference in D/H ratio between the two hydrogens attached to the C6 position of glucose (designated C6H^R and C6H^S) from C3 plants varies by c. 200‰, changing systematically with environmental pCO$_2$/pO$_2$ (Ehlers *et al.* 2015). This is believed to result from the fact that varying pCO$_2$/pO$_2$ changes the relative rates of photosynthesis and photorespiration, which changes the mechanism of hydrogen addition to the relevant carbon site of 3-phosphoroglycerate – a sugar precursor. It is unknown whether this or other similar site-specific hydrogen isotope variations can survive diagenesis and catagenesis to appear in petroleum or gas compounds. Nevertheless, it provides some sense of the possible amplitudes and environmental controls of the hydrogen isotope structures of biomolecules.

Despite the fact that the bulk oxygen isotope composition of cellulose is well explored as a palaeoclimate archive, we are not aware of any previous observations of the oxygen isotope structures of natural carbohydrates.

Terpenes. Martin *et al.* (2006) present a summary and interpretation of NMR measurements of the hydrogen isotope structures of geraniol and α-pinene of various origins, showing that site-specific D/H variations are very large (a factor of c. 4) and differ systematically between C3 and C4 plants. These findings can be interpreted as consequences of inheritance of hydrogens from metabolic precursors (pyruvate, glyceraldehyde 3-phosphate, dimethylallyl diphosphate and isopentenyl diphosphate) accompanied by large KIEs. The resulting hydrogen isotope structures differ from common patterns of equilibrium hydrogen isotope distribution by a factor of two or more and so present another instance of strong violations of equilibrium in molecular-scale isotopic distributions in biomolecules. The terpenoids are abundant in petroleum, and, if their biosynthetic hydrogen isotope structures survive diagenesis and catagenesis, they may present an opportunity to isotopically fingerprint the biosynthetic origins of an oil component. Alternatively, if isotopic exchange and fractionation during burial overprint these high-amplitude biosynthetic signatures, it is imaginable that the extent of preservation v. homogenization of such effects could serve as a proxy for oil maturity.

Lignin. The isotopic structures of biopolymers and other large biomolecules present great challenges to existing technologies for site-specific and clumped-isotope analyses (see Appendix B). Nevertheless, there have been significant efforts to observe at least some features of the isotopic anatomy of lignin, which has been the subject of a measurement technique involving chemical extraction of CH$_3$

from methoxy groups, followed by hydrogen and/ or carbon isotope analysis (Keppler *et al.* 2007; Feakins *et al.* 2013). This technique reveals that methoxy group hydrogen has a D/H ratio that is similar to the non-exchangeable hydrogens in cellulose; i.e. it is c. 150–200‰ lower than but correlated with the δD of environmental water. This relationship is of interest primarily because of its potential as a palaeoclimate archive. The carbon isotope compositions of methoxy groups from lignin are less well explored because they do not obviously add any information beyond a bulk analysis of the global δ^{13}C of woody materials, and there is evidence that they are vulnerable to changes during biological degradation of plant litter (Anhauser *et al.* 2015). We revisit this issue below, in our discussion of the carbon isotope structures of lignites and coals that have been subjected to diagenetic and catagenetic changes.

General observations

We conclude this discussion with a few general observations: The carbon isotope structures of fatty acids are perhaps the firmest ground on which we can stand when trying to predict the isotopic structures of petroleum and gas components, because lipids are relatively abundant and well preserved after early diagenesis, and because they exhibit relatively simple systematic isotopic variations (i.e. the 'even/odd ordering' carbon isotope motif). However, the differences in isotopic structures between fatty acids from bacteria and Eukarya reveal complexity that complicates our task. It will not be possible to develop a predictive framework for the isotopic structures of petroleum compounds derived from fatty acids until we have a more complete understanding of the diversity in isotopic structures of these precursors, and some sense of how that diversity maps onto different forms of sedimentary organic matter. The same could be said for the amino acids, with their systematic differences between autotrophs and heterotrophs, and carbohydrates, with their systematic differences among C3, C4 and CAM plants. Perhaps more problematically, the isotopic structures of primary biological straight-chain hydrocarbons are unknown, despite their abundance and obvious relevance to interpreting structurally similar petroleum compounds. It seems plausible that they are closely related to the fatty acids (i.e. simply through decarboxylation); nevertheless, observations of the isotopic structures of these important compounds should be a high priority. Similarly, little is known about the isotopic structures of hopanoids; these compounds represent a diverse and potentially information-rich target for future studies of isotopic structure. Finally, we note that there are no observations of any kind that constrain the abundances of any of the multiply substituted forms of any of the abundant classes of biomolecules (excepting biogenic methane; see Douglas *et al.* (2017) and Stolper *et al.* (2017)).

Kerogen

Essentially all of the work summarized in this review concerns the isotopic structures of biomolecules (above) or components of oil and natural gas (below). It is natural to wish to explain the compositions of the latter through derivation from the former, but there is a third major class of natural organic compounds that stand between these two: the humins, lignins and kerogens in sedimentary rocks (Fig. 1). These structurally complex organic polymers are derived from buried organic matter, they contain components that are recognizable biomolecules, and their thermal degradation produces most petroleum and natural gas components. It will probably be impossible to reach a complete, detailed and coherent understanding of the isotopic structures of petroleum and gas components until we come to terms with the isotopic structures of the materials from which many of them are derived. The key question is whether these organic polymers formed during diagenesis and catagenesis mostly preserve the isotopic structures of biomolecules, or are instead dominated by isotopic fractionations during decarboxylation, reduction, deamination, polymerization, condensation (ring closure) and other diagenetic and catagenetic reactions.

There are vanishingly few direct observations of isotopic structures of kerogens and other sedimentary organic polymers that can guide us, but our understanding of kerogen synthesis (Tissot & Welte 1984; Burnham & Sweeney 1989; Hedges *et al.* 2000; Mao *et al.* 2010) leads to several expectations. First, the increasing proportions of straight-chain hydrocarbons, fatty acids, lignins and cellulose (relative to amino acids and soluble sugars) during remineralization and early diagenesis (Wakeham *et al.* 1997; Hedges *et al.* 2000; Freeman 2001; Marshall & Rodgers 2008) should lead us to expect that kerogens have isotopic structures more similar to these preserved compounds. Specifically, it is reasonable to predict the type I kerogens (rich in aliphatic hydrocarbons) should be most similar in isotopic structure to the fatty acids, the type II kerogens (rich in aromatic hydrocarbons) should be more similar to terpinoids and hopanoids (unfortunately uncharacterized for their isotopic structures at present), and type III kerogens (derived from sources rich in woody land plants) should more closely resemble the isotopic structures of lignin and cellulose.

The most obvious caveat to these predictions is that the chemical transformations that occur during formation and thermal maturation of kerogen may be associated with KIEs having significant amplitudes, potentially modifying the isotopic structures in biomolecules into new patterns with which we are currently unfamiliar (Schimmelmann *et al.* 1999, 2001). Furthermore, it is known that at least some of the hydrogen in kerogens is exchangeable with water during burial diagenesis (Schimmelmann *et al.* 2006). This veil of diagenetic and catagenetic processes that separates the biomolecules from the geological kerogens may be the greatest hurdle to really characterizing and understanding the isotopic structures of kerogen-derived petroleum hydrocarbons.

What is known about these processes is that they do not result in dramatic differences in bulk $\delta^{13}C$ between recently deposited sedimentary organic matter and more deeply buried kerogens (Schimmelmann *et al.* 1999, 2001; Nabbefeld *et al.* 2010). Even relatively mature kerogens generally have bulk $\delta^{13}C$ and δD values in the range of bulk organic-rich sediment. Thus, it seems reasonable to expect that fractionations and exchange reactions accompanying diagenesis and catagenesis are likely to be restricted to particular sites that are relatively reactive, labile and/or exchangeable.

We know of only two related observation that speak directly to these issues: a study of the C and H isotope compositions of methoxyl groups in lignin subjected to microbial attack during burial in leaf litter (Anhauser *et al.* 2015), and a more recent study of the same moieties in lignites from the Japan margin, sampled by cruise IODP-337, as described by Lloyd *et al.* (2016) as well as Inagaki *et al.* (2015). Anhauser *et al.* showed that moderate amounts (*c.* 50%) of removal of methoxyl groups from lignin through microbial attack have no effect on δD, but drive $\delta^{13}C$ of the residual methoxyl groups up by *c.* 5‰. Lloyd *et al.* (2016) found that the trace of residual methoxyl groups in ancient, deeply buried lignites (estimated as *c.* 1% of the initial methoxy inventory remains) have δD_{SMOW} values of −220 to −230‰ (similar to modern plant lignins) but extraordinary $\delta^{13}C_{PDB}$ values of +25−+40‰, which are among the highest $\delta^{13}C$ values known for terrestrial geological materials and *c.* 50‰ enriched relative to methoxyl groups in common plant lignins. One explanation for these findings is that the methoxyl groups are degraded during deposition, diagenesis and earliest catagenesis (microbially and/or possibly by other mechanisms), and that this loss is rate limited by cleavage of the C—O bond in the methoxyl group, which is accompanied by a *c.* 10‰ primary KIE for the methoxyl carbon but no or only negligible secondary KIEs for the adjacent methoxyl-group hydrogens. (One prediction of this hypothesis is that the oxygens in the residual methoxy groups should be similarly enriched by tens of per mille in ^{18}O and ^{17}O, although it is not obvious at present how they might be measured for their isotopic composition.) These findings, sparse as they are, suggest that the site-specific isotope geochemistry of kerogens hold potential as a record of the mechanisms and extent of kerogen maturation.

KIEs associated with 'cracking' reactions

The formation of geological bitumen, oil and hydrocarbon gas is mostly due to thermal degradation of kerogen; formation of light oil and gas components by thermal degradation of oil and bitumen components is widely suspected but more controversial (Tissot & Welte 1984; Lewan 1985, 1993; Quigley & Mackenzie 1988; Mango 1991; Price & Schoell 1995; Seewald *et al.* 1998). The liquid products (bitumens and oils) are dominated by naphthenes and paraffins – that is, aromatic and straight- or branched-chain hydrocarbons, respectively – whereas the gas products are overwhelmingly composed of low-molecular weight (C1–5) alkanes. These reactions generally proceed by cleavage of kerogen components, either at the sites of functional groups or at C—C bonds in chain- or ring-structures, producing diverse petroleum compounds, abundant H_2O and CO_2, and leaving residues that are depleted in H and O and condense, or aromatize, towards a progressively more graphitic structure.

If we are to predict the isotopic structure of a particular petroleum compound that is a product of these processes, the key questions are: what specific components of kerogen are its precursors; what are the isotopic structures of those precursors; what specific reactions occurred (e.g. in a simple case, what bond in a precursor broke to liberate the compound of interest); and what are the KIEs associated with that reaction? This list is straightforward and necessary, but probably cannot be answered rigorously for any petroleum or gas compound. Kerogens are structurally complex, containing many compounds and sites that potentially serve as reactants for any given product (particularly the relatively small, structurally simple hydrocarbons that are most amenable to isotopic studies). The mechanisms and stoichiometries of 'cracking' reactions have been the subject of much prior study, but are generally described only at a schematic level intended to represent some average across a broad family of reactants and products.

Two ways of making progress with this challenging problem have been suggested. The first way is to conduct experimental or theoretical studies that examine KIEs associated with simplified model systems (e.g. production of methane by breaking the CH_3—CH_{2a} bond in an *n*-alkane). These studies have the advantage of constraining potentially generalizable atomistic mechanisms, and the disadvantage that they obviously oversimplify the reality of geological petroleum and gas formation. The second way is to conduct experimental studies that empirically relate the isotopic composition of a petroleum or gas product to the isotopic content of a kerogen reactant. These studies have the advantage of being conceptually straightforward and obviously relevant to at least the kerogens and experimental conditions that were studied; their disadvantage is that they provide few mechanistic insights that might let one generalize their findings to other source materials, conditions and products.

Idealized models

The first widely used mechanistic model for the KIEs associated with oil and gas formation was presented by Chung *et al.* (1988). (Note our discussion is also generally consistent with the elaboration of this model to include Rayleigh effects in Rooney *et al.* (1995).) This model is inspired by the observation that the $\delta^{13}C$ values of individual molecular components of natural gases (methane, ethane, propane, butanes) often display a linear negative correlation with the inverse of their carbon number (i.e. one-quarter for butane, one-third for propane, etc.), with a vertical intercept similar to the $\delta^{13}C$ values of inferred source kerogens for those gases (Fig. 2). One interpretation of such trends is that these gas components are derived from pyrolysis of an isotopically homogeneous pool of kerogen carbon, with a carbon isotope fractionation that affects only one carbon atom in the product molecule. In this case, the methane–kerogen fractionation equals the KIE (usually inferred to be *c.* 20‰), the ethane–kerogen fractionation half this value (i.e. because only one of the two carbons expresses the fundamental KIE), and so forth. Chung *et al.* (1988) conceptualized this model by suggesting the source substrate can be thought of as a straight-chain hydrocarbon, and that the 'cracking' reaction involves severing one C—C bond to release the product alkane (e.g. breaking CH_3—CH_{2a}, where CH_3 is the terminal methyl group and CH_{2a} is the methylene group immediately adjacent to it, yields methane; breaking CH_{2a}—CH_{2b} yields ethane, etc.). It follows that the carbon isotope composition of the carbon adjacent to the broken bond (e.g. one of the two terminal carbons in propane) should have a $\delta^{13}C$ equal to that for kerogen minus the KIE, and that all other carbons in the product molecule should have a $\delta^{13}C$ equal to that for source kerogen.

Although Chung *et al.* (1988) do not address site-specific or clumped-isotope variations, their model can be extrapolated to predict the possible consequences for such indices. Site-specific

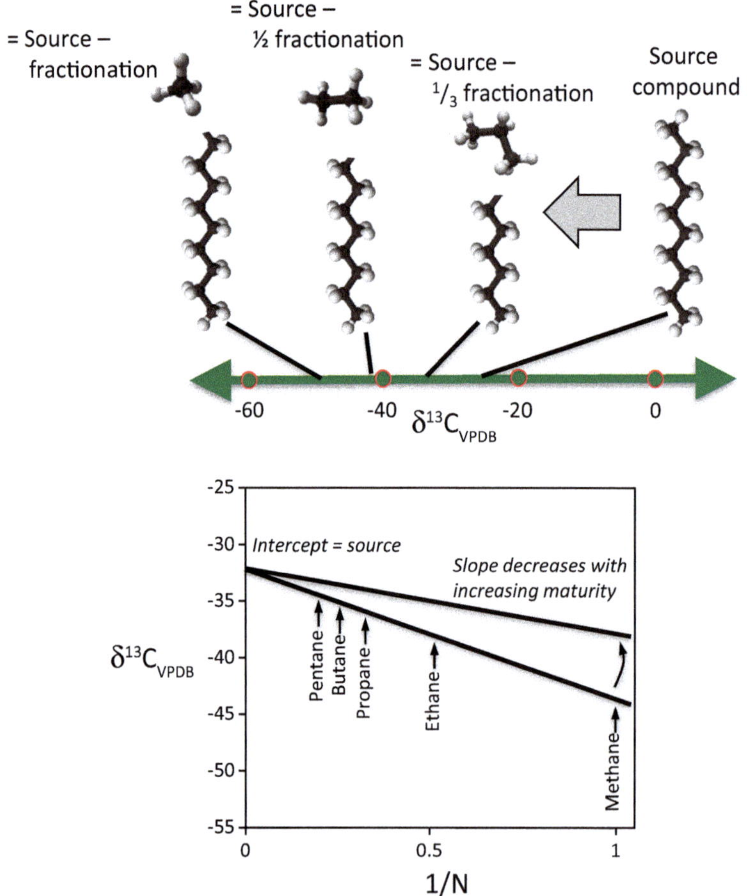

Fig. 2. Conceptual model for the carbon isotope compositions of natural gas components formed by cracking a larger organic molecule (represented here by an *n*-alkane), adapted from Chung *et al.* (1988). Each product molecule is created by cleaving a single C—C bond in the source compound, with a fractionation that may vary with temperature but is assumed to be a constant for all reactions having different product compounds. The smallest product (methane) expresses this full fractionation, and larger products dilute that fractionation by combining the fractionated site with adjacent unfractionated sites. The δ^{13}C$_{VPDB}$ of the products will exhibit a negative linear correlation with $1/N$, where N is the number of carbon atoms in each product molecule, with a slope that flattens with increasing temperature (and thus diminishing fractionation) and with increasing reaction progress. δ^{13}C$_{VPDB} = (R^{13}_{sample}/R^{13}_{VPDB} - 1) \times 1000$, where R^{13} is the ^{13}C/^{12}C abundance ratio and VPDB stands for the Vienna Pee Dee Belemnite reference standard.

compositions of *n*-alkanes produced according to this model are relatively easy to predict: in all cases, the average δ^{13}C of the two terminal CH$_3$ groups should be equal to that of the source kerogen (assumed for the time being to be homogeneous) minus half of the KIE; e.g. if kerogen has a δ^{13}C$_{PDB}$ of −25‰ and C—C bond cleavage involves a 20‰ normal KIE, then product *n*-butane should have an average δ^{13}C$_{PDB}$ of −35‰ for the two indistinguishable terminal methyl groups, and of −25‰ for the two indistinguishable interior CH$_2$ groups. We are not aware of any effort to extend the Chung *et al.* (1988) model to consider hydrogen

isotope compositions of natural gas components, or as a means of predicting isotopic compositions of oil compounds.

The clumped-isotope effects of the Chung *et al.* (1988) model are also easy to predict, though somewhat counterintuitive. Consider the case of a precursor straight-chain hydrocarbon having a random distribution of ^{13}C throughout its structure, 'cracking' at one of the CH$_{2a}$–CH$_{2b}$ bonds to produce ethane, with a KIE of 20‰. Immediately prior to the cracking reaction, the probabilities of ^{13}C being present on each of the two carbon sites that will eventually be transferred to ethane are equal to each other and to the bulk ^{13}C concentration in the precursor ('$[^{13}C]_i$'). Thus, in this case the probability that both of those two sites contain ^{13}C is $[^{13}C]_i^2$, and the Δ_{13C2H6} value (enrichment of the clumped-isotope species relative to a random distribution) of that population of sites is 0‰. An infinitesimal progress of the cracking reaction will produce a population of product ethane molecules, which will consist of one carbon atom inherited from the terminal CH_3 site of the precursor and one carbon atom inherited from the CH_{2a} site of that precursor. According to the Chung model, the first of these atoms will have a probability of inheriting ^{13}C exactly equal to the bulk $[^{13}C]_i$ of the precursor, but the second will have a probability equal to that for the precursor, $[^{13}C]_i$, multiplied by a number closely related to the fractionation factor for the cleavage reaction (if the $\alpha_{KIE} = 0.98$, then this multiplier is 0.9802; the difference is a simple result of the difference between $^{13}C/^{12}C$ ratios and ^{13}C concentrations). Once ethane forms, these two carbon atoms are symmetrically equivalent, and the concentration of ^{13}C in the ethane pool as a whole, averaged across the molecule, will be ($[^{13}C]_i \times (0.5 + [0.9802]/2)$). If that ^{13}C were randomly distributed among all carbon atoms, then the probability of finding an ethane with two ^{13}Cs would just be this number squared: ($[^{13}C]_i \times (0.5 + [0.9802]/2)^2$. However, these two carbon atoms come from two different sources and have two different probabilities of containing ^{13}C: one of them is unfractionated and has the composition of the precursor, whereas the other is fractionated and contains less than the substrate's $[^{13}C]_i$ value. As a result, the actual probability of encountering a $^{13}C_2H_6$ molecule is: ($[^{13}C]_i^2 \times 0.9802$), which is lower than suggested by the expression for the random distribution, above. That is, even without considering the distinctive fractionation behaviours of the clumped-isotope species, or any possible non-random isotopic structure of the precursor, the Chung et al. (1988) model predicts that cracking reactions should produce ethane with a negative Δ_{13C2H6} value. This is just one example of many possible clumped-isotope effects that could arise from cracking reactions that subsample organic molecules accompanied by KIEs.

Several subsequent models add significant sophistication to the treatment of stable isotope fractionation during natural gas formation but the researchers in question do not clearly comment on the expected molecular-scale isotopic structures of products or residual reactants (e.g. Berner et al. 1995). An exception that does add new details regarding isotopic structure is the work of Tang et al. (2000, 2005), who present the most sophisticated treatment of KIEs associated with hydrocarbon cracking published to date. Similar to the Chung et al. (1988) model, Tang et al. (2000) consider a relatively large n-alkane ($C_{16}H_{34}$, hexadecane) that undergoes cleavage reactions at or near one of its ends to produce methane (breaking the CH_3—CH_{2a} bond), ethane (breaking the CH_{2a}—CH_{2b} bond) or propane (breaking the CH_{2b}—CH_{2c} bond). However, rather than simply making a first-order assumption about the sites and magnitudes of the associated KIEs, Tang et al. (2000) performed a density functional theory (DFT) model of the vibrational energetics of the reactants and products and calculated the relative reaction rates of the singly ^{13}C-substituted isotopologues. Their findings are loosely consistent with the assumptions of Chung et al. (1988) (a 'normal' KIE, which favours transfer of ^{12}C to products, and which is strongest at the carbon sites adjacent to the breaking bond, and c. 10–20‰ in amplitude). However, this more rigorous treatment also reveals two features that could be common to the KIEs associated with cracking reactions. First, 'secondary isotope effects', or isotope effects associated with ^{13}C substitution at sites not adjacent to the bond being broken (e.g. a difference between the rate of forming propane from $^{12}CH_3$—$^{12}CH_2$—$^{12}CH_2$—$^{12}CH_2$... and $^{13}CH_3$—$^{12}CH_2$—$^{12}CH_2$—$^{12}CH_2$, where the bold line indicates the bond to be broken). These secondary isotope effects have amplitudes c. one-quarter to one-half of those of the primary isotope effects (i.e. associated with substitution adjacent to the breaking bond), which are surprisingly large, and significant if one is to understand the isotopic compositions of petroleum compounds at the level of a few per mille or better. Second, the amplitudes of KIEs (both primary and secondary) depend on reaction temperature; in the cases Tang et al. (2000) consider, fractionations generally decrease in amplitude with increasing temperature. This second finding raises a first-order problem with the empirical interpretation of gas stable isotope data: Chung et al. (1988) emphasize the importance of reaction *progress* in controlling the δ^{13}C of a given gas component (with products approaching compositions of substrates as reaction progress increases). Tang et al. (2000) demonstrate that the same changes can reflect reaction to similar extents but at higher *temperature*.

Tang et al. (2005) experimentally examine the isotopic compositions of residues of pyrolysis of oil-fraction compounds (C13–21 n-alkanes) and extend the model of Tang et al. (2000) to explain their findings. The most important new points made by this second paper are (1) larger alkanes crack more rapidly than smaller ones at a given thermal maturity, meaning the progress of pyrolysis reactions with thermal maturity can vary markedly from small to large compounds; and (2) oil-fraction compounds of intermediate size are both the products of cracking larger compounds (tending to reduce their δ^{13}C and δD) and residues of their own cracking reactions (tending to raise their δ^{13}C and δD values). These arguments could be extended to encompass condensate and gas fraction compounds (though Tang *et al.* (2005) do not present detailed arguments regarding these species).

The treatment by Tang *et al.* (2000, 2005) of alkane-cracking reactions is an advance on that presented by Chung *et al.* (1988), both because of the use of first-principles chemical physics to describe KIEs and because of the consideration of the consequences of simultaneous cracking of a family of related compounds. However, for our purposes these two approaches are at least qualitatively similar in their predictions regarding isotopic structures of hydrocarbons, because they both examine the relatively narrow case of *n*-alkanes that react only be cleaving C—C bonds to produce smaller *n*-alkanes. While this is probably an important process in natural petroleum systems, it provides little insight into the expected consequences of cracking branched alkanes, aromatic compounds, fatty acids and other organic acids, kerogen components, or any of the many other organic components of petroleum and its source rocks. (In the interests of brevity, we do not explore in detail the quantitative differences between the Tang and Chung model approaches in the predicted isotopic structures of oil and gas compounds, but see Piasecki *et al.* (2018) for further discussion.)

Experimental simulations

There have been numerous previous studies of bulk stable isotope fractionations (i.e. averaged across all isotopologues of a given molecule) associated with pyrolysis and hydropyrolysis of kerogen, bitumen and oil (e.g. Lewan 1983; Berner *et al.* 1995; Schimmelmann *et al.* 1999, 2001; Tang *et al.* 2000, 2005). These demonstrate that carbon isotope effects are generally 'normal' in direction (favouring transfer of light isotopes from reactants to products), have amplitudes of *c.* 10–20‰ for $\delta^{13}C$ and *c.* 100‰ for δD, and decrease in amplitude with increasing size of the product molecule, temperature of reaction, and integrated progress of the reaction. All of these trends are broadly consistent with the theoretical expectations reviewed in the previous section.

We are aware of only three previous studies that document the clumped-isotope and position-specific isotope effects associated with experimental pyrolysis of kerogen to produce hydrocarbons larger than methane (see Stolper *et al.* (2017), for a review of experimental constraints on methane clumped-isotope compositions from such experiments). Results from these studies are currently in press, in review or available only in abstract form, but we consider them here in the interests of making this review as current and complete as possible.

Ethane $\Delta^{13}C_2H_6$. Clog *et al.* (2013, 2014) and Clog & Eiler (2014) present analyses of the bulk and clumped-isotope ($\Delta^{13}C_2H_6$) composition of ethane produced by hydropyrolysis of two kerogen-rich sedimentary rocks: Woodford Shale and Albian/Aptian lacustrine shale from the Araripe basin, Brazil (Fig. 3a). The $\Delta^{13}C_2H_6$ index is currently reported relative to an arbitrary reference standard – a commercial ethane gas that has a $\delta^{13}C$ and δD broadly similar to that of common thermogenic gases (below). For this reason, it is not known whether ethane produced by kerogen cracking has a $^{13}C_2H_6$ abundance more or less than expected for a random distribution (i.e. $\Delta^{13}C_2H_6$ above or below zero in an absolute reference frame). Nevertheless, these experimental products are 1–2‰ lower in $\Delta^{13}C_2H_6$ than the reference gas and *c.* 1–3‰ lower than *c.* 75% of the natural ethanes analysed to date. Thus, kerogen cracking produces ethane that is at least relatively low in $\Delta^{13}C_2H_6$.

Increasing the temperature and time of hydropyrolysis of Woodford Shale increases the $\Delta^{13}C_2H_6$ of product ethane subtly but significantly (by a few tenths of per mille) – consistent with our expectations based on extrapolation of the Chung *et al.* (1988) model (above). However, increasing temperature and duration of hydropyrolysis of Araripe shale has the opposite effect, further decreasing $\Delta^{13}C_2H_6$ by several tenths of per mille. The differences between these two experimental series could have several explanations. Perhaps the two source rocks in question contain ethane precursors with different carbon isotope structures, or perhaps the relative contributions of kerogen, bitumen and oil cracking over the temperature ranges of these experiments were not the same (presuming ethane formed by cracking these different precursors could differ in $\Delta^{13}C_2H_6$). It would be premature to prefer one of these or other plausible ideas before conducting more experiments of this kind. Nevertheless, these findings demonstrate that kerogen cracking can produce ethane with diverse clumped-isotope composition, varying by source rock and maturity.

Clog *et al.* (2013, 2014) and Clog & Eiler (2014) also present results of experiments in which ethane is destroyed by pyrolysis, either exposed to glass at 600°C or in the presence of glass and Ni catalyst at 200°C. Analysis of the residues of these experiments show them to be higher in $\delta^{13}C$ relative to starting materials (as expected for kinetically controlled pyrolysis that is faster for ^{12}C than ^{13}C), but substantially reduced in Δ_{13C2H6} (Fig. 3b). The fact that Δ_{13C2H6} values decrease as $\delta^{13}C$ increases is not surprising – this is the sign of the clumped-isotope effect anticipated for residues of molecular diffusion in most cases (Eiler 2007; 2013), and simple models of chemical-KIEs can follow mass laws similar to molecular diffusion (Bigeleisen 1949). However,

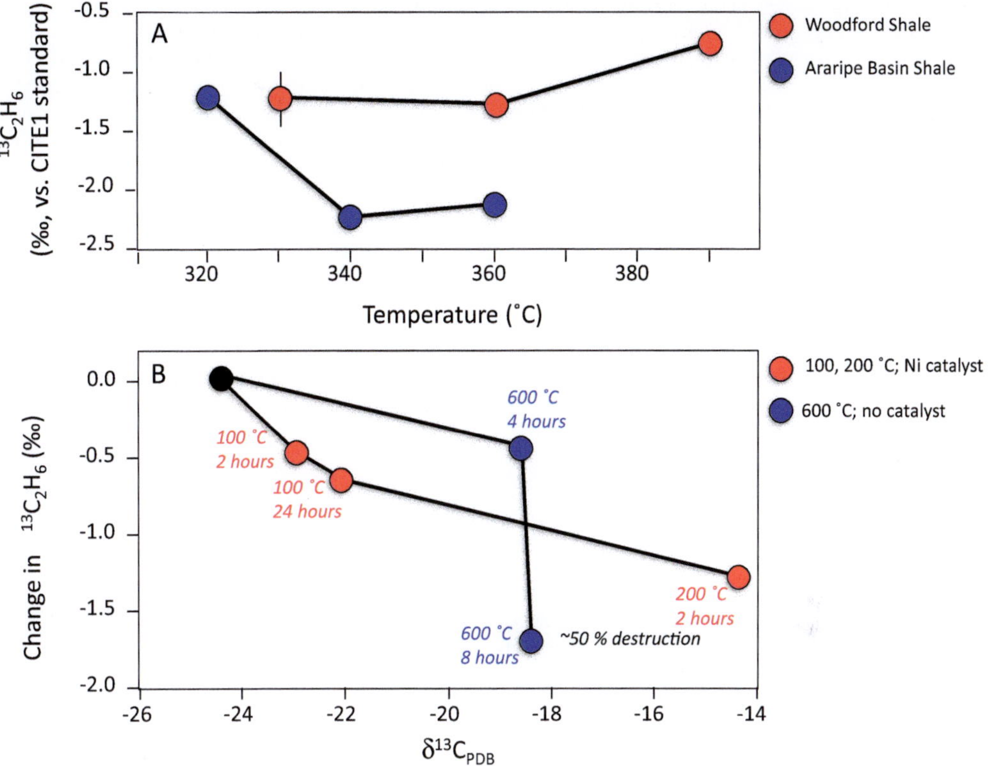

Fig. 3. $2 \times {}^{13}C$ clumped-isotope compositions ($\Delta^{13}C_2H_6$ values) of ethane: **(a)** produced by hydropyrolysis of shales; or **(b)** residual to ethane cracking. In (a), Δ_i values are reported relative to an arbitrary ethane standard; in (b) the vertical axis reflects the change in Δ_i from the starting composition. The highest $\delta^{13}C$ points in panel (b) reflect *c.* 50% ethane consumption. Data from Clog *et al.* (2014).

the magnitude of the decrease in $\Delta^{13}C_2H_6$ observed in ethane-cracking experiments is greater than can be explained by these sorts of transport-limited mass laws (which are generally restricted to *c.* 0.5‰ or less for tens of per mille ranges in $\delta^{13}C$). One possibility is that the KIE for ethane pyrolysis violates simple diffusion-like behaviour (e.g. if fractionations are controlled by differences in vibrational energy between reactant and transition states, or if the reaction mechanism has several steps). Another possibility is that the reactant has a large excess in $\Delta^{13}C_2H_6$ and that ethane consumption is slightly reversible, and thus is accompanied by a re-equilibration process that reduces that excess.

Propane and higher order hydrocarbon site-specific ^{13}C. Piasecki (2015) and Piasecki *et al.* (2016*b*, 2018) present analyses of the site-specific ^{13}C composition of propane (i.e. difference in $\delta^{13}C$ between terminal CH_3— and central —CH_2— groups) for products of hydropyrolysis of Woodford Shale (Fig. 4a). Their findings indicate that terminal carbon

positions increase in $\delta^{13}C$ with increasing extent and temperature of pyrolysis, as expected by both the Chung *et al.* (1988) and Tang *et al.* (2000, 2005) models of isotope effects associated with *n*-alkane cracking to form propane. However, these experiments also reveal that the central carbon position also rises over the course of cracking, by several per mille. This finding clearly disagrees with the simplified assumptions implicit in the Chung *et al.* (1988) treatment; it is consistent in direction with the predictions of Tang *et al.* (2000, 2005), though different in magnitude (*c.* 100% greater than the predicted change in central position $\delta^{13}C$). There are several possible explanations of this result. First, the various temperature steps of this hydropyrolysis experiment produce propane from different proportions of precursors – kerogen dominating early and oil dominating late – and these precursors could differ in the $\delta^{13}C$ for the carbon that is transferred to central position of propane (this is the explanation favoured by Piasecki (2015)). Second, propane formation may involve secondary KIEs that are

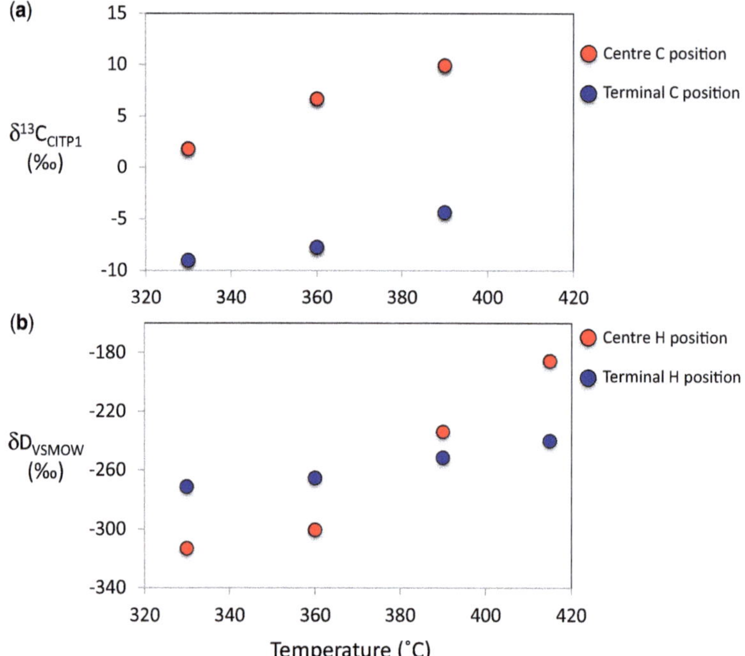

Fig. 4. Site-specific isotopic composition of propane produced by hydropyrolysis of Woodford Shale: (**a**) presents carbon isotope compositions, reported relative to the intralaboratory standard, CITP1 (data from Piasecki (2015)); (**b**) presents hydrogen isotope compositions reported relative to VSMOW, meaning the plotted centre-terminal difference reflects the actual value of the site-specific fractionation (data from Ponton *et al.* (2016)).

qualitatively similar to those predicted by Tang *et al.* (2000) but quantitatively stronger (possible, though perhaps unlikely given that these data imply the secondary isotope effect on the central C position is similar in strength to the primary isotope effect on the terminal C position). Third, it may be that propane forms not by cleavage of the CH_{2b}—CH_{2c} bond of an *n*-alkane, as assumed in the Chung and Tang models, but rather by decomposition of a branched or aromatic compound, involving a primary isotope effect on the site that donates carbon to the central position of product propane.

Julien *et al.* (2015) present the results of a study of site-specific carbon isotope fractionations associated with evaporation of several potential organic environmental contaminants. Although they find significant site-specific signatures of evaporation for ethanol and other small, polar compounds, the two naturally significant hydrocarbons they examined (toluene and *n*-heptane) show only insignificant or subtle site-specific differences in effective vapour pressure isotope effect (*c.* 1‰ or less). These findings suggest isotopic fractionations associated with volatility may be significant for some oil compounds (e.g. organic acids) but are unlikely to be large for the major components of oil and natural gas.

Propane site-specific D. There are no published details regarding the site-specific H isotope fractionations associated with pyrolysis of kerogen or other natural hydrocarbon precursors, and it is not obvious that they can be reasonably guessed at by extrapolation of the proposed models for ^{13}C effects. Thus, our only guide to this subject is the long-standing finding that δD values of natural gas and petroleum compounds often rise with increasing maturity in products from the same or similar source rocks, implying some sort of 'normal' KIE (i.e. H-bearing species react faster than D-bearing species, and increasing the progress of the reaction slowly diminishes this effect as products approach the composition of original substrates).

A recent conference presentation (Ponton *et al.* 2016) includes preliminary results documenting the site-specific D/H variation in propane generated by hydropyrolysis of Woodford Shale. These experiments (reproduced in Fig. 4b) show that the δD of both the terminal and central positions of propane rise by tens of per mille over the course of heating from 330 to 415°C (roughly corresponding to the transition from primary to secondary cracking in natural petroleum-generating systems), and that the difference in δD between these two sites also rises by tens of per mille, from a non-equilibrium fractionation where the central position is lower in δD than the ends at low temperature/maturity, to an approximately equilibrated signature where the central position δD is greater than the terminal at high temperature/maturity (Fig. 4b). One interpretation of this finding is that increasing thermal stress is associated with an approach to equilibrium intramolecular hydrogen isotope structure, either through internal exchange of hydrogens among different sites in the same molecule or through heterogeneous exchange between the molecule of interest and some other compound (water, clay, etc.).

Natural bitumen and oil-fraction compounds

A single study has been published documenting the isotopic structures of samples of compounds that are major components of natural bitumens and oils: Gilbert *et al.*'s (2013) study of the carbon isotope compositions of the CH_3, CH_{2a} and CH_{2b} positions (i.e. terminal and adjacent two carbon sites) of C11–31 *n*-alkanes. Before we discuss

these measurements, one key point must be made first: the sample set examined in this study consists of pure compounds purchased from chemical supply companies. Thus, while it seems possible that they are related in some way to natural biological and/ or petroleum sources, there is little concrete information that ties them to any particular natural material or process. While we will attempt to relate Gilbert *et al.*'s (2013) findings to our understanding of the isotopic structures of fatty acids and isotope effects associated with kerogen and oil cracking (as those authors did), it is possible that these measurements are not directly relevant to any natural petroleum compounds.

Gilbert *et al.* (2013) find three patterns of carbon isotope variation in the C11–31 *n*-alkanes (Fig. 5a; note these measurements were made by NMR, and that they do not constrain differences between the three plotted sites and the bulk molecules – i.e. only site-to-site differences in each molecule are constrained): (1) all compounds having an odd number of carbons and 17 or more carbons in all exhibit a pattern where both CH_3 and CH_{2b} are several per mille lower in $\delta^{13}C$ than the CH_{2a} position; (2) all compounds having an even number of carbons and

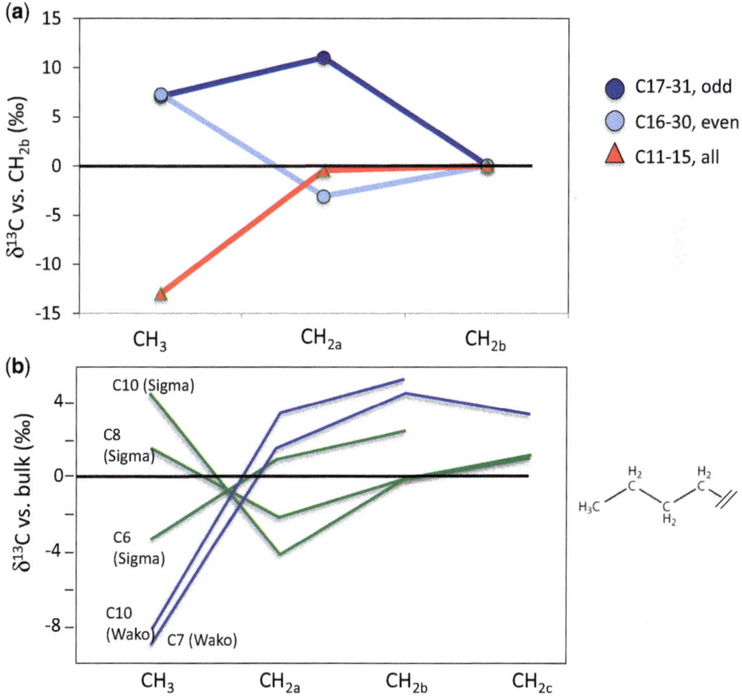

Fig. 5. Site-specific carbon isotope structures of the terminal 3 or 4 sites of *n*-alkanes from (**a**) oil-fraction compounds (from Gilbert *et al.* (2013)) or (**b**) condensate-fraction compounds (from Gilbert *et al.* (2016*b*)). In (a) data are reported relative to the CH_{2b} site (third from the end) and it is not known how any of these sites relate to the bulk $\delta^{13}C_{PDB}$; in (b), data are reported relative to the known molecule-average ('bulk').

16 or more carbons in all have a pattern where both CH_3 and CH_{2b} are several per mille higher in $\delta^{13}C$ than the CH_{2a} position (curiously, the difference between CH_3 and CH_{2b} is the same, on average, for these two groups); and (3) all compounds having 15 or fewer total carbons (whether even- or odd-carbon number) display a pattern where the $\delta^{13}C$ of the CH_{2a} and CH_{2b} sites are the same as one another but the CH_3 site is substantially ^{13}C-depleted relative to both CH_{2a} and CH_{2b}.

Gilbert *et al.* (2013) suggest that the carbon isotope patterns of the C16+ *n*-alkanes can be understood to be products of decarboxylation of fatty acids, which are known to exhibit even/odd ordering of alternately high and low $\delta^{13}C$ values (though in detail it is not clear how the two patterns they observe in *n*-alkanes relate to the diverse carbon isotope structures of fatty acids from *E. coli* and yeast; see above). The C11–15 *n*-alkanes could be understood to be products of cracking longer-chain alkanes and fatty acids, where mixing between fragments derived from even- and odd-carbon number precursors averages out to produce a mixture with no significant 'even/odd' isotopic ordering in the interiors of carbon chains, but a KIE associated with the cracking reaction leaves its imprint as ^{13}C depletion of the terminal carbons (at least one of which was presumably adjacent to the bond that broke in the longer compound).

Gilbert *et al.* (2013) clearly demonstrate the existence of systematic, high-amplitude carbon isotope variations in the *n*-alkanes; and, although one could propose other explanations of their findings, the hypotheses they present are easily understood and sensibly derived from an understanding of fatty acid synthesis. For these reasons, this study suggests large alkanes inherit their isotopic structures from their fatty acid precursors, whereas small ones mostly express 'mean reversion' through mixing, plus the fractionations associated with cracking (Gilbert *et al.* 2013). But, we must recall that none of this argument is based on measurements of recognized natural petroleum compounds. Studying natural geochemical materials and processes will be much more technically challenging because the NMR techniques used by Gilbert *et al.* (2013) currently require *c.* 0.1–1 g of pure samples. Each of the *n*-alkanes generally make up only *c.* 0.1–0.5 wt % of natural crude oil (*Oil in the Sea*, National Research Council 2003), meaning one analysis would require isolation out of one or more litres of crude oil, hundreds of litres of oil-contaminated waters, or kilograms of oil-containing rock. This is feasible but presents a so-far unaddressed technical challenge. Alternatively, it may be possible to study isotopic structures of oil-fraction compounds by pyrolysis followed by mass spectrometry (Gilbert *et al.* 2016*a, b*) or direct mass spectrometry (Piasecki

et al. 2016*b*), with significantly reduced sample sizes. However, the instruments and methods capable of such measurements are still in development (see Appendix B).

Natural condensate-fraction compounds

Gilbert *et al.* (2016*b*) present (in abstract form) the initial results of an NMR study of the carbon isotope structures of the C6,7,8 and C10 *n*-alkanes – species that are most strongly concentrated into the condensate fraction of petroleum systems (i.e. species less volatile than natural gas constituents, found in light oils, and more volatile than most components of bitumen and heavy oils). Their findings (Fig. 5b) loosely resemble observations for the C11–15 *n*-alkanes in their previous study (Fig. 5a; Gilbert *et al.* 2013): the interior CH_2 sites are relatively uniform in $\delta^{13}C$ and conform to a single, relatively simple pattern (CH_{2a} 1–2‰ lower in $\delta^{13}C$ than CH_{2b}, which is within *c.* 1‰ of CH_{2c} when present), whereas the terminal sites are strongly fractionated relative to the average of all CH_2 sites. However, there are two discrepancies between this pattern and that for the average C11–15 pattern in Figure 5a: (1) terminal CH_3 groups may be strongly ^{13}C depleted *or* enriched relative to CH_2 sites; and (2) the same compound obtained from different sources (C10 from Waco or Sigma Aldrich suppliers) can have opposite signs of the CH_3—CH_{2a} fractionation. It is difficult to imagine how these findings can be reconciled with the concept that these compounds are formed by mixing the products of cleaving fatty acid or longer alkane chains, superimposed on a normal primary KIE at the terminal sites. Furthermore, as for the C11–31 alkanes discussed above, these data were generated on pure alkanes purchased from chemical suppliers that generally do not provide detailed information about sources and formation mechanisms. For this reason, we can only speculate as to how these findings might speak to the carbon isotope structures of natural low-molecular-weight hydrocarbons.

Natural gas constituents

Up to this point, our focus on our subject has gradually blurred as we have moved from the long-standing body of work on relatively well-understood biosynthetic compounds, through the veil of barely known isotopic structures of humins, lignins and kerogens, to a sparse sampling of compounds that may or may not be related to components of natural bitumens and oils. However, there is some promise at the end of our story, as the isotopic structures of constituents of natural gases have been the subject of a relatively large, rapidly growing and diverse body of analytical work. The largest part by far concerns the abundances of doubly substituted forms of methane ($^{13}CH_3D$ and, more recently, $^{12}CH_2D_2$). This subject has grown so large in the last year that it warrants its own review (Douglas *et al.* 2017; Stolper *et al.* 2017). We focus here on the smaller but still significant body of recent work on the carbon and hydrogen isotope structures of natural propane and ethane.

Site-specific ^{13}C in natural propane

Currently, four independent research groups using four different analytical technologies are exploring the difference in $\delta^{13}C$ between the central and terminal C positions in propane, including: (1) high resolution gas source mass spectrometry (Piasecki *et al.* 2016*b*, 2018); (2) pyrolysis followed by gas chromatography (GC) separation and combustion of the products and IRMS of the resulting CO_2 peaks (Gilbert *et al.* 2016*a*); (3) sequential chemical degradation of propane followed by separate combustion of the products and IRMS on the resulting CO_2 (Gao *et al.* 2016); and (4) NMR on propane condensed in high pressure tubes (Liu *et al.* 2015). Only the first of these techniques has been applied to suites of natural gases having known origins, but all four provide important complementary constraints: NMR data and selective chemical degradations yield site-specific data that can be directly anchored to absolute scales (e.g. Pee Dee Belemnite (PDB)) but their sample sizes make it challenging to create large datasets on natural materials, whereas the mass spectrometric and pyrolitic techniques permit smaller sample sizes but are essentially relative methods, constraining only differences in site-specific $\delta^{13}C$ between samples and reference propane. Taken together, these various methods provide the following picture of intramolecular ^{13}C distribution in natural propane.

Gao *et al.* (2016) show that propane in the widely used natural gas reference standard, 'NG3', has a difference: $\delta^{13}C_{centre} - \delta^{13}C_{ends}$ of 19.2‰, where both quantities are known on the PDB scale. This is in the same direction and has the same order of magnitude as the equilibrium carbon isotope fractionation predicted by first-principles theory at geologically relevant temperatures (above). Liu *et al.* (2015) report an NMR measurement showing a similar centre–end fractionation of 15.8‰ for an unknown (or at least unreported) propane gas; considering that nearly all readily available propane is from natural thermogenic gas, it seems likely that this is a material generally similar to propane in NG3. Gilbert *et al.* (2016*a*) report the centre–end carbon isotope fractionation for three propanes: NGS2 (another widely available natural gas standard) = 12.8‰, whereas two other commercial propanes of uncertain origin have much lower values of 8.2 and 1.8‰. This

result is compromised by the fact that the pyrolysis method used in this study involves unknown fractionations that potentially effect accuracy; thus, the fact that they find smaller absolute fractionations than Gao *et al.* (2016) and Liu *et al.* (2015) may not be significant. Nevertheless, it is significant that Gilbert *et al.* (2016*a*) report a large (11‰) range in the centre–end fractionation.

Piasecki (2015) and Piasecki *et al.* (2016*b*, 2018) report the site-specific carbon isotope compositions of ten natural propanes from the Antrim Shale (Michigan), Eagle Ford Shale (Texas), and Potiguar Basin (Brazil). These data were measured by mass spectrometry and (as for the Gilbert *et al.* (2016*a*) observations) constrain *relative* differences in $\delta^{13}C_{centre}$ and $\delta^{13}C_{ends}$ between samples but not the *absolute* centre–end fractionation. The finding of this work (Fig. 6a) is that, as one moves from lower maturity wet gases (Antrim Shale and lower maturity oil-associated Potiguar gases) to higher maturity gases (higher maturity Potiguar and Eagle Ford suites), the $\delta^{13}C$ of both terminal and centre-position carbons rise. Assuming that the rise in bulk $\delta^{13}C$ is a generally useful proxy for increasing thermal maturity (Prinzhofer & Huc 1995), the data trend in Figure 6a suggests that this rise is first concentrated exclusively in the terminal C position for the first half of the maturity range (a trend consistent with the Chung *et al.* (1988) model), after which the centre position becomes the dominant site of further increases in $\delta^{13}C$. One interpretation of this finding is that the first half of the trend reflects primary kerogen cracking following the expectations of the Chung *et al.* model, whereas the second half of the trend marks a shift to another precursor, such as bitumen or oil cracking (so-called secondary cracking). This interpretation is consistent with the findings for Woodford Shale cracking experiments (above; Piasecki *et al.* 2018), but remains speculative and should be examined with further experiments and studies of closely related natural gases that vary in thermal maturity.

Site-specific D in natural propane

Liu *et al.* (2015) report NMR data documenting site-specific hydrogen isotope fractionation in propane, $\delta D_{centre} - \delta D_{ends} = -26 \pm 10$‰ for an undescribed propane. This contrasts with the fractionation of *c.* +50–80‰ theoretically predicted for equilibrium at relevant geological temperatures, and, taken at face value, suggests that propane can be generated with an unequilibrated hydrogen isotope structure.

Ponton *et al.* (2016) present preliminary results of mass spectrometric measurements of the site-specific hydrogen isotope composition of 12 natural propanes from several petroleum-forming systems:

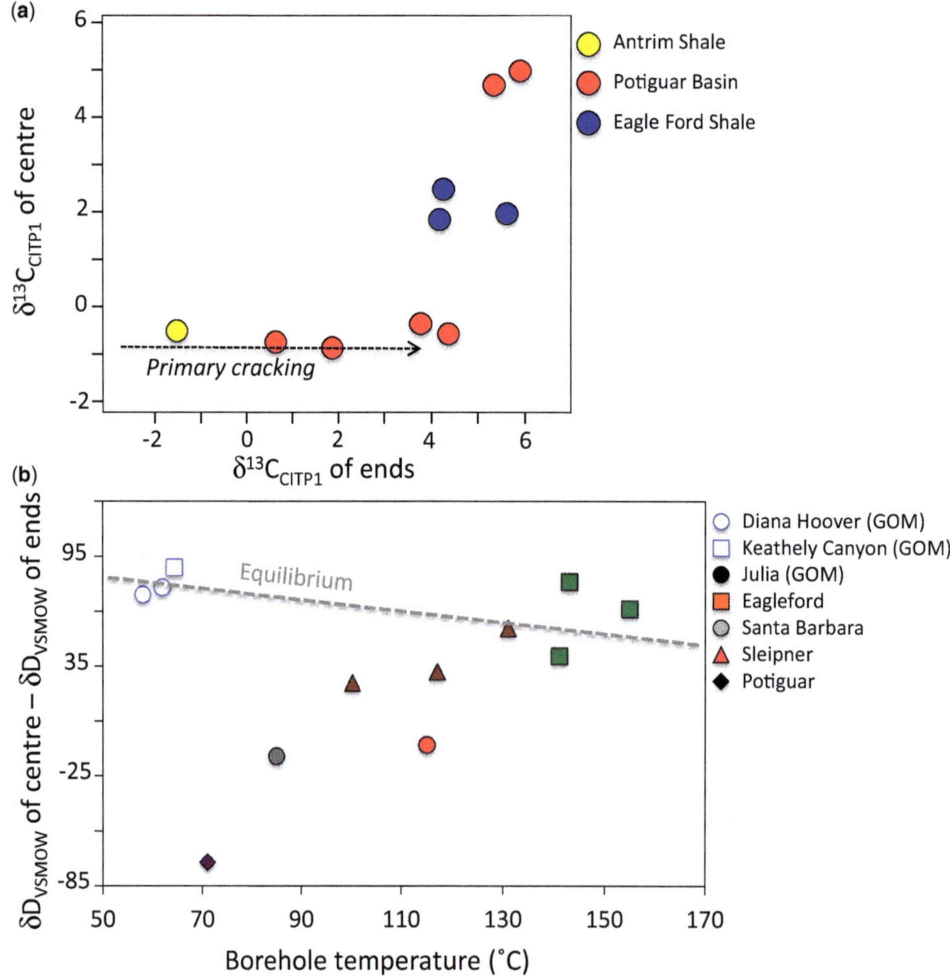

Fig. 6. Site-specific isotopic structures of natural propane. (**a**) carbon isotopes, from Piasecki (2015) (reported v. the intralaboratory standard, CITP1); (**b**) hydrogen isotopes, from Ponton *et al.* (2016) (on an absolute reference frame scale). In (a) the arrow represents the path through this composition space predicted by the model of Chung *et al.* (1988). In (b), the dashed line represents the equilibrium site-specific hydrogen isotope fractionation predicted by Piasecki *et al.* (2016*a*) (similar to that in Webb & Miller (2014)).

three wells in the Gulf of Mexico, the Eagle Ford Shale (Texas), Sleipner (North Sea), Potiguar Basin (Brazil) and Sacate field (coastal California). These data were generated by mass spectrometry but, unlike the ^{13}C data described above, can be tied to an absolute δD_{VSMOW} scale because the same techniques were applied to propane that had been experimentally driven to an equilibrium isotopic structure by exposure to Pd catalyst (i.e. propane can be standardized using a 'heated gas' or 'absolute reference frame' of the kind previously used for clumped-isotope and/or site-specific stable isotope measurements of CO_2, O_2, CH_4 and N_2O; Eiler 2007, 2011). The commercially obtained propane

gas cylinder used by Ponton *et al.* (2016) as an intralaboratory standard (a gas with an unknown origin, but probably purified from thermogenic natural gas) yields a site-specific hydrogen isotope fractionation, $\delta D_{centre}-\delta D_{ends} = -25 \pm 6‰$, which is essentially identical to the finding by Liu *et al.* (2015) on their intralaboratory reference material. It is tempting to imagine that commercial suppliers provided the same or similar propane to both labs, although this is not known to be the case and could equally well be a coincidence (or conspiracy of errors!).

The results of Ponton *et al.* (2016) for natural propanes should be considered preliminary, but define a trend where the bulk molecular δD rises from the lowest to highest values commonly seen in thermogenic propane (-170 to $-115‰$) while the centre–end hydrogen isotope fractionation rises from -72 to $+89$ (±6)$‰$. This finding is similar to results for propane generated by hydropyrolysis of Woodford Shale (centre–end fractionation rises from -52 to $+67‰$ over a $69‰$ rise in δD; above). One interpretation of this result is that the site-specific hydrogen isotope fractionation in propane is a maturity indicator, analogous to common interpretations of the bulk $\delta^{13}C$ or δD of natural gas components. However, a different interpretation is suggested by two observations: (1) the high end of the range of centre–end fractionations is similar to the expected equilibrium fractionation at shallow crustal temperatures; and (2) the centre–end hydrogen isotope fractionation exhibits a relationship with the borehole temperature of the well from which each sample was collected (Fig. 6b). In particular, 9 of the 12 studied samples exhibit a positive trend, the high end of which is indistinguishable from the predicted equilibrium relationship, whereas the remaining three samples (from the Dianna Hoover and Keathley Canyon wells in the Gulf of Mexico) also lie on the predicted equilibrium trend but at lower temperatures (60 v. 150°C). This finding suggests the possibility that thermogenic propane may be formed out of equilibrium with respect to its site-specific hydrogen isotope structure but then evolve towards equilibrium.

The most obvious process that might promote such equilibration is intra- or intermolecular hydrogen isotope exchange, perhaps catalyzed by some coexisting material (e.g. water, clay or oil). This process would probably proceed more quickly at higher temperature and thus would provide an explanation for the fact that the gases from the highest temperature wells approach the equilibrium distribution. It is less obvious why the group of three Gulf of Mexico gases also have an equilibrium hydrogen isotope structure despite their low borehole temperatures. Four possibilities occur to us. First, there could have been prior storage of propane at greater temperatures in some deeper reservoir. Second, some material may be present in the reservoirs for these samples (that is absent in the other reservoirs) and particularly effective at promoting re-equilibration down to low temperatures. Third, reservoir microbial ecosystems (known to be present and actively generating methane in the Keathley Canyon field) may be able to catalyze hydrogen isotope exchange of petroleum compounds. This idea is inspired by prior work of Valentine et al. (2004), who showed that the enzymatic pathway responsible for biological methanogenesis operates nearly reversibly when the partial pressure of H_2 is low; thus, microbes could act as a kind of specialized catalyst for hydrogen isotope exchange that short-circuits the

thermally-activated, abiological pathways. Fourth, perhaps low maturity gases are generated by a mechanism that promotes intramolecular equilibrium, whereas higher temperature cracking of propane precursors is kinetically controlled (perhaps counterintuitive, but we would say possible). All four of these suggestions are speculative at this stage, but illustrate the sorts of processes that may be revealed by further exploration of this proxy.

$^{13}C_2H_6$ in natural ethane

Clog et al. (2013, 2014) and Clog & Eiler (2014) present $\Delta^{13}C_2H_6$ values for ethane from 25 natural gases from the Marcellus Shale (Pennsylvania), Haynesville Shale (Texas and Louisiana), Eagleford Shale (Texas) and the Potiguar and Sergipe Alagoas petroleum fields (Brazil). To the best of our knowledge, this constitutes the largest and most diverse set of observations constraining the intramolecular isotopic structure of a hydrocarbon larger than methane recovered from natural samples. The most striking finding of this work is the relatively large range of variability in $\Delta^{13}C_2H_6$ (5 ‰, or a factor of c. 10–20 greater can be easily explained by equilibrium clumped-isotope effects or simple physical processes like diffusion or mixing). The measured range is also c. five-fold greater than that observed in products of shale pyrolysis experiments (above). This suggests that most of the variance in this isotopic proxy must be driven by something other than common kerogen and oil-cracking reactions, temperature-dependent equilibrium or simple processes like diffusive transport. One process that might explain these findings is ethane cracking: the experiments summarized above produced a 1.8‰ range in $\Delta^{13}C_2H_6$; if this process operated as a Rayleigh distillation, the residue left over after c. 90% ethane loss could manifest the full range in $\Delta^{13}C_2H_6$ values Clog et al. (2014) observed in natural samples. If this process is in fact responsible for most of the measured range, then $\Delta^{13}C_2H_6$ might be most useful as a proxy for secondary cracking of wet gas components. Two observations offer support for this possibility: (1) relatively dry shale gases are consistently in the lower half of the range of measured $\Delta^{13}C_2H_6$ values; and (2) a plot of gas wetness v. $\Delta^{13}C_2H_6$ (Fig. 7) reveals a positive overall trend, and better-defined positive trends for each of three suites of related samples – the Potiguar, Eagle Ford and Sergipe Alagoas. These trends are broadly consistent with the relationship between wetness and $\Delta^{13}C_2H_6$ predicted based on the ethane-cracking experiments (solid curve in Fig. 7).

A word on the doubly substituted methanes

The clumped-isotope compositions of natural methanes are reviewed in Stolper et al. (2017); see also

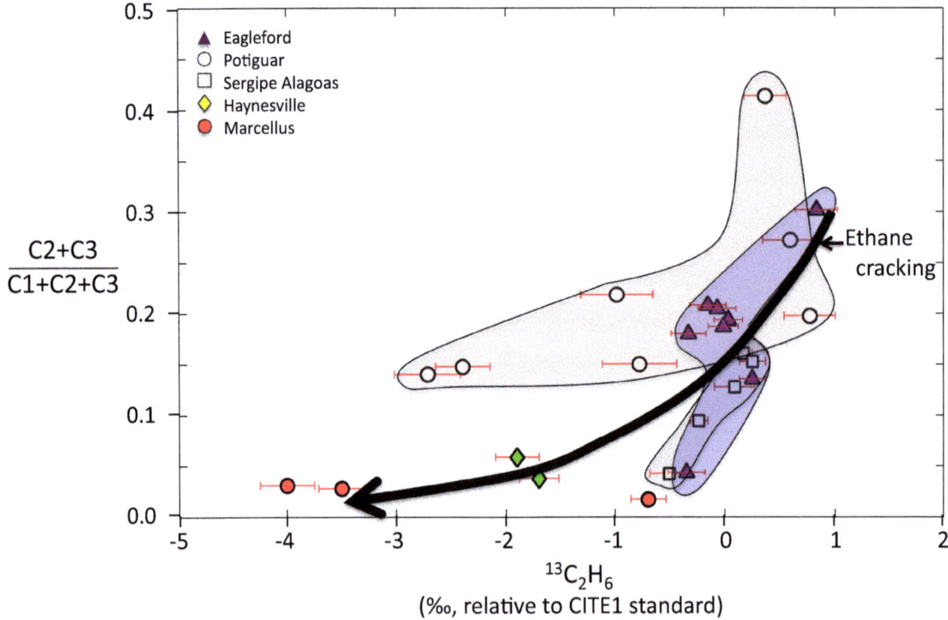

$$\frac{C2+C3}{C1+C2+C3}$$

$^{13}C_2H_6$
(‰, relative to CITE1 standard)

Fig. 7. Clumped-isotope composition of ethane from natural gas samples, plotted v. the 'gas wetness' (proportion of ethane + propane to the total of methane + ethane + propane) in those samples; data from Clog *et al.* (2013, 2014) and Clog & Eiler (2014) (reported v. the intralaboratory standard, CITE1). For comparison, the heavy black arrow shows the trend predicted for ethane cracking, assuming Rayleigh distillation and isotopic fractionation based on experiments depicted in Figure 3b.

the review by Douglas *et al.* 2017). We do not discuss this topic in detail here, but point out one finding that is more striking and meaningful when contrasted with the subjects covered in this paper: methane is commonly, even typically, in clumped-isotope equilibrium at its temperature of formation, across a wide range of conditions and formation mechanisms. Recent papers on methane clumped-isotope geochemistry have focused attention on non-equilibrium clumped-isotope compositions. However, these findings have been observed only in a few settings and formation mechanisms: a subset of biogenic and serpentine-related methanes; experimental thermogenic methane formed by alkyl cracking; and possibly a subset of oil-associated unconventional gases. The more common finding (estimated at *c.* 80+ % of natural samples analysed to date) is that methane apparent temperatures are similar to known or expected temperatures of methanogenesis. For this reason, the methane clumped-isotope proxies may be a rare instance where Galimov's (1974, 1985) hypothesis of an organic stable isotope geochemistry dominated by equilibrium thermodynamics seems to be usually true. Why does methane stand in contrast to the other groups of organic compounds that have been explored for their isotope structures – lipids, sugars, amino acids, oil and condensate alkanes,

propane, ethane? The question is particularly puzzling when one considers that the bulk stable isotope fractionations associated with methanogenesis have been successfully described through KIEs (Tang *et al.* 2000, 2005). A reasonable conclusion is that methanogensis often occurs in chemical environments that promote local reversible H exchange among the moieties that will become methane (and/or with some other, unrecognized, hydrogen pool, such as H radicals in kerogen structures). It is striking that this seems to be the case both for most natural biogenic methanogenesis in subsurface environments and for thermogenic methanogenesis involving kerogen cracking. An important question for future research is whether this is also true of the hydrogen isotope structures of larger organic compounds, which are essentially unknown outside of sugars and propane.

The end game: graphitization

Burial or subduction of organic matter to conditions where rocks undergo macroscopic metamorphic reactions (*c.* 300+ °C) leads to dramatic reorganization of the H/C/N/O/S volatiles from organics into just a few stable forms: graphite (or, at high pressure, diamond), pyrite or pyrrhotite, ammonium ion in feldspar or sheet silicates, and simple molecular gases (Harrison 1976). However, this transformation is gradual and severely limited by kinetics, preserving carbonaceous material that is structurally intermediate between kerogen and crystalline graphite well into high metamorphic grades (Mao *et al.* 2010). (There is also evidence for traces of C2+ hydrocarbons in metamorphic rocks (Mango 1991), although one might ask whether some such evidence reflects contamination rather than persistence of complex organic molecules.) Nevertheless, as temperature rises through the metamorphic range, eventually NH_3, H_2S, CO_2 and even CH_4 become unstable, leaving crystalline graphite and the inorganic molecular gases (CO_2, CO, H_2O, H_2) as the only abundant remnants of buried organic matter.

These processes raise three questions about the stable isotope contents of organics in the rock record. (1) How long do vestiges of the original isotopic structures of biomolecules or fingerprints of catagenetic 'cracking' reactions persist? (2) At what stage can one no longer distinguish between the products of metamorphosing biomolecules and non-biological sources of graphite and simple molecular gases? (3) Do any organic materials have thermodynamically controlled isotopic structures under such extreme conditions? There is a long history of approaching these questions using bulk stable isotope data (e.g. the $\delta^{13}C$ value of graphite, or the hydrogen isotope fractionation between methane and H_2; Dunn & Valley 1992; Wada *et al.* 1994; Horibe & Craig 1995; Schimmelmann *et al.* 2001). Such constraints are useful, but must contend with ambiguities that arise from the wide range of geological processes that can raise or lower bulk isotopic contents. We ask: what insights could molecular isotopic structure (site-specific and clumped-isotope compositions) bring to these questions?

Graphitization of kerogen – transformation of its non-periodic polymerized structure into crystalline graphite, accompanied by loss of H, N, O and S – begins in the 'gas window' (*c.* 200°C) and occurs very gradually, with increases in crystallinity and coarsening in grain size persisting well into the amphibolite facies (Harrison 1976). The $\delta^{13}C$ and δD values of kerogens do not evolve radically over the oil and gas windows (*c.* 100–250°C), but studies of the bulk carbon isotope fractionation between graphitic carbon and coexisting carbonate in metasediments show that subsequent graphitization is accompanied by carbon isotope exchange (Dunn & Valley 1992). This exchange starts around 400°C (Wada *et al.* 1994) but is slow and does not reach equilibrium until temperatures reach *c.* 500–600°C. A key insight is that even at amphibolite-facies metamorphic conditions, small graphite grains are further from equilibrium with coexisting carbonate than are

large graphite grains (Dunn & Valley 1992); this implies that carbon isotope exchange requires grain growth – i.e. little or no exchange occurs by swapping carbon atoms in and out of the sites of existing graphitic lattices. Taken together, these findings suggest the carbon isotope structures of components of kerogens may be preserved to great depths and temperatures in the Earth. If so, preservation of large site-specific carbon isotope variations in kerogens and 'graphitic' carbon might provide a more conclusive proof of biogenicity than bulk $\delta^{13}C$ values – a widely used but controversial and often ambiguous biomarker (e.g. Van Zuilen *et al.* 2002; Ohtomo *et al.* 2014). On the other hand, it is also possible that graphitization involves little heterogeneous carbon isotope exchange but significant amounts of molecular- and lattice-scale isotope rearrangement that could erase original biogenic isotopic structures.

Conclusions and prospects

We have attempted to provide a comprehensive and up-to-date overview of what is known about the molecular-scale isotopic structures of natural organic compounds, from their origins in biosynthesis through their transformation by catagenesis, to their destruction by metamorphism. This is a topic of tremendous complexity and with great potential to record information of scientific and practical importance. However, the technologies that enable site-specific and clumped-isotope analyses of organic compounds have only recently emerged (though they are advancing rapidly; Appendix B), there has been little inter-calibration of the diverse instruments and labs, and the application of these technologies has been sparse and in some cases difficult to relate to environmental and geological samples. For these reasons, the most lasting value of this review may be to organize and illuminate the scattered elements of an emerging field, in hopes that we can lend the field structure, pose initial hypotheses about its most important processes and phenomena, and make clear some of its most pressing needs.

The work summarized in this paper, and in the accompanying paper by Stolper *et al.* (2017) provide evidence that the isotopic structures of geological organic compounds can reflect any combination of the following factors: inheritance from primary biomolecules (e.g. the carbon isotope structures of aliphatic hydrocarbons, presumably derived from lipids); KIEs associated with maturation of refractory organics into humin, lignin and kerogen (e.g. extreme ^{13}C enrichment in methoxyl groups of lignins); KIEs during kerogen cracking (e.g. ^{13}C depletion in terminal sites of propane and other alkanes); thermodynamic equilibria at conditions of thermogenic reactions (the best example of which may be the clumped-isotope compositions of the most thermogenic methanes); KIEs during secondary cracking (e.g. the $\Delta^{13}C_2H_6$ value of ethane, particularly in dry gases); equilibration after formation due to protracted time at geological temperatures (e.g. the site-specific hydrogen isotope compositions of some propanes); KIEs during biological production and consumption (e.g. clumped-isotope compositions of some biogenic methanes in earth-surface environments); and biologically catalyzed equilibration (e.g. clumped-isotope compositions of methanes in subsurface environments). This rich set of processes has been revealed by just the last several years of exploratory work, promising much more to come as our analytical tools mature and see broader application.

We close our discussion with a list of conclusions and suggestions regarding future work that will be particularly important for near-term progress,

(1) *The interplay of chemical kinetics and equilibria*: Our first explorations of the isotopic structures of geological organics indicate that both chemical kinetics and thermodynamic equilibria are capable of controlling intramolecular isotopic properties. For example, carbon isotope structures of common biomolecules are clearly dominated by kinetic factors and the same seems to be true of hydrogen isotope structures of freshly synthesized sugars and at least some geological propanes. Yet methanes often have equilibrated distributions of isotopologues, propanes from some natural environments seem to have equilibrated hydrogen isotope structures, and at least some moieties of kerogens undergo hydrogen isotope exchange with their environments during burial diagenesis. The fact that both kinetic and equilibrium processes seem to be important in natural organics is sure to lead to confusion, at least in the near term, and raising the following questions. (1) When should an isotopic structure be interpreted as a biosignature v. a thermometer. (2) What are we to make of samples that have only partially equilibrated some initially kinetically controlled structure (or kinetically modified an initially equilibrated structure). (3) How will we parse the isotopic structures of molecules that have some properties controlled by kinetics and other properties controlled by thermodynamic equilibria? However, a longer-term view suggests these phenomena will be a rich archive rather than a problem. The combined site-specific and clumped-isotope properties of organic molecules include a large number of analysable species, and it seems plausible that we will learn to read complex multi-property measurements as records of sources, formation mechanism and environmental conditions. Furthermore, once the kinetics controlling the approach to equilibrium are better understood, it is possible that molecular isotopic structures will present opportunities for reconstructing temperature–time histories (i.e. a form of 'geospeedometry'). The most promising opportunities of this last sort may be the hydrogen isotope structures of relatively volatile and structurally simple compounds, such as the low-molecular-mass alkanes.

(2) *Defining isotopic equilibrium*: Considering the geological, environmental and economic importance of hydrocarbons, and the major role stable isotope analysis plays in their study, it is surprising how little concrete information constrains isotope exchange equilibria involving these compounds. This generalization is true both for the homogeneous equilibria that potentially control molecular isotopic structure and for the more familiar heterogeneous isotope exchange reactions between different chemical species. Other than vapour pressure isotope effects measured on low-temperature liquids (Jansco & Van Hook 1974), there are few concrete data that confidently constrain isotope exchange equilibria involving hydrocarbons larger than methane (Julien *et al.* 2015). This is an extraordinary gap in our understanding of the fundamentals of this subject, and should serve as a call for future research.

(3) *First-principles models*: Given the importance of irreversible reactions in petroleum formation and the great number and diversity of petroleum compounds, it is clear that the conceptual and theoretical models summarized in section 'KIEs associated with 'cracking' reactions' are too simple and narrow to serve as the basis for a rigorous interpretation of the stable isotope compositions of oil and gas constituents. The highest-priority needs include: extension of rigorous models of KIEs to encompass precursors other than *n*-alkanes and products larger than propane (particularly condensate and oil-fraction compounds); consideration of hydrogen isotope effects associated with diagenetic and catagenetic reactions; and a theoretical exploration of KIEs for clumped-isotope species (i.e. beyond the simple sampling-statistics effects discussed above with reference to ethane).

(4) *The kerogen problem*: The greatest weakness of our subject is that we have little insight into the detailed isotopic structures of kerogens, and it is not obvious how any of the existing

analytical technologies (Appendix B) could be adapted to study them. We suggest that the best way to approach this problem may be by analysis of the isotopic structures of oil and gas compounds produced by controlled heating of kerogens – that is, to approach it as an inverse problem rather than by direct measurements. Such studies might be made more insightful by studying the products of cracking experiments performed on compounds that are model systems for kerogenous material, with sites of interest isotopically labelled to aid characterization of reaction mechanisms.

(5) *Technology development*: The subject of this review is an emerging field that has come into being as a result of technology experiments aimed at expanding the capabilities of stable isotope geochemistry (Appendix B). Most of those technologies are relatively young and rapidly evolving, such that we should expect the next several years to result in dramatic increases in the numbers of active laboratories, the volume and quality of observations, and the numbers and types of molecular isotopic properties that can be observed. The most pressing need is for technological innovations that will permit analyses of small samples (of the order of micromols and less) of large molecules (of the order of ten and more atoms per molecule), preferably constraining diverse site-specific and clumped-isotope properties. NMR is the most mature of the relevant techniques, but it requires a dramatic reduction in sample size and the development of the ability to quantify multiply substituted species – something that will only become possible with a very large improvement in sensitivity. Mass spectrometry seems to have the greatest promise as a highly sensitive technique that is applicable to diverse properties of diverse compounds, but demonstrated instruments lack the mass resolution or number of detectors that would be required for a truly general approach to the problem. Fourier transform mass spectrometry holds promise as a way past these hurdles, but is as yet little understood as a quantitative precise tool for measuring natural isotope abundances.

The authors are grateful to the editor and two anonymous reviewers for their insightful comments and suggestions on an earlier draft of this work. Work presented in this paper was supported by the National Science Foundation programmes in Petrology and Geochemistry and EAR Instrumentation, by sponsored research support from Exxon Mobil and Petrobras, and by the California Institute of Technology. We thank Nami Kitchen for her contributions to analytical and experimental work presented here.

Appendix A

Nomenclature

The molecular-scale isotopic structures of organics is an emerging, relatively specialized field, and therefore uses a language and quantitative units that may be unfamiliar to some readers. A summary of this subject is challenging because its nomenclature and units are currently mixed (and in some respects confused). However, the following synopsis should provide a useful guide to most recent papers.

The isotopic forms of molecules have been described using the words: 'isotopologue', which we take to mean a version of a molecule that is unique in its number and/or location of rare-isotope substitutions; and 'isotopomer', which we take to mean one of some number of isotopologues that share a common isotopic stoichiometry (e.g. the same number of ^{13}C atoms) but differ from one another in the sites of those rare-isotope substitutions. Thus, by the usage adopted here (and in many recent papers) all isotomomers are also unique isotopologues of the same molecule. At least one recent study suggests the term 'isotopocule' (Toyoda *et al.* 2015), which is effectively equivalent to our use of the term isotopologue.

Any isotopologue that contains two or more rare isotopes (^{13}C, D, ^{15}N, ^{34}S, etc.) may be termed a 'multiply substituted' or (colloquially) 'clumped' isotopologue. Thus, $^{13}C_2H_6$ is a clumped isotopologue of ethane. 'Site-specific' or 'position-specific' isotopic differences generally refer to differences in proportions of two or more isotopologues that share a common number and type of rare isotopes but differ in the molecular sites of isotopic substitution. Thus, $^{12}CH_2D—^{12}CH_2—^{12}CH_3$ and $^{12}CH_3—^{12}CHD—^{12}CH_3$ are two isotopologues of propane that differ in their sites of deuteration; thus they are also isotopomers of one another. Samples that differ in proportions of these two species are said to exhibit site-specific isotopic fractionation or variation.

The units used to report clumped-isotope variations are relatively uniform and easily explained, though to date they have only been applied to relatively simple molecules (CO_2, N_2O, CO_3^{-2}, N_2, CH_4, C_2H_6), and it is not clear that current practice will be appropriate for discussion of the multiply substituted isotopologues of larger, more complex molecules (particularly those where clumped-isotope effects are themselves site-specific). Nevertheless, there is only one common practice in this field: clumped-isotope compositions are reported as Δ_i values, where i refers to an isotopologue of interest (or sometimes a collection of isotopologues that all share the same cardinal mass; thus i can be either a chemical and isotopic formula or a cardinal number). Δ_i values are calculated as:

$$\Delta_i = \left(\frac{R_i}{R_i^*} - 1 \right) \times 1000 \qquad (A1)$$

where R_i is the ratio of the isotopologue of interest to the unsubstituted isotopologue of that same molecule, and R_i^*

is the value that ratio would have if all isotopes were randomly distributed among all isotopologues of the compound of interest. A less commonly used alternate to this nomenclature defines Δ_i values to equal the deviation, in per mille, of the equilibrium constant for an isotope exchange reaction from the value of that equilibrium constant in the case of random distribution of isotopes among all isotopologues (generally equal to the high temperature limit of that equilibrium constant; Wang *et al.* 2004; Piasecki *et al.* 2016*a*).

The nomenclature used to describe site-specific isotopic variations is more variable, in many cases unique to each paper (even when considering multiple papers from the same research group). Previously proposed units can be broadly defined as falling into two categories.

First, reports of the isotopic composition of a single indicated site are presented as: a concentration of a rare isotope (e.g. [^{13}C]); or, an isotope ratio (e.g. ^{13}C/^{12}C, or R^{13}); or, a 'delta' value relative to an absolute reference scale (e.g. δ^{13}C$_{VPDB}$); or, a 'delta' value relative to an arbitrary standard (δ^{13}C$_{STD}$). Second, reports of an isotopic difference between two non-equivalent sites, A and B, are usually given as: a ratio of ratios (e.g. R$_A^{13}$/R$_B^{13}$, or equivalently α_{A-B}^{13C}); or, as an ε value (where $\varepsilon_{A-B} = 1000 \ln (\alpha_{A-B}^i)$; or, as a difference between two 'delta' values (e.g. δ^{13}C$_{VPDB}$ of site A minus δ^{13}C$_{VPDB}$ of site B). A dizzying diversity of superscripts, subscripts and other symbols are used to indicate which molecular site is being discussed, meaning one generally must approach the units and terms of each paper as a unique kind of jargon.

Clearly, there is a need to improve the consistency of units used to report site-specific isotopic variations. We suggest the best nomenclature would avoid arithmetic artifacts arising from non-linearities in δ^iX$_{STD}$ scales, and would focus on the unique information recorded by site-specific variations as distinct from that recorded by the more familiar indices of bulk stable isotope composition; ε_{A-B}^i values may be best suited to this purpose (i.e. a fractionation factor, expressed in per mille, of isotope ratio R^i between sites A and B).

Appendix B

Analytical technologies

There can be little question that the isotopic structures of organic molecules record a rich archive constraining sources, conditions and geochemical evolution. The question is, how can this archive be read? The technologies of measurements of molecular isotopic structure were recently reviewed by Eiler (2013); here we update this review and briefly summarize the technologies that have been applied to the study of petroleum and natural gas compounds.

(1) *Chemical degradation*: The oldest well-demonstrated approaches to measuring the site-specific isotopic compositions of organic molecules involve selective chemical attack, such as decarboxylation (Abelson & Hoering 1961), oxidative cleaving by ozonolysis followed by decarboxylation (Monson & Hayes 1980, 1982*a*, *b*), or release of methyl groups from methoxy-bearing compounds through reaction with iotic acid (Keppler *et al.* 2007; Feakins *et al.* 2013). We are aware of only one instance where such methods have been applied to a petroleum or natural gas compound: Gao *et al.* (2016) demonstrate that propane can be converted to propanol, then to acetone and finally to acetic acid. Combustion followed by IRMS of the acetic acid yields the average δ^{13}C of the terminal and central carbon sites, whereas combustion followed by IRMS of the original propane constrains the quantity: $(2 \times \delta^{13}$C$_{end} + \delta^{13}$C$_{centre})/3$. Thus, these two measurements can be used to solve for the difference, δ^{13}C$_{centre} - \delta^{13}$C$_{end}$.

(2) *NMR:* NMR spectroscopy permits quantification of the abundances of nuclides having net nuclear spin (including H, D and ^{13}C), and in many instances discriminates between signals from non-equivalent molecular sites. Its most common use is characterizing structures of unknown organic molecules, but it has been used to observe the relative abundances of stable isotopes at natural abundances and per mille precisions since the early 1980s (Martin & Martin 1981; Caer *et al.* 1991). However, only in the last 10 years has NMR emerged as a broadly applied technique. These methods have been primarily applied to food science forensics and studies of natural products (Martin *et al.* 2006). Three studies use NMR to examine the carbon or hydrogen isotope structures of petroleum compounds (Gilbert *et al.* 2013, 2016*a*, *b*; Liu *et al.* 2015). A significant challenge of these techniques is that they generally require pure samples in approximately tens to hundreds of milligram quantities. For this reason, all work on hydrocarbons published or presented so far has examined the isotopic structures of commercial chemicals (and/or a few widely distributed reference standards), making it difficult to understand how the results might relate to natural geological materials. Nevertheless, NMR is currently the most productive and well-demonstrated technique for site-specific analysis of organic compounds larger than propane, and, even if other techniques are better suited to natural samples, it is likely to remain an essential tool for characterizing reference standards. It has not yet been shown that NMR can observe 'clumped' isotope species at their natural abundances and useful precision.

(3) *Pyrolytic degradation*: Pyrolysis of organic molecules can yield products that selectively sample specific molecular sites, such that a site-specific isotopic analysis can be performed by separating the products of pyrolysis (such as by GC) and then analysing them separately for an isotopic property of interest. Past efforts to do this have generally involved the passing of the products of pyrolysis through a

gas chromatographic column, followed by combustion and continuous flow IRMS of each eluted peak (Corso & Brenna 1997, 1999). The primary use of such methods is to analyse the CH_3 site of ethanol (which is sampled by CH_4 produced by pyrolysis). This technique has been applied to propane to discriminate between the $\delta^{13}C$ value of the terminal, methyl carbon site and the $\delta^{13}C$ of the bulk molecule, thus constraining the difference between the terminal and central positions (Gilbert *et al.* 2016*a*, *b*).

(4) *Mass spectrometry*: Mass spectrometric measurements of clumped-isotope compositions can be made simply by quantifying proportions of singly and multiply substituted ion beams, where the highest precisions can be obtained by simultaneous detection in a multi-collector sector mass spectrometer (Eiler 2007, 2013). It is also well established that site-specific isotopic variations can be constrained by mass spectrometric analysis of the isotopic compositions of two or more ion species that differ in their proportions of the sites of interest (e.g. the ^{15}N fractionation between α v. β nitrogen in N_2O can be distinguished by comparing the isotopic compositions of NO^+ and N_2O^+ ions; Yoshida & Toyoda 2000). Until recently, such measurements were restricted to simple C—N—O gases because of the presence of complex interferences in the mass spectra of other compounds. However, with the advent of high resolution isotope ratio mass spectrometry (Eiler *et al.* 2013), these measurements can be made on a variety of low-molecular-weight organic and inorganic molecular gases (Magyar *et al.* 2016; Piasecki *et al.* 2016*b*; Stolper *et al.* 2017). The recent exploration of stable isotope ratio analysis using the exceptionally high mass resolution of Fourier transform mass spectrometry is expanding this capability into diverse higher molecular weight organics (Eiler *et al.* 2017).

(5) *Infrared spectroscopy*: Infrared spectroscopy should be an ideal basis for site-specific and clumped-isotope analyses because the vibrational spectra of isotopologues are effectively unique, and sensitivity can be exceptional. Such techniques are demonstrated for site-specific ^{15}N analysis of N_2O (Waechter *et al.* 2008) and clumped-isotope analysis of methane (Ono *et al.* 2014). However, no infrared spectroscopic measurements have been made constraining isotopic structures of organic molecules more complex than methane, so no such data appear in this review. It is possible that spectroscopic measurements of clumped-isotope compositions of ethane and clumped- and position-specific measurements of propane could be developed, though it is unlikely they will ever be applied more broadly than this due to the complexity of vibrational spectra of larger molecules, and the difficulty of performing such measurements on species that are not room-temperature gases.

References

ABELSON, P.H. & HOERING, T.C. 1961. Carbon isotope fractionation in formation of amino acids by photosynthetic organisms. *Proceedings of the National Academy of Sciences*, **47**, 623–632.

ANHAUSER, T., GREULE, M., ZECH, M., KALBITZ, K., McROBERTS, C. & KEPPLER, F. 2015. Stable hydrogen and carbon isotope ratios of methoxyl groups during plant litter degradation. *Isotopes in Environmental and Health Studies*, **51**, 143–154.

AUGUSTI, A. 2007. *Monitoring climate and plant physiology using deuterium isotopomers of carbohydrates*. PhD thesis, Umea University, Sweden.

AUGUSTI, A., BETSON, T.R. & SCHLEUCHER, J. 2008. Deriving correlated climate and physiological signals from deuterium isotopomers in tree rings. *Chemical Geology*, **252**, 1–8.

BERNER, U., FABER, E., SCHEEDER, G. & PANTEN, D. 1995. Primary cracking of algal and landplant kerogens: kinetic models of isotope variations in methane, ethane and propane. *Chemical Geology*, **126**, 233–245.

BETSON, T.R., AUGUSTI, A. & SCHLEUCHER, J. 2006. Quantification of deuterium isotopomers of tree-ring cellulose using nuclear magnetic resonance. *Analytical Chemistry*, **78**, 8406–8411.

BIGELEISEN, J. 1949. The relative reaction velocities of isotopic molecules. *Journal of Chemical Physics*, **17**, 675–678.

BILLAULT, I., GUIET, S., MABON, F. & ROBINS, R. 2001. Natural deuterium distribution in long-chain fatty acids is nonstatistical: a site-specific study by quantitative ^2H NMR spectroscopy. *ChemBioChem*, **2**, 425–431.

BURNHAM, A.K. & SWEENEY, J.J. 1989. A chemical kinetic model of vitrinite maturation and reflectance. *Geochimica et Cosmochimica Acta*, **53**, 2649–2657.

CAER, V., TRIERWEILER, M., MARTIN, G.J. & MARTIN, M.L. 1991. Determination of site-specific carbon isotope ratios at natural abundance by carbon-13 nuclear magnetic resonance spectroscopy. *Analytical Chemistry*, **63**, 2306–2313.

CHACKO, T., COLE, D.R. & HORITA, J. 2001. Equilibrium oxygen, hydrogen and carbon isotope fractionation factors applicable to geologic systems. *In*: VALLEY, J.W. & COLE, D.R. (eds) *Stable Isotope Geochemistry*. Reviews in Mineralogy and Geochemistry, **43**. Mineralogical Society of America, Chantilly, VA, 1–81.

CHUNG, H.M., GORMLY, J.R. & SQUIRES, R.M. 1988. Origin of gaseous hydrocarbons in subsurface environments: theoretical considerations of carbon isotope distribution. *Chemical Geology*, **71**, 97–104.

CLOG, M. & EILER, J. 2014. C-H and C-C clumping in ethane by high-resolution mass spectrometry. Abstract presented at the 2014 Fall Meeting of the American Geophysical Union, San Francisco, USA.

CLOG, M., EILER, J., GUZZO, J.V.P., MORAES, E.T. & SOUZA, I.V.A. 2013. Doubly ^{13}C-substituted ethane. Abstract presented at the 2013 Goldschmidt Meeting, Florence, Italy. *Mineralogical Magazine*, **77**, 897.

CLOG, M.D., FERREIRA, A.A., SANTOS NETO, E.V., EILER, J.M. 2014. Ethane C-C clumping in natural gas: a proxy for cracking processes? Abstract presented at the 2014 Fall Meeting of the American Geophysical Union, San Francisco, USA.

CORSO, T.N. & BRENNA, J.T. 1997. High-precision position-specific isotope analysis. *Proceedings of the National Academy of Sciences*, **94**, 1049–1053.

CORSO, T.N. & BRENNA, J.T. 1999. On-line pyrolysis of hydrocarbons coupled to high-precision carbon isotope ratio analysis. *Analytica Chimica Acta*, **397**, 217–224.

DENIRO, M.J. & EPSTEIN, S. 1977. Mechanism of carbon isotope fractionation associated with the lipid synthesis. *Science*, **197**, 261–363.

DOUGLAS, P., STOLPER, D. *ET AL.* 2017. Methane clumped isotopes: progress and potential for a new isotopic tracer. *Organic Geochemistry*. First published online 16 August, 2017, https://doi.org/10.1016/j.orggeochem.2017.07.016

DUAN, J.-R., BILLAULT, I., MABON, F. & ROBINS, R. 2002. Natural deuterium distribution in fatty acids isolated from peanut seed oil: a site-specific study by quantitative 2H NMR spectroscopy. *Chembiochem*, **3**, 752–759.

DUNN, S.R. & VALLEY, J.W. 1992. Calcite-graphite thermometry: a test for polymetamorphism in marble, Tudor gabbro, Ontario. *Journal of Metamorphic Geology*, **10**, 487–501.

EHLERS, I., AUGUSTI, A., BETSON, T.R., NILSSON, M.B., MARSHALL, J.D. & SCHLEUCHER, J. 2015. Detecting long-term metabolic shifts using isotopomers: CO_2-driven suppression of photorespiration in C-3 plants over the 20th century. *Proceedings of the National Academy of Sciences*, **112**, 15585–15590.

EILER, J.M. 2007. 'Clumped-isotope' geochemistry – The study of naturally-occurring, multiply-substituted isotopologues. *Earth and Planetary Science Letters*, **262**, 309–327.

EILER, J.M. 2011. Paleoclimate reconstruction using carbonate clumped isotope thermometry. *Quaternary Science Reviews*, **30**, 3575–3588.

EILER, J.M. 2013. The isotopic anatomies of molecules and minerals. *Annual Reviews of Earth and Planetary Sciences*, **41**, 411–441.

EILER, J.M., CLOG, M. *ET AL.* 2013. A high-resolution gas-source isotope ratio mass spectrometer. *International Journal of Mass Spectrometry*, **335**, 45–56.

EILER, J., CESAR, J. *ET AL.* 2017. Analysis of molecular isotopic structures at high precision and accuracy by Orbitrap mass spectrometry. *International Journal of Mass Spectrometry*, **422**, 26–142, https://doi.org/10.1016/j.ijms.2017.10.002

ELSNER, M., JOCHMANN, M.A., HOFSTETTER, T.B., HUNKELER, D., BERNSTEIN, A., SCHMIDT, T.C. & SCHIMMELMANN, A. 2012. Current challenges in compound-specific stable isotope analysis of environmental organic contaminants. *Analytical and Bioanalytical Chemistry*, **403**, 2471–2491.

EPSTEIN, S., YAPP, C.J. & HALL, J.H. 1976. Determination of D/H ratio of non-exchangeable hydrogen in cellulose extracted from aquatic and land plants. *Earth and Planetary Science Letters*, **30**, 241–251.

FEAKINS, S.J., ELLSWORTH, P.V. & STERNBERG, L.da S.L. 2013. Lignin methoxyl hydrogen isotope ratios in a coastal ecosystem. *Geochimica et Cosmochimica Acta*, **121**, 54–66.

FERREIRA, A.A., SANTOS NETO, E.V., SESSIONS, A.L., SCHIMMELMANN, A. & NETO, F.R.A. 2012. $^2H/^1H$ ratio of hopanes, tricyclic and tetracyclic terpanes in oils and source rocks from the Potiguar Basin, Brazil. *Organic Geochemistry*, **51**, 13–16.

FREEMAN, K.H. 2001. Isotopic biogeochemistry of marine organic carbon. *In*: VALLEY, J.W. & COLE, D.R. (eds) *Stable Isotope Geochemistry*. Reviews in Mineralogy and Geochemistry, **43**. Mineralogical Society of America, Chantilly, VA, 579–605.

GALIMOV, E.M. 1973. *Izotopy ugleroda v heftegazovoy geologii* [Carbon Isotopes in Oil-Gas Geology]. NASA Technical Translation F-682.

GALIMOV, E.M. 1974. Organic geochemistry of carbon isotopes. *In*: TISSOT, B. & BLENNER, F. (eds) *Advances in Organic Geochemistry, 1973; Proceedings of the 6th International Meeting on Organic Geochemistry*. Éditions Technip, Paris, 439–452.

GALIMOV, E.M. 1985. *The Biological Fractionation of Isotopes*. Academic Press, London.

GALIMOV, E.M. & SHIRINSKII, V.G. 1975. Ordered distribution of carbon isotopes in individual compounds and components of lipid fraction of organisms. *Geokhimiya*, **4**, 503–528.

GAO, L., HE, P., JIN, Y., ZHANG, Y., WANG, X., ZHANG, S. & TANG, Y. 2016. Determination of position-specific carbon isotope ratios in propane from hydrocarbon gas mixtures. *Chemical Geology*, **435**, 1–9.

GILBERT, A., SILVESTRE, V., ROBINS, R.J., REMAUD, G.S. & TCHERKEZ, G. 2012*a*. Biochemical and physiological determinants of intramolecular isotope patterns in sucrose from C3, C4 and CAM plants accessed by isotopic ^{13}C NMR spectrometry: a viewpoint. *National Product Reports*, **29**, 476–486.

GILBERT, A., ROBINS, R.J., REMAUD, G.S. & TCHERKEZ, G.G.B. 2012*b*. Intramolecular ^{13}C pattern in hexoses from autotrophic and heterotrophic C3 plant tissues. *Proceedings of the National Academy of Sciences*, **109**, 18204–18209.

GILBERT, A., YAMADA, K. & YOSHIDA, N. 2013. Exploration of intramolecular ^{13}C isotope distribution in long chain n-alkanes (C_{11}–C_{31}) using isotopic ^{13}C NMR. *Organic Geochemistry*, **62**, 56–61.

GILBERT, A., YAMADA, K., SUDA, K., UENO, Y. & YOSHIDA, N. 2016*a*. Measurement of position-specific ^{13}C isotopic composition of propane at the nanomole level. *Geochimica et Cosmochimica Acta*, **177**, 205–216.

GILBERT, A., YAMADA, K. & YOSHIDA, N. 2016*b*. Evaluation of on-line pyrolysis coupled to isotope ratio mass spectrometry for the determination of position-specific ^{13}C isotope composition of short chain n-alkanes (C6-C12). *Talanta*, **153**, 158–162.

GUY, R.D., FOGEL, M.L. & BERRY, J.A. 1993. Photosynthetic fractionation of the stable isotopes of oxygen and carbon. *Plant Physiology*, **101**, 37–47.

HARRISON, W.E. 1976. Laboratory graphitization of a modern estuarine kerogen. *Geochimica et Cosmochimica Acta*, **40**, 247–248.

HAYES, J.M. 2001. Fractionation of carbon and hydrogen isotopes in biosynthetic processes. *In*: VALLEY, J.W. & COLE, D.R. (eds) *Stable Isotope Geochemistry*. Reviews in Mineralogy and Geochemistry, **43**. Mineralogical Society of America, Chantilly, VA, 225–277.

HEDGES, J.I., EGLINGTON, G. *ET AL.* 2000. The molecularly-uncharacterized component of nonliving organic matter in natural environments. *Organic Geochemistry*, **31**, 945–958.

HORIBE, Y. & CRAIG, H. 1995. D/H fractionation in the system methane-hydrogen-water. *Geochimica et Cosmochimica Acta*, **59**, 5209–5217.

INAGAKI, F., HINRICHS, K.U. *ET AL.* 2015. Exploring deep microbial life in coal-bearing sediment down to *c.* 2.5 km below the ocean floor. *Science*, **349**, 420–424.

IVLEV, A.A., LOROLEVA, M.Y. & KALOSHIN, A.G. 1974. Possible mechanisms of carbon isotope effect appearance in autotrophic organisms. *Doklady Akademii Nauk SSSR*, **217**, 224–227.

JANSCO, G. & VAN HOOK, W.A. 1974. Condensed phase isotope effects (especially vapor pressure isotope effects). *Chemical Reviews*, **74**, 689–750.

JULIEN, M., PARINET, J., NUN, P., BAYLE, K., HOHENER, P., ROBINS, R.J. & REMAUD, G.S. 2015. Fractionation in position-specific isotope composition during vaporization of environmental pollutants measured with isotope ratio monitoring by C-13 nuclear magnetic resonance spectrometry. *Environmental Pollution*, **205**, 299–306.

KEPPLER, F., HARPER, D.B. *ET AL.* 2007. Stable hydrogen isotope ratios of lignin methoxyl groups as a paleoclimate proxy and constraint on the geographic origin of wood. *New Phytologist*, **176**, 600–609.

KUBICKI, J.D., LACROCE, M.V. & TROUT, C.C. 2016. H-D fractionation factors at individual sites on model petroleum compounds. Abstract presented at the 2016 San Diego Meeting of the American Chemical Society, April, San Diego, USA.

LESOT, P., BAILLIF, V. & BILLAULT, I. 2008. Combined analysis of C-18 unsaturated fatty acids using natural abundance deuterium 2D NMR spectroscopy in chiral oriented solvents. *Analytical Chemistry*, **80**, 2963–2972.

LEWAN, M.D. 1983. Effects of thermal maturation on stable organic carbon isotopes as determined by hydrous pyrolysis of Woodford Shale. *Geochimica et Cosmochimica Acta*, **47**, 1471–1479.

LEWAN, M.D. 1985. Evaluation of petroleum generation by hydrous pyrolysis experimentation. *Philosophical Transactions of the Royal Society of London*, **315**, 123–134.

LEWAN, M.D. 1993. Laboratory simulation of petroleum formation – hydrous pyrolysis. *In*: ENGEL, M.H. & MACKO, S.A. (eds) *Organic Geochemistry*. Plenum Press, New York, 419–442.

LIU, C., MCGOVERN, G.P. & HORITA, J. 2015. Position-specific hydrogen and carbon isotope fractionations of light hydrocarbons by quantitative NMR. Abstract presented at the 2015 Fall Meeting of the American Geophysical Union, December, San Francisco, USA.

LLOYD, M., SESSIONS, A., SCHIMMELMANN, A., FEAKINS, S. & EILER, J. 2016. Determination of clumped ^{13}C-2H-H2 compositions of methoxyl groups in wood, lignin and simple organic monomers. Abstract presented at the 2016 Organic Geochemistry Gordon Conference, June, Tokyo.

MAGYAR, P.M., ORPHAN, V.J. & EILER, J.M. 2016. Measurement of rare isotopologues of nitrous oxide by high-resolution multi-collector mass spectrometry. *Rapid Communications in Mass Spectrometry*, **30**, 1923–1940.

MANGO, F. 1991. The stability of hydrocarbons under the time-temperature conditions of petroleum genesis. *Nature*, **352**, 146–148.

MAO, J., FANG, X., LAN, Y.Q., SCHIMMELMANN, A., MASTALERZ, M., XU, L. & SCHMIDT-ROHR, K. 2010. Chemical and nanometer-scale structure of kerogen and its change during thermal maturation investigated by advanced solid-state ^{13}C NMR spectroscopy. *Geochimica et Cosmochimica Acta*, **74**, 2110–2127.

MARKAI, S., MARCHAND, P.A., MABON, F., BAGUET, E., BILLAULT, I. & ROBINS, R.J. 2002. Natural deuterium distribution in branched-chain medium-length fatty acids is nonstatistical: a site-specific study by quantitative ^{2}H NMR spectroscopy of the fatty acids of capsaicinoids. *ChemBioChem*, **3**, 212–218.

MARSHALL, A.G. & RODGERS, R.P. 2008. Petroleomics: chemistry of the underworld. *Proceedings of the National Academy of Sciences*, **105**, 18090–18095.

MARTIN, G.J. & MARTIN, M.L. 1981. Isotopic labeling in natural abundance – application of high-resolution deuterium NMR to the study of vinyl compounds. *Comptes Rendus de L'Academie des Sciences Serie II*, **293**, 31–33.

MARTIN, G.J., MARTIN, M.L. & REMAUD, G. 2006. SNIF-NMR – Part 3: From mechanistic affiliation to origin inference. *In*: WEBB, G.A. (ed.) *Modern Magnetic Resonance*. Springer, 1669–1680.

MARTINI, A.M., WALTER, L.M., KU, T.C., BUDAI, J.M., MCINTOSH, J.C. & SCHOELL, M. 2003. Microbial production and modification of gases in sedimentary basins: a geochemical case study from a Devonian shale gas play, Michigan basin. *AAPG Bulletin*, **87**, 1355–1375.

MONSON, K.D. & HAYES, J.M. 1980. Biosynthetic control of the natural abundance of carbon 13 at specific positions within fatty acids in Escherichia Coli: evidence regarding the coupling of fatty acid and phospholipid synthesis. *Journal of Biological Chemistry*, **255**, 11435–11441.

MONSON, K.D. & HAYES, J.M. 1982a. Carbon isotopic fractionation in the biosynthesis of bacterial fatty acids. Ozonolysis of unsaturated fatty acids as a means of determining the intramolecular distribution of carbon isotopes. *Geochimica et Cosmochimica Acta*, **46**, 139–149.

MONSON, K.D. & HAYES, J.M. 1982b. Biosynthetic control of the natural abundance of carbon 13 at specific positions within fatty acids in Saccharomyces cerevisiae: isotopic fractionations in lipid synthesis as evidence for peroxisomal regulation. *Journal of Biological Chemistry*, **257**, 5568–5575.

NABBEFELD, B., GRICE, K., SCHIMMELMANN, A., SAUER, P.E., BOTTCHER, M.E. & TWITCHETT, R. 2010. Significance of $\delta D_{kerogen}$, $\delta^{13}C_{kerogen}$ and $\delta^{34}S_{pyrite}$ from several Permian/Triassic (P/Tr) sections. *Earth and Planetary Science Letters*, **295**, 21–29.

NATIONAL RESEARCH COUNCIL 2003. *Oil in the Sea*. National Academies Press, Washington, DC.

OHTOMO, Y., KAKEGAWA, T., ISHIDA, A., NAGASE, T. & ROSING, M.T. 2014. Evidence for biogenic graphite in early Archean Isua metasedimentary rocks. *Nature Geoscience*, **7**, 25–28.

O'LEARY, M.H., RIFE, J.E. & SLATER, J.D. 1981. Kinetic and isotope effect studies of maize phosphoenolpyruvate carboxylase. *Biochemistry*, **20**, 7308–7314.

ONO, S., WANG, D.T. *ET AL.* 2014. Measurement of a doubly substituted methane isotopologue, ^{13}CH$_3$D, by tunable infrared laser direct absorption spectroscopy. *Analytical Chemistry*, **86**, 6487–6494.

PIASECKI, A. 2015. *Site specific isotopes in small organic molecules.* PhD thesis, California Institute of Technology, Pasadena, CA, USA.

PIASECKI, A., SESSIONS, A., PETERSON, B. & EILER, J. 2016*a*. Prediction of equilibrium distributions of isotopologues for methane, ethane and propane using density functional theory. *Geochimica et Cosmochimica Acta*, **190**, 1–12.

PIASECKI, A., SESSIONS, A., LAWSON, M., FERREIRA, A.A., NETO, E.V.S. & EILER, J.M. 2016*b*. Analysis of the site-specific carbon isotope composition of propane by gas source isotope ratio mass spectrometer. *Geochimica et Cosmochimica Acta*, **188**, 58–72.

PIASECKI, A., SESSIONS, A. *ET AL.* 2018. Position-specific ^{13}C distributions within propane from experiments and natural gas samples. *Geochimica et Cosmochimica Acta*, **220**, 110–124. First published online October 6, 2017, https://doi.org/10.1016/j.gca.2017.09.042

PONTON, C., XIE, H. *ET AL.* 2016. Experiments constraining blocking temperatures of H isotope exchange in propane and ethane. Abstract presented at the 2016 Meeting of the International Clumped Isotope Workshop, January, Saint Petersburg USA.

PRICE, L.C. & SCHOELL, M. 1995. Constraints on the origins of hydrocarbon gas from compositions of gases at their site of origin. *Nature*, **378**, 368–371.

PRINZHOFER, A.A. & HUC, A.Y. 1995. Genetic and postgenetic molecular and isotopic fractionations in natural gases. *Chemical Geology*, **126**, 281–290.

QUIGLEY, T.M. & MACKENZIE, A.S. 1988. The temperatures of oil and gas formation in the sub-surface. *Nature*, **333**, 549–552.

ROONEY, M.A., CLAYPOOL, G.E. & CHUNG, H.M. 1995. Modeling thermogenic gas generation using carbon isotope ratios of natural gas hydrocarbons. *Chemical Geology*, **126**, 291–232.

ROSSMANN, A., BUTZENLECHNER, M. & SCHMIDT, H.-L. 1991. Evidence for a nonstatistical carbon isotope distribution in natural glucose. *Plant Physiology*, **96**, 609–614.

RUSTAD, J.R. 2009. Ab initio calculation of the carbon isotope signatures of amino acids. *Organic Geochemistry*, **40**, 720–723.

SCHIMMELMANN, A., LEWAN, M.D. & WINTSCH, R.P. 1999. D/H isotope ratios of kerogen, bitumen, oil, and water in hydrous pyrolysis of source rocks containing kerogen types I, II, IIS, and III. *Geochimica et Cosmochimica Acta*, **63**, 3751–3766.

SCHIMMELMANN, A., BOUDOU, J.-P., LEWAN, M.D. & WINTSCH, R.P. 2001. Experimental controls on D/H and ^{13}C/^{12}C ratios of kerogen, bitumen and oil during hydrous pyrolysis. *Organic Geochemistry*, **32**, 1009–1018.

SCHIMMELMANN, A., SESSIONS, A.L. & MASTALERZ, M. 2006. Hydrogen isotopic (D/H) composition of organic matter during diagenesis and thermal maturation. *Annual Reviews of Earth and Planetary Sciences*, **34**, 501–533.

SEEWALD, J.S., BENITEZ-NELSON, B.C. & WHELAN, J.K. 1998. Laboratory and theoretical constraints on the generation and composition of natural gas. *Geochimica et Cosmochimica Acta*, **62**, 1599–1617.

STOLPER, D.A., LAWSON, M., FORMOLO, M.J., DAVIS, C.L., DOUGLAS, M.J. & EILER, J.M. 2017. The utility of methane clumped isotopes to constrain the origins of methane in natural gas accumulations. *In*: LAWSON, M., FORMOLO, M.J. & EILER, J.M. (eds) *From Source to Seep: Geochemical Applications in Hydrocarbon Systems*. Geological Society, London, Special Publications, **468**. First published online December 14, 2017, https://doi.org/10.1144/SP468.3

TANG, Y., PERRY, J.K., JENDEN, P.D. & SCHOELL, M. 2000. Mathematical modeling of stable carbon isotope ratios in natural gases. *Geochimica et Cosmochimica Acta*, **64**, 2673–2687.

TANG, Y., HUANG, Y. *ET AL.* 2005. A kinetic model for thermally induced hydrogen and carbon isotope fractionation of individual n-alkanes in crude oil. *Geochimica et Cosmochimica Acta*, **69**, 4505–4520.

TISSOT, B.P. & WELTE, D.H. 1984. *Petroleum Formation and Occurrence*. Springer Verlag, Berlin.

TOYODA, S., YOSHIDA, N. & KOBA, K. 2015. Isotopocule analysis of biologically produced nitrous oxide in various environments. *Mass Spectrometry Reviews*, https://doi.org/10.1002/mas.21459

VALENTINE, D.L., CHIDTHAISONG, A., RICE, A., REEBURGH, W.S. & TYLER, S.C. 2004. Carbon and hydrogen isotope fractionation by moderately thermophilic methanogens. *Geochimica et Cosmochimica Acta*, **68**, 1571–1590.

VAN ZUILEN, M., LEPLAND, A. & ARRHENIUS, G. 2002. Re-assessing the evidence for the earliest traces of life. *Nature*, **418**, 627–630.

WADA, H., TOMITA, T., MATSUURA, K., IUCHI, K., ITO, M. & MORIKIYO, T. 1994. Graphitization of carbonaceous matter during metamorphism with references to carbonate and politic rocks of contact and regional metamorphisms, Japan. *Contributions to Mineralogy and Petrology*, **118**, 217–228.

WAECHTER, H., MOHN, J., TUZSON, B., EMMENEGGER, L. & SIGRIST, M.W. 2008. Determination of N$_2$O isotopomers with quantum cascade laser based absorption spectroscopy. *Optics Express*, **16**, 9239–9244.

WAKEHAM, S.G., LEE, C., HEDGES, J.I., HERNES, P.J. & PETERSON, M.L. 1997. Molecular indicators of diagenetic status in marine organic matter. *Geochimica et Cosmochimica Acta*, **61**, 5363–5369.

WANG, Y. & SESSIONS, A.L. 2009*a*. Equilibrium ^{2}H/^{1}H fractionations in organic molecules. I. Experimental calibration of ab initio calculations. *Geochimica et Cosmochimica Acta*, **73**, 7060–7075.

WANG, Y. & SESSIONS, A.L. 2009*b*. Equilibrium ^{2}H/^{1}H fractionations in organic molecules. II. Linear alkanes, alkenes, ketones, carboxylic acids, esters, alcohols and ethers. *Geochimica et Cosmochimica Acta*, **73**, 7076–7086.

WANG, Z., SCHAUBLE, E.A. & EILER, J.M. 2004. Equilibrium thermodynamics of multiply-substituted isotopologues of molecular gases. *Geochimica et Cosmochimica Acta*, **68**, 4779–4797.

WEBB, M.A. & MILLER, T.F. 2014. Position-specific and clumped stable isotope studies: comparison of the Urey and path-integral approaches for carbon dioxide, nitrous oxide, methane, and propane. *The Journal of Chemical Physics A*, **118**, 467–474.

WEBB, M.A., WANG, Y., BRAAMS, B.J., BOWMAN, J.M. & MILLER II, T.F., III 2017. Equilibrium clumped-isotope effects in doubly substituted isotopologues of ethane. *Geochimica et Cosmochimica Acta*, **197**, 14–26, https://doi.org/10.1016/j.gca.2016.10.001

WHITICAR, M.J., FABER, E. & SCHOELL, M. 1986. Biogenic methane formation in marine and freshwater environments: CO_2 reduction vs acetate fermentation – Isotope evidence. *Geochimica et Cosmochimica Acta*, **50**, 693–709.

YOSHIDA, N. & TOYODA, S. 2000. Constraining the atmospheric N_2O budget from intramolecular site preference in N_2O isotopomers. *Nature*, **405**, 330–334.

ZHANG, B.-L., BILLAULT, I., LI, X., MABON, F., REMAUD, G. & MARTIN, M.L. 2002. Hydrogen isotopic profile in the characterization of sugars. Influence of the metabolic pathway. *Journal of Agricultural and Food Chemistry*, **50**, 1574–1580.

Vanadium isotope composition of crude oil: effects of source, maturation and biodegradation

YONGJUN GAO[1]*, JOHN F. CASEY[1], LUIS M. BERNARDO[2], WEIHANG YANG[1] &
K. K. (ADRY) BISSADA[1]

[1]*Department of Earth and Atmospheric Sciences, University of Houston, Houston, TX 77584, USA*

[2]*Repsol, USA, 2455 Technology Forest Blvd., The Woodlands, TX 77381, USA*

**Correspondence: ygao@central.uh.edu*

Abstract: We present a study of vanadium (V) isotope compositions for 17 crude oils spanning a wide range of concentrations and formation ages. Bulk organic geochemical and biomarker compositions are investigated for 11 co-genetic crude oils from the Barbados oil field. About 2‰ V isotope fractionation was observed and they are primarily correlated with V/Ni ratios and most likely reflect the depositional environment of the petroleum source rocks. Factors such as the lithology of the source rocks, Eh and pH of the depositional environment, and possibly the V isotope composition of seawater could all play a role in the V isotope composition. The Eh and pH conditions determine the speciation and coordination of V ions in fluids, whereas the lithology of the source rock defines the competing phases for available V ions in solution. The V isotopes are significantly modified by maturation and biodegradation. The V isotope fractionation during biodegradation has most probably resulted from the microbial activity-induced changes in the species and coordination geometries of V ions in fluids. The progressive decrease of $\delta^{51}V$ with the increase of maturation might suggest a preferential loss of ^{51}V during de-metalation and/or a preferential incorporation of ^{50}V in the newly formed V-organometallic compounds.

The inorganic component in crude oil is generally less than 1 wt%, but it can be important because it influences both the refining process and investigations into oil–oil or oil–source rock correlations in exploration (Filby 1994). Among the inorganic elements in petroleum, vanadium (V) is recognized as the most abundant trace metal constituent whose concentration can range upwards to more than 2000 $mg\ kg^{-1}$, although some crude oils contain as little as 0.1 $mg\ kg^{-1}$ (Amorim *et al.* 2007). The abundances of V and nickel (Ni), as well as the V/Ni ratio in crude oils, vary widely depending on the source-rock composition, depositional environment and the nature of organic matter (Lewan & Maynard 1982; Lewan 1984; Barwise 1990; Filby 1994). The concentrations of these metals can be influenced by secondary processes, such as thermal alteration, deasphalting (a process in which the asphalt content of crude oil is removed or reduced), biodegradation and water washing or during migration. However, the V/Ni ratio itself tends to be nearly constant due to the structural similarities among organometallic compounds that contain V and Ni (Galarraga *et al.* 2008). Vanadium occurs in crude oil predominantly as vanadyl ions (VO^{2+}) in the form of organometallic pyrole complexes of porphyrins in which V substitutes for Mg in plant chlorophylls or Fe in haem. Vanadium also occurs as other largely unknown non-porphyrins or as the cation of organic acids (e.g. Curiale 1987; Filby 1994; Amorim *et al.* 2007; Zhao *et al.* 2014). It has been shown that VO (II) porphyrins tend to predominate over Ni (II) in more anoxic marine environments, and that Ni (II) porphyrins predominate over VO (II) under more oxic marine and non-marine conditions (Lewan 1984). Hence, the V/Ni ratio has been successfully used to characterize the redox conditions of the depositional environment of petroleum source rocks (Moldowan *et al.* 1986; Barwise 1990; Sundararaman *et al.* 1993; Galarraga *et al.* 2008).

Stable isotope studies of V could provide important information additional to elemental studies, which are prone to more uncertainties such as initial source concentration and H_2O content. Mass-dependent stable isotope fractionation is fundamentally driven by the differences in chemical and physical properties of the atoms arising from the relative mass difference among different atoms of a given element, the so-called isotope effect (Hoefs 2008). The magnitude of isotope fractionation is inversely related to the temperature and diminishes to zero at very high temperatures (Urey 1947). Mass-dependent isotope fractionation between different phases/species is mainly affected by the valence states and the chemical binding environment, where the heavier isotope generally preferentially accumulates in the

From: LAWSON, M., FORMOLO, M. J. & EILER, J. M. (eds) 2018. *From Source to Seep: Geochemical Applications in Hydrocarbon Systems*. Geological Society, London, Special Publications, **468**, 83–103.
First published online December 14, 2017, https://doi.org/10.1144/SP468.2
© 2018 The Author(s). Published by The Geological Society of London. All rights reserved.
For permissions: http://www.geolsoc.org.uk/permissions. Publishing disclaimer: www.geolsoc.org.uk/pub_ethics

species or compound with higher bonding strength or higher oxidation state (e.g. Urey 1947; Schauble *et al.* 2001). Vanadium exists in multiple valence states (V^{2+}, V^{3+}, V^{4+} and V^{5+}). First-principle calculations predict that the heavy V isotope (^{51}V) is enriched in a more oxidized valance state with an equilibrium isotope fractionation factor of 6.3‰ at 25°C between $[V^{5+}O_2(OH)_2]^-$ and $[V^{3+}(OH)_3(H_2O)_3]$ (Wu *et al.* 2015). Theoretical calculations also show that V isotope fractionation occurs among different species with the same valance state caused by the differences in bond lengths and coordination numbers (Wu *et al.* 2015). This makes the V isotope system particularly important in tracking depositional or post-sedimentary environments, which can lead to a better understanding of the biogeochemical cycles and the near-surface and subsurface pathways that lead to oil formation and possibly the oil degradation processes. This is especially true when hydrocarbon biomarker indices using steranes, hopanes or terpanes have been altered by biodegradation. It has been proposed that biological processing of V by marine life prior to diagenesis may also introduce fractionation (Premović *et al.* 2002). Vanadium has been identified as an essential trace element for plants and animals (Amorim *et al.* 2007) and is thought to be important for bacterial nitrogen fixation (Anbar & Knoll 2002; Bellenger *et al.* 2008; Kraepiel *et al.* 2009). Hence, V isotopes might also have the potential to track the source of biomass within the depositional environment and eventually be developed as a regular biomarker in low-maturity crude oils.

The first attempts to resolve the systematics of the distribution of V isotopic compositions in petroleum systems was conducted on crude oils and kerogens from the Cretaceous La Luna Formation of Venezuela, but using thermal ionization mass spectrometry (TIMS) (Premović *et al.* 2002). This work revealed that the $\delta^{51}V$ values of petroleum asphaltenes and source-rock kerogens appeared to be significantly lighter than that of the inorganic matrix of the La Luna Formation, and was postulated to result from the biological processing of seawater V (Premović *et al.* 2002). In contrast, TIMS analyses of the V isotope compositions of sea squirts that take up significant V from seawater suggested that no notable isotope fractionation between sea squirt and seawater occurred during the biological V uptake process, which had been indicated by their overlapping $^{50}V/^{51}V$ ratios within the analytical uncertainty of ±0.04 (Nomura *et al.* 2012). However, these earlier TIMS studies all suffer unfortunately from low precision with errors up to ±0.3% (2 SE) of the absolute $^{51}V/^{50}V$ ratios due to the incomplete removal of polyatomic or isobaric interferences and the lack of a well-characterized reference standard (Zhang 2003).

Thus, high precision analysis on V isotopes in crude oil is required to correctly assess the significance of its fractionation. Precise V isotope analyses in geological materials have been only reported recently because of major analytical challenges that occurred prior to the recent advancements in multi-collector inductively coupled plasma mass spectrometry (MC-ICP-MS) and the development of more specific chemical separation techniques for V isotopes (Nielsen *et al.* 2011; Prytulak *et al.* 2011, 2013; Nielsen 2015; Ventura *et al.* 2015; Wu *et al.* 2016). The $^{51}V/^{50}V$ ratio of the only two stable isotopes of V, ^{51}V (99.76%) and ^{50}V (0.24%), is *c.* 420 and the minor ^{50}V isotope has direct isobaric interferences from isotopes of chromium (Cr) and titantium (Ti) – specifically ^{50}Cr and ^{50}Ti. Improvements in both MC-ICP-MS and separation protocols have greatly improved analytical precision and made it possible to investigate V isotopes to precisions of about 0.10‰ (2σ) in silicate matrices (Nielsen *et al.* 2011, 2014, 2016; Prytulak *et al.* 2011, 2013; Wu *et al.* 2016). However, crude oils are more challenging for precise analysis of V isotopes because of their heavy organic matrix, high sulphur (S) content and wide variations in V content. The complex organic matrix may not only introduce various polyatomic interferences that could form isobars on target analytes, but also cause significant instability of the plasma. Sulphur is abundant in many marine-sourced crude oils and its concentration can vary from below 0.05 wt% up to *c.* 6 wt% or more (Khuhawar *et al.* 2012). Removal of isobaric interferences on V isotopes from sulphur oxide molecules using ion exchange chemical separation techniques described below have been investigated here. Sulphur interferences were initially reported by us when describing a $\delta^{51}V$ value for the NIST 8505 crude oil standard value (Casey *et al.* 2015); they have also recently been reported using higher resolution modes in MC-ICP-MS analyses (Nielsen 2015, 2016; Wu *et al.* 2016).

Ventura *et al.* (2015) recently published V isotope compositions for a suite of 11 crude oils from various basins based on the analytical method established by Nielsen *et al.* (2011) and aided by microwave-assisted digestion (Ventura *et al.* 2015). The V isotope compositions of these crude oils with V concentrations ranging from 5 to 335 ppm spanned a range of −1.64‰ to −0.22‰, somewhat less than the range reported here. Based on the first-order source lithological dependence of $\delta^{51}V$, Ventura *et al.* (2015) proposed that the isotopic compositions in crude oils are primarily inherited from the initial sedimentary organic matter or metal-bearing fluids present during the time of sediment deposition. According to these authors, V stable isotopic composition of crude oils were unlikely to have resulted from mass-dependent fractionation associated with the generation, expulsion or migration of petroleum (Ventura *et al.* 2015). Nonetheless, the lack of a co-genetic sample set, and the absence of organic geochemical parameters (e.g. the maturation index and biodegradation index) to help constrain processes related to fractionation, may have hindered the detection of other controlling factors and various processes that can result in V isotope fractionation in crude oils.

Here we present a detailed study of V isotope compositions coupled with certain trace and/or major element abundances (V, Ni, S) in 17 crude oils using newly developed procedures. The methods used are capable of yielding external precisions in the V isotope ratio measurement of better than ±0.3‰ (2σ) for crude oils. The crude oils were selected from eight localities from the worldwide crude oil library of the Center for Petroleum Geochemistry at the University of Houston. They were chosen because of their highly variable V content, ranging from as little as 0.71 µg g^{-1} up to 391 µg g^{-1} (Table 1) and strongly contrasting V/(V + Ni) ratios; this ratio constitutes a well-known redox index for crude oils (Lewan 1984). The selected samples were used to investigate the first-order primary controlling factor for V isotope variability of crude oils from various depositional environments. Among the 17 crude oils analysed here, a subset of 11 samples have been chosen from a single restricted location and represent a co-genetic set of marine crude oils from multiple wells from the Barbados oil field. Detailed bulk organic geochemical and biomarker analyses had been conducted earlier on the Barbados set of samples in connection with a study to assess the process of hydrocarbon generation, expulsion, migration and preservation in Barbados (Repsol, pers. comm.). The Barbados oils are interpreted to be generated within siliciclastic marine source rocks that reached the early part of the oil-generation window, although small differences in the maturity level can be recognized between the oils. Other differences in the Barbados oil properties appear to be related to strong post-accumulation processes such as biodegradation and possibly some evaporative fractionation in some of the samples. The sample set was therefore deemed useful to further evaluate the potential isotopic fractionation associated with depositional and post-depositional maturation and biodegradation processes. This restricted field locality approach for high precision V isotope measurements coupled with organic geochemistry studies to assess maturity, biodegradation and depositional environment effects on V fractionation has not been previously attempted.

We aim to evaluate the methodology and the validity of using V isotope as a fingerprint for oil–oil and oil–source rock correlation and geochemical inversion studies (Bissada *et al.* 1993) for a range of crude oils including those variably biodegraded or of variable maturity. The results of this work may contribute to a host of current postulates involving interaction of source, maturation, and migration parameters, and their influences on oil composition. They could ultimately provide insight into some of the biogeochemical cycles, the importance of redox conditions and the variations in pathways that lead to oil formation. The importance of minerals, transition metal catalysis and water in petroleum generation is yet not completely understood (Goldstein 1983; Seewald 1994; Mango 1996; Lewan 1997).

Method

Samples

The basic geological origin, formation age and chemical compositions of all samples are given in Table 1. The detailed information for the studied samples is as follows.

Sample TP-1 is a heavy crude oil (20°API gravity) from the Timan Pechora Province, Russia. This high S crude is generally thought to be generated from Paleozoic anoxic marine carbonate source rocks at low to moderate maturity levels (Requejo *et al.* 1995).

NIST RM8505 is a Venezuelan heavy crude oil (15°API gravity) generated from the Cretaceous La Luna marly source rocks that were deposited under marine anoxic conditions (Galarraga *et al.* 2008).

Sample BFW-1 is a relatively light crude oil (35° API gravity) from the northwestern part of the Bend Arch, Fort Worth Basin in Texas, USA. The oils in this region are thought to have been generated from Pennsylvanian clastic rocks deposited in settings ranging from oxic marine to marginal marine to continental deltaic systems (Thompson 1982; Pollastro *et al.* 2003, 2007).

Sample BFW-2 (36°API gravity) and sample BFW-3 (38°API gravity) are relatively light oils from the northeastern part of the Bend Arch, Fort Worth Basin in Texas, USA. These oils are believed to be generated primarily from marine shales deposited during the Late Mississippian on a continental shelf under dysoxic conditions (Jarvie *et al.* 2001; Pollastro *et al.* 2003, 2007; Hill *et al.* 2007).

Sample PB-1 is a relatively light oil (32°API gravity) from the eastern margin of the Permian Basin adjacent to the Eastern Shelf, in Texas, USA. The source rocks for the oils in this region could be the Lower Mississippian Barnett Shale or the Upper Devonian Woodford Shale or a composite Woodford–Barnett source deposited under marine to marginal marine conditions (Comer 1991; Pollastro *et al.* 2003).

Table 1. *Geochemical properties and V isotope compositions of crude oil samples*

Sample	Origin	Source-rock age*	Isotope composition				Trace element				API gravity	Biomarker		
			$\delta^{51}V$ (‰)	2 sd	n	N	S (wt%)	V (mg/g)	Ni (mg/g)	V/ (V + Ni)	API°	Pr/ nC17	Ts/ (Ts + Tm)	Tricyclic Index
RM 8505	Venezuela	Cretaceous/Tertiary	−0.02	0.30	17	6	2.17	391	48.6	0.89	15	NA	NA	NA
TP-1	Timan Pechora Province, Russia	Paleozoic	−0.46	0.31	4	2	1.67	119	73.4	0.62	20	NA	NA	NA
PB-1	Tom Green Co. TX, USA	Lower Mississippian or Upper Devonian	−0.62	0.25	6	2	0.18	0.71	0.95	0.43	32	NA	NA	NA
BFW-1	Foard Co. TX, USA	Pennsylvanian	−1.76	0.04	3	1	0.26	1.48	11.3	0.12	35	NA	NA	NA
BFW-2	Grayson Co. TX, USA	Late Mississippian	−0.44	0.13	3	1	0.30	18.3	7.44	0.71	36	NA	NA	NA
BFW-3	Grayson Co. TX, USA	Late Mississippian	−0.11	0.11	3	1	0.20	5.41	2.24	0.71	38	NA	NA	NA
BBD-01	Barbados	Cretaceous	−0.25	0.09	3	1	1.25	80.1	74.2	0.52	19	ND	0.36	0.14
BBD-02	Barbados	Cretaceous	−0.84	0.30	1	1	0.75	82.4	76.0	0.52	20	ND	0.37	0.14
BBD-03	Barbados	Cretaceous	−0.63	0.30	1	1	0.80	44.0	41.9	0.51	25	38.6	0.39	0.14
BBD-04	Barbados	Cretaceous	−1.26	0.27	3	1	0.92	51.6	49.1	0.51	25	10.8	0.38	0.14
BBD-05	Barbados	Cretaceous	0.40	0.30	1	1	0.37	23.3	21.6	0.52	25	29.0	0.42	0.15
BBD-06	Barbados	Cretaceous	−0.67	0.30	1	1	0.98	179	172.8	0.51	19	1.03	0.37	0.13
BBD-07	Barbados	Cretaceous	−0.66	0.30	1	1	0.70	40.1	39.0	0.51	27	0.77	0.38	0.14
BBD-08	Barbados	Cretaceous	−1.60	0.30	1	1	0.21	5.54	7.39	0.43	35	0.83	0.46	0.18
BBD-09	Barbados	Cretaceous	−0.56	0.30	1	1	0.80	57.0	55.8	0.50	24	0.87	0.37	0.13
BBD-10	Barbados	Cretaceous	−1.10	0.30	1	1	0.33	18.6	20.1	0.48	31	0.73	0.43	0.17
BBD-11	Barbados	Cretaceous	−0.95	0.30	1	1	0.48	17.8	18.9	0.49	30	0.78	0.43	0.16

*Source for ages: Thompson 1982; Larue *et al.* 1985; Comer 1991; Speed *et al.* 1991*a, b*; Babaie *et al.* 1992; Requejo *et al.* 1995; Jarvie *et al.* 2001, 2007; Pollastro *et al.* 2003, 2007; Hill & Schenk 2005; Hill *et al.* 2007; Galarraga *et al.* 2008.
Elemental concentrations were analysed by ICP-OES and ICP-MS. Biomarker data were analysed with GC-MS, and V isotope compositions were analysed with a Nu Plasma II MC-ICP-MS at University of Houston.
NA: not available for publication; ND: not detectable.

Sample BD01 through BD11 are a set of 11 crude oils from the island of Barbados, part of the forearc accretionary prism in the eastern Caribbean, which is underlain by sedimentary rocks and mélanges that include those rocks accreted to the hanging wall of the Caribbean plate, which have been detached from the lower footwall South Atlantic oceanic plate with its Mesozoic and Cenozoic cover sequences. The source-rock ages for these crude oils are a subject of active debate. The crude oils have been suggested to be generated either from Cretaceous marine shales deposited under normal salinity and dysoxic conditions (Burggraf *et al.* 2002; Lawrence *et al.* 2002; Leahy *et al.* 2004; Hill & Schenk 2005) or from marine turbiditic shales of Tertiary age deposited under similar marine conditions (Larue *et al.* 1985; Speed *et al.* 1991*a*; Babaie *et al.* 1992). The Tertiary interpretation is based, in part, on similarities between whole oil gas chromatograms, carbon isotopes and other geochemical data of Barbados Tertiary shale extracts and crude oils (Babaie *et al.* 1992). Cellulose higher plant debris in wells suggest Tertiary sources were derived from the ancestral Oronoco deltaic systems (Speed *et al.* 1991*a*; Babaie *et al.* 1992) accreted to the front of the Caribbean arc as it migrated eastwards. Evidence to suggest a Cretaceous source rock for Barbados oils includes: (1) that most of the potential source rocks sampled at the surface or in drill holes on Barbados did not reach the oil window with Ro < 0.6 implying deeper, more mature sources (Babaie *et al.* 1992; Hill & Schenk 2005); (2) that a Barbados mud diapir contains not only Paleogene lithic clasts, but a Cretaceous clast (Joes River Formation), which probably indicates a deeper, more mature Cretaceous source (Speed *et al.* 1991*a*; Deville *et al.* 2003); (3) that direct oil–oil correlations yielded similarities between Barbados oils and some oils from eastern Venezuela and Trinidad that are from Upper Cretaceous La Luna-like source rocks based on their organic geochemistry (Zumberge 1987; Lawrence *et al.* 2002; Hill & Schenk 2005); and (4) that in the Barbados oils there is an absence of the higher plant biomarkers oleanane and lupanes typically associated with Tertiary terrestrial organic matter (Hill & Schenk 2005; our unpublished data). It is also well documented that the basal décollement in Trinidad has penetrated into the Cretaceous section (Wood 2000) and similarly Cretaceous fossils have been sampled from mud-diapir clasts of Trinidad (Speed *et al.* 1991*b*; Deville *et al.* 2003). Like Trinidad, Barbados is underlain by multiple accreted thrust slices that can repeat stratigraphy and effectively structurally thicken and bury Tertiary and/or Tertiary/Cretaceous stratigraphic sections, with Tertiary only of Tertiary/Cretaceous sections dependent on the depth of the basal décollement.

Barbados crude oil could be interpreted to be co-genetically formed at multiple stratigraphic/structural levels of strictly Tertiary or Tertiary/Cretaceous sections creating the variable thermal maturities down-section indicated by the Barbados crude oil geochemistry (e.g. Hill & Schenk 2005; this study). Drilling has only penetrated immature Paleocene-aged sediments and/or mélanges in Barbados (Speed *et al.* 1991*a*; Babaie *et al.* 1992), but similar deeper and/or older stratigraphic sections in the oil window underlying Barbados could be possible. Deeper penetration of the basal décollement into the Cretaceous at the time when Eocene–Paleocene subsurface assemblages of Barbados were accreted cannot be precluded at this time. Because there is no measured age-related geochemical biomarker in the crude oil datasets available from various geochemical studies of Barbados oils that could reliably distinguish Tertiary from Cretaceous sources (Hill & Schenk 2005; J.M. Moldowan pers. comm. 2015), we regard Tertiary and Cretaceous source-rock models as both permissible with the current published data and assessment of the internal data available.

The Barbados crude oils have API gravities ranging from 19° to 35°. In spite of the large variations in the elemental concentrations of V and Ni, these oils have a relatively narrow range of $V/(V + Ni)$ ratio of 0.50 ± 0.03 (Table 1), suggesting a small range of source-rock characteristics. Some of these crude oils have also been subjected to variable degrees of biodegradation. Oils range from severely altered in shallow reservoirs to completely unaffected in deeper reservoirs (Hill & Schenk 2005). All of these characteristics of varying biodegradation and maturity are reflected in the compositional data of the 11 wells sampled in this study (Table 1).

Mineralization to remove organic matrix

Crude oils are very complex in composition, viscosity and phase relationships. Direct injection of crude oils into ICP-MS may influence the stability of the plasma or even cause plasma extinction. The analytical challenge for both elemental and isotopic composition of metal elements in crude oils also comes from the various polyatomic and isobaric interferences due to their heavy organic matrix. A mineralization procedure has been developed to destroy the organic structures of crude oils with strong oxidizing acids in pressurized vessels with either the Parr high-pressure acid digestion bomb (Parr 4749) or the single reaction chamber microwave digestion system (Milestone Ultrawave®). These techniques have proved effective to extract metal elements from crude oils with recovery of better than $98 \pm 5\%$ (Yang 2014; Casey *et al.* 2016). Special precautions have been taken to minimize sampling heterogeneity of high viscosity crude oils by shaking in capped glass vessels while heating at *c.* 80°C on a hot plate. For Parr bomb digestion, precisely weighed crude oil of *c.* 100 mg or *c.* 400 mg was loaded into the Teflon cup of the digestion bomb; 3 ml of 16N HNO_3 was then added to the cup which was then capped and heated on a hot plate at 150°C for 2 h. The cap was then removed and its contents evaporated down to incipient dryness. This pre-oxidizing treatment can destroy some part of the organic matrix and thus help reduce the pressure that can develop inside the Parr bomb during the following heating step. After dry-down, 3 ml of 16N HNO_3 and 1 ml of 12N HCl were added to the Teflon cup; the cup was loaded into the bomb, which, after assembly, was heated at 160°C inside a Lindberg/Blue M oven for 12 h. After the apparatus had cooled down, the inner Teflon cup was taken out of the bomb and the digested solution was then transferred into a Savillex PFA beaker. The Teflon cup was repeatedly washed with 2% HNO_3 to ensure 100% transfer. The solution inside the PFA beaker was then dried down on a hot plate at 150°C. Finally, the digested sample was re-dissolved with 5 ml of 0.2N HNO_3. For microwave digestion, *c.* 400 mg of precisely weighed crude oils were loaded into 40 ml quartz digestion tubes and then 5 ml of 16N HNO_3, 1 ml of 12N HCl and 1 ml of 30% H_2O_2 were added to the digestion tubes for digestion. A five-step microwave irradiation programme was used for digestion: (i) 5 min to 120°C; (ii) 10 min to 160°C; (iii) 10 min to 180°C; (iv) 8 min to 260°C; and (v) hold at 260°C for 15 min. After digestion the resultant solutions were transferred into PFA beakers, dried down to incipient dryness on a hot plate and then re-dissolved with 5 ml of 0.2N HNO_3. Total carbon content of the resultant sample solution was no more than 0.1% (g/g) analysed by a LECO® CS230 Carbon/Sulfur Analyzer. Near-complete recovery of V in crude oils by this mineralization procedure has been demonstrated by repeated digestion experiments on two NIST reference materials for oil (NIST RM 8505 and NIST 1634c) The analysed V concentrations are 390 ± 0.8 µg g^{-1} ($2\sigma, \underline{n} = 25$) for NIST RM 8505 and 28.9 ± 2.3 µg g^{-1} ($2\sigma, \underline{n} = 10$) for NIST 1634c, both being in good agreement with the recommended values by NIST of 390 ± 10 and 28.2 ± 0.4 µg g^{-1}, respectively (Yang 2014; Casey *et al.* 2016). The complete recovery of V during the digestion steps is important to eliminate a potential cause of isotopic fractionations of V. To achieve adequate V (minimum = 1000 ng) for isotope measurements in the method described below, solutions from multiple digestions of low-abundance crude oils were combined before purification, with the number of digestions dependent on V abundance measurements.

Chemical purification

We use a modified approach from that of Nielsen *et al.* (2011) to isolate V from other elements, especially from S, Ti and Cr. To achieve near-quantitative removal of S for successful V isotope analysis, our chemical purification protocol for crude oils uses a multi-step liquid chromatography process with combined cation and anion exchange columns. It has been demonstrated that S molecules can interfere with both V isotopes and ^{49}Ti – which is a monitor for Ti levels in solution for isobar correction (Nielsen *et al.* 2011, 2016; Nielsen 2015). To minimize these interferences, our changes to the existing methods include an added column for removal of S, as well as other, more specific, slight modifications to previous column procedures involving somewhat different resins, and achieving reduction in some reagents and modest increases in efficiencies of some column steps, as described below.

First column chemistry. First, 5 ml of wet Bio-Rad AG 50W X-12 (200–400 mesh) cation exchange resin was packed into a quartz glass column with an internal diameter of 10 mm. Before loading the resin, all the columns were repeatedly acid-leached with 8N HNO_3 and 6N HCl. The resin was subsequently washed with 6N HCl and Milli-Q H_2O, and conditioned with 0.2N HNO_3. Next, a 5 ml sample solution in 0.2N HNO_3 was loaded onto the resin. The resin was then washed with Milli-Q H_2O and 1N HNO_3 to remove S and major and trace elements such as Na and Li. The V was then completely washed out of the column with 40 mL of 1N HNO_3. The analysis of the collected solution for the V fraction using ICP-MS indicated that V was $100 \pm 5\%$ recovered from the column together with Ti, K and minor amounts of Na. Other cations, such as Mg, Ca, Fe, Al, Si and Cr were all effectively separated from the V fraction by being held in the resin due to them having higher distribution coefficients than V coefficients (Strelow *et al.* 1965). Sulphur (mainly existing as the anion SO_3^{2-}) was almost completely removed from the sample matrix. To achieve a complete removal of S, a second pass through this cation exchange column was required for samples with high S content. The removal of Mg, Ca and the majority of other cations by this step makes it possible to use small columns and also to use further separation columns that involve HF to remove Ti and other cations (avoiding the formation of Mg and Ca insoluble fluorides). The collected V fraction in 1N HNO_3 was then evaporated and refluxed with 6N HCl to reduce V^{5+} to V^{4+}; the V was then picked up with 1 ml 0.3N HCl in 1.0N HF mixed acids for the second column.

Second column chemistry. Initially, 1 ml of anion exchange resin AG1 X8 (200–400 mesh) was packed in a polypropylene column (Bio-Rad, USA). This column was used to separate Ti from V based on the method reported by Makishima *et al.* (2002) and Nielsen *et al.* (2011). Before loading the sample, the columns were cleaned with 6N HCl, 2N HF and Milli-Q H_2O and then conditioned with eluent (0.3N HCl in 1.0N HF mixed acid); a 1 ml sample solution in mixed 0.3N HCl and 1.0N HF prepared from the first column was then loaded onto the column. Finally, the V was eluted with 20 ml of eluent, which left Ti in the column.

Third column chemistry. As with the second polypropylene columns, the third set of columns were repeatedly acid-leached with 8N HNO_3, 6N HCl and 10N HF before filling with 1 ml AG1 X8 resin. These columns were designed to further purify V from other cations remaining in the V fraction after the first and second columns, i.e. mainly K and minor Cr. For this column procedure, V remained absorbed strongly in the band at the top of the column due to the formation of a polyvanadate complex inside the anion exchange resin particles with a matrix of weak acid containing H_2O_2, whereas all the cations have no absorption to the resin (Strelow & Bothma 1967; Kiriyama & Kuroda 1983). The V fraction obtained from the second column was dried down and then refluxed with 15N HNO_3 – to oxidize all V to the V^{5+} form – and then re-dissolved in 1 ml of a mixture of 0.01N HNO_3 and 1.0% H_2O_2. The resin was washed with 2N HNO3 and Milli-Q H_2O and conditioned with a mixed solution of 0.01N HNO_3 and 1.0% H_2O_2. The sample solution was then loaded onto the column and the resin washed with 10 ml of the mixed solution (0.01N HNO_3 and 1.0% H_2O_2) to remove K, Cr and possibly other cations. Finally, the V was collected with 14 ml of 1N HNO_3.

The chemical separation procedure described above resulted in $^{50}Ti/^{50}V$, $^{50}Cr/^{50}V$ and S/V ratios of less than 0.06, 0.04 and 0.1, respectively. The full procedural blank for V from this separation technique was less than 10 ng.

Elemental analysis

Concentrations of V, Ni and S in the crude oils were determined by ICP-OES (Inductively Coupled Plasma Optical Emission Spectrometry, Agilent 725) or triple quadrupole (QQQ) ICP-MS (Inductively Coupled Plasma Mass Spectrometry, Agilent 8800), depending on their concentrations after digestion, at the University of Houston. The long-term precision of this method in terms of reproducibility of V, Ni and S concentrations is generally about or better than 1% (1σ) evaluated by repeated analysis on multiple full replicates of the NIST RM8505 standard. As to accuracy, the V concentration on NIST RM8505 determined by this method is 391 ± 2 $\mu g\ g^{-1}$

$(2\sigma, n = 17)$, which is within 1% of the NIST recommended value of 390 ± 10 ppm.

Mass spectrometry for V isotope analysis

For V isotope analysis, we utilized the high resolution Nu Plasma II MC-ICP-MS at the University of Houston. The samples were introduced through a Cetac Aridus II desolvating nebulizer at a concentration of 1000 ng g^{-1} of V in 2% HNO_3. Sample analyses were bracketed by a 1000 ng g^{-1} in-house V single-element standard ('UH') prepared from the V ICP standard from Inorganic Ventures, USA (IVU). Isotope compositions in δ notation were first calculated relative to this University of Houston in-house standard ($\delta^{51}V_{UH}$) in parts per 1000 ($\delta^{51}V$ per mille):

$$\delta^{51}V_{UH} = 1000 \times [(^{51}V/^{50}V_{sample}/^{51}V/^{50}V_{IVU}) - 1] \quad (1)$$

The $\delta^{51}V_{UH}$ compositions were all converted and reported relative to the Alfa Aesar (AA) V standard of Nielsen *et al.* (2011) by:

$$\delta^{51}V = \delta^{51}V_{UH} - \Delta \quad (2)$$

where Δ is the average measured per mil difference between the IVU and the AA standards of $-0.1 \pm 0.02‰$ (SE; $n = 82$).

Masses of 49, 50, 51, 52 and 53 were collected in five Faraday cups (L4, L1, H2, H5 and H7, respectively) fitted with $10^{11}\ \Omega$ resisters. For a solution with 1000 ng g^{-1} V and an uptake rate of 40 μl min^{-1}, the typical intensity of ^{51}V is about 40 volts. Isobaric interferences from ^{50}Ti and ^{50}Cr on ^{50}V were corrected by monitoring masses 49 and 53, respectively. Values for $^{50}Ti/^{49}Ti$ and $^{50}Cr/^{53}Cr$ used for the isobaric interference corrections are 0.9725 (Niederer *et al.* 1985) and 0.4574 (Shields *et al.* 1966), respectively. These values were corrected for instrumental mass fractionation using an exponential law. An exponential mass fractionation factor (β value) was applied to correct for instrumental mass fractionation of the $^{50}Ti/^{49}Ti$ and $^{50}Cr/^{53}Cr$ ratios used for isobaric corrections according to the equation:

$$R_M = R_T/(m_1/m_2)^\beta \quad (3)$$

where R_T and R_M are the true and measured isotope ratios for a given element and m_1/m_2 is the mass ratio of $^{50}Ti/^{49}Ti$ and $^{50}Cr/^{53}Cr$. The calculated ratio R_M is used for isobaric interference corrections. Dynamic determination of β values for Ti and Cr during an analysis is impossible even with the wide collector array of the Nu Plasma II. Therefore, prior to an analytical session, the β values for Cr and Ti were independently determined and these fixed values were applied to the online data correction protocol. Overall, $\beta_{(Ti)}$ and $\beta_{(Cr)}$ were *c.*−1.7 and drift during an analytical session (typical of about 20 to 24 hours) was less than ±0.01. Maximum $^{50}Ti/^{50}V$ and $^{50}Cr/^{50}V$ ratios for samples analysed were 0.02 and 0.03, respectively, and the magnitude of offset in the corrected $\delta^{51}V$ derived from such a small variation of β values was less than 0.002‰. Doping experiments with Ti, Cr and S confirm that the relative levels of these particular elements in the purified V solutions result in no systematic bias in the V isotope ratio measurements (Fig. 1). Our tests on natural crude oils with high and low V and S abundances indicate that this method yields a 2σ precision of ±0.30‰ and is applicable to, and suitable for, crude oils with a wide range of V concentrations (Table 1). It is worth noting that this precision is comparable to that obtained by Ventura *et al.* (2015) but with five times less material (1 ppm v. 5 ppm), mostly owing to the high sensitivity of the Nu Plasma II used for this study. The analysed V isotope compositions of the BDH V standard solution over a 6-month period yielded a mean value of −1.15‰ ±0.3‰ (2σ) which is comparable to the values of −1.19 ± 0.12 (2σ) published by Nielsen *et al.* (2011). The accuracy on the V isotope analysis was further checked by repeated analyses on two NIST reference materials – NIST RM 8505 and NIST SRM 1634c. Repeated analyses (six analyses for two digestions and separations) of NIST SRM 1634c, which is a residual fuel oil with 28.2 µg g^{-1} V and 2 wt% of S, yielded a mean $\delta^{51}V$ value of −1.41 ± 0.11‰ (2σ). Repeated analyses (17 runs from six digestions and separations) of NIST RM 8505, which is a Venezuelan crude oil with 390 µg g^{-1} V and 2.2 wt% of S, yielded a mean $\delta^{51}V$ value of −0.02 ± 0.3‰ (2σ). Both values are in good agreement within the quoted uncertainties with their reported values of − 1.49‰ and −0.33 ± 0.36‰ for NIST SRM 1634c and NIST RM 8505, respectively (Ventura *et al.* 2015). It is interesting to note that our $\delta^{51}V$ values for NIST RM 8505 and NIST 1634c, although within uncertainties, were both slightly heavier than those obtained by Ventura *et al.* (2015), which might be related to the S effect on the V isotope analysis as stated in the above section.

Results

The V isotope compositions of the 17 crude oils are compiled in Table 1 and Figure 2. The $\delta^{51}V$ compositions of these oils showed a wide variation, ranging from −1.76‰ to +0.40%, which is wider than the reported range from −0.27‰ to −1.29‰ in mantle-derived mafic and ultramafic rocks from diverse localities and of various degrees of alteration (Prytulak *et al.* 2013). The V isotope composition of these

crude oils indicated that they have been fractionated to be both heavier and lighter compared to that of Bulk Silicate Earth (BSE) ($\delta^{51}V_{BSE} = -0.7 \pm 0.2‰$, Prytulak *et al.* 2013). The crude oils analysed here define a slightly broader range compared with the previous results (Ventura *et al.* 2015), though the two datasets largely overlap (Fig. 2). Most strikingly in this study, the 11 co-genetic crude oils from Barbados have almost the same degree of V isotope fractionation defined by the globally distributed sample sets used in both studies, but surprisingly very restricted ratios of V/Ni and V/(V = Ni) compared with the global values (Table 1; Fig. 2).

Discussion

Source and reservoir

The crude oils in this study have a broad range of V isotope compositions (−1.76‰ to +0.40‰, Table 1). The observed total V isotope fractionation of about 2.2‰ is smaller than the predicted equilibrium fractionation at 25°C of 6.3‰ between V(V) and V(III) species and of 3.9‰ between V(V) and V(IV) species for inorganic V species in an aqueous system (Wu *et al.* 2015). However, it is somewhat close to the predicted equilibrium fractionation at 25°C between V(IV) and V(III) species of 2.5‰, and that of 1.5‰ between two V (V) species in an aqueous system (Wu *et al.* 2015). There is a first-order correlation between the V isotope composition and the V/(V + Ni) ratio in the regionally distributed datasets (Fig. 3). As shown in Figure 3, isotopically heavy crude oils are generally associated with high V/(V + Ni) ratios, whereas isotopically light crude oils tend to have low V/(V + Ni) ratios. Such an overall positive correlation between $\delta^{51}V$ values and the V/(V + Ni) ratio in crude oils has also been reported previously (Ventura *et al.* 2015). We, however, also observed significant variation in $\delta^{51}V$ values at a given V/(V + Ni) ratio when examining the Barbados dataset from a single locality (Fig. 3) where a large number of more regionally discrete samples from different wells were available for study.

The primary prevalence and proportionality of Ni and V in crude oils are believed to be controlled by the Eh–pH conditions, the availability of S and the lithology of the source rock in the depositional environment (Lewan & Maynard 1982; Lewan 1984; Barwise 1990). The metallo-porphyrins of V and Ni are formed during early diagenesis of source rocks through the metalation of free-based porphyrins which form during a series defunctionalization and conversion of chlorophyll (Baker & Louda 1986; Filby & Van Berkel 1987; Keely *et al.* 1990; Filby 1994). In general, oils of marine origin in more anoxic environments have higher concentrations of trace elements predominated by V porphyrins

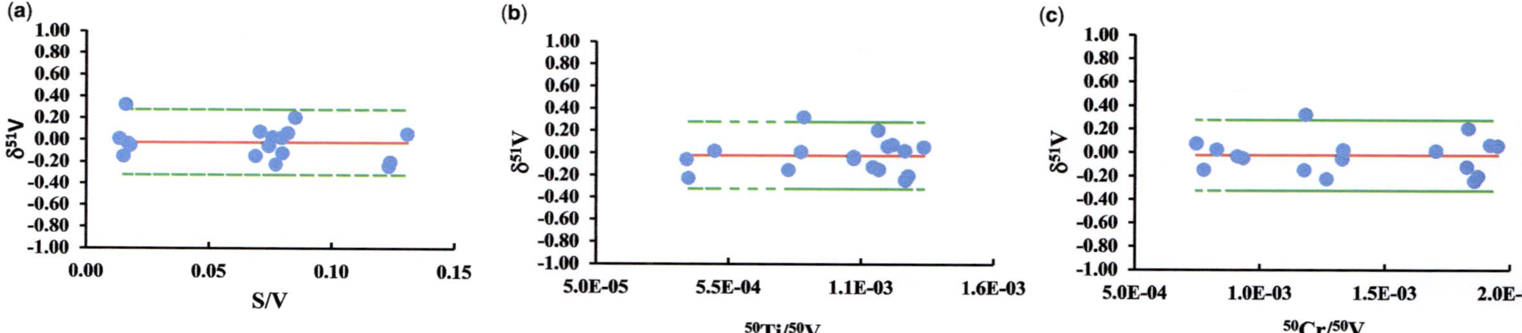

Fig. 1. Measured V isotope compositions for NIST Crude Oil Reference Material RM 8505, reported relative to the AA V standard solution defined as 0‰ (Nielsen *et al.* 2011). Dashed green lines represent the 2s error of ±0.3‰, and the solid red line represents the average value of −0.02‰. This shows that multi-step column chemistry can efficiently remove S (**a**), Ti (**b**) and Cr (**c**), and their presence in the purified V solutions results in no systematic bias in the V isotope ratio measurements.

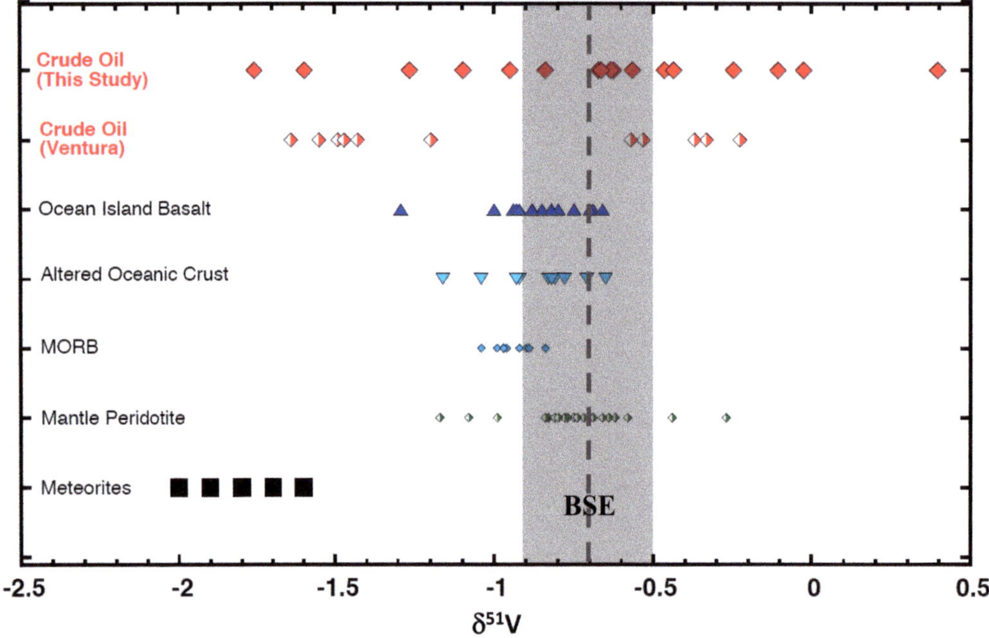

Fig. 2. V isotopic compositions of crude oils (top two rows) as reported here and by Ventura *et al.* (2015) compared with all other data available from terrestrial and meteorite reservoirs (Prytulak *et al.* 2011, 2013; Nielsen *et al.* 2014). Values of $\delta^{51}V$ are reported relative to the AA V solution standard defined as 0‰ (Nielsen *et al.* 2011). Shaded area is the reported range for Bulk Silicate Earth (BSE): (-0.7 ± 0.2‰; Prytulak *et al.* 2013).

over Ni, whereas oils derived from terrestrial sources under more oxic conditions have lower concentrations of trace elements predominated by Ni porphyrins over V (Lewan 1984; Barwise 1990; Filby 1994). Hence, both the $V/(V + Ni)$ and V isotopic compositions of crude oils might be primarily controlled by the depositional environment of the petroleum source rocks where the V was incorporated into organic matter during sediment deposition and early diagenesis. It is worth noting that the observed correlation trend between V isotope composition and the oxidizing conditions indicated by the $V/(V + Ni)$ ratio (Fig. 3) seems to somewhat contradict the theoretical predictions that the heavy V isotope (^{51}V) should be enriched in a more oxidized valance state (Wu *et al.* 2015). However, it should also be noted that validation of the observed first-order correlation between the V isotope compositions and the $V/(V + Ni)$ ratios for globally distributed crude oils (Ventura *et al.* 2015; this study) requires further detailed organic geochemical data to support their pristine nature (i.e. these oils were not altered by any secondary process). As discussed in the following sections, secondary processes, such as maturation and biodegradation, can significantly shift the primary V isotope compositions, because in crude oil V mainly occurs as vanadyl ions (VO^{2+}) in the form of organometallic complexes (Branthaver &

Filby 1987; Filby & Van Berkel 1987; Filby 1994). The observed overall correlation between the V isotope compositions and the $V/(V + Ni)$ ratios, if proven, might indicate that the V isotope compositions in crude oils are primarily controlled by factors other than the valance status of V in crude oils. It has been shown that the V isotope could fractionate between different species with the same valence status due to the differences in bond lengths and coordination numbers (Wu *et al.* 2015). However, this could also indicate that the dominant speciation and/or the valance status of the V-organometallic complexes could be different in crude oils generated under different depositional environments. For instance, the V in crude oils ordinated from strongly reducing, H_2S-rich marine, carbonaceous, sedimentary environments mainly exists as vanadyl (VO^{2+}) porphyrins, whilst the V in crude oils from less anoxic, H_2S-poor, non-marine, clay-rich siliciclastic environments might dominantly exist as V (III) in non-porphyrins or as the cation of organic acids. The V in clay minerals that is present mainly as V (III) (Maylotte *et al.* 1981; Premović 1984; Wanty *et al.* 1990) could potentially be leached into interstitial water and be incorporated into crude oils. Further detailed work on the relationship between V isotopes and its speciation in crude oils will be required to resolve this question.

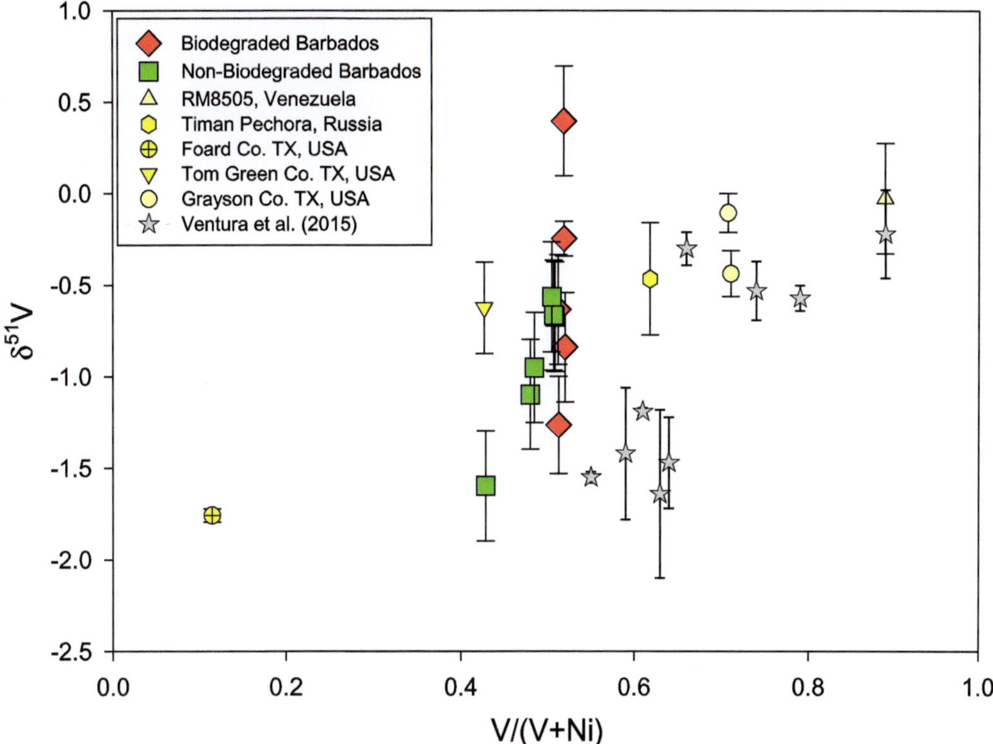

Fig. 3. V isotopic compositions of crude oils from this study and Ventura *et al.* (2015) plotted against V/(V + Ni) ratios. Error bars are 2 SD listed in Table 1. Values of δ^{51}V are reported relative to the AA V solution standard defined as 0‰ (Nielsen *et al.* 2011).

The source of V in organic-rich sediments is a matter of controversy and is far from being solved. Two plausible sources have been postulated: (i) the buried remains of V-rich marine organisms (Premović *et al.* 1986); or (ii) preferential adsorption and reduction of V from the water column, from the water/sedimemt depositional surface, or from subsurface pore water (Szalay & Szilágyi 1967; Bloomfield & Kelso 1973; Cheshire *et al.* 1977; McBride 1978; Premović 1978).

Despite being toxic in humans, V is categorized as an essential trace element for plants and other animals. It stimulates the synthesis of chlorophyll and promotes the growth of young animals (Amorim *et al.* 2007). It is found in marine algal haloperoxidase enzymes (Butler 1998). It also accumulates in Tunicata (Crans *et al.* 2004) that first appear in the fossil record in the early Cambrian period (Fedonkin *et al.* 2012). Biological activity may produce measurable V isotopic fractionation because the metabolic processing of V involves a number of steps, such as transport across membranes and uptake by enzymes (Maréchal *et al.* 1999; Premović *et al.* 2002). Premović *et al.* (2002) proposed that the

difference in the ^{50}V/^{51}V values between the La Luna petroleum asphaltenes/source kerogens and inorganic sources can be ascribed to biological processing of seawater V. In contrast, direct comparison of the V isotope composition of sea squirts and that of seawater appears to show that no notable isotope fractionation occurs during the biological V uptake process, admittedly at the given analytical precision (errors up to ±5% for ^{51}V/^{50}V ratio, Nomura *et al.* 2012). Obviously, better and more precise investigations are required to resolve the observed controversy. Wedepohl (1971) demonstrated by mass balance calculation that the mechanism for the enrichment of V by V-rich marine organisms must have been of only minor importance. This would indicate that the large V enrichments observed in organic material/crude oil are largely the consequence of absorption and reduction during deposition and early diagenesis and lithification. This is consistent with V being preferentially concentrated in sediments overlain by anoxic waters through diffusion across the sediment–water interface and immobilization within sediments (Emerson & Huested 1991).

Normal seawater has a V content of only *c.* 2 ppb (Wedepohl 1971), but enrichment of V at the sea bottom may occur under highly reducing conditions associated with organic-matter scavenging action within enriched V solutions (Premović *et al.* 1986). The ultimate sources of V in seawater include hydrothermal solutions related to submarine volcanism (Boström & Fisher 1971), and continental runoff (Wedepohl 1964). Laboratory experiments suggest that V is not leached from basalt during hydrothermal alteration (Seyfried & Mottl 1982). As with serpentinization and seafloor weathering, there is little evidence for elemental V addition or removal during low-temperature hydrothermal alteration (Kelley *et al.* 2003). It has also been reported that the hydrothermal fluids show little overall change in concentration in comparison to seawater (Jeandel *et al.* 1987), which suggests that high-temperature hydrothermal processes cannot be very important in determining the concentration of V in the ocean.

The oceanic residence time of V has been calculated as *c.* 50 kyr (Tribovillard *et al.* 2006), which is significantly longer than the mixing time of the global oceans (*c.* 2000 years; Palmer *et al.* 1988). This would be sufficient for the ocean to have an isotopically homogenous V composition at any given geological time. Hence, one possible explanation of the observed variations in V isotope composition of crude oil would be a simply inheritance from a variable V isotope composition of seawater at the time of source-rock deposition. Theoretical calculations suggest that the V isotope composition of seawater might vary with change of ambient oxygen levels (Wu *et al.* 2015). However further work is required to establish such a speculative control mechanism.

Due to the limitation on high precision age constraints on the crude oils investigated in this study, it is not possible or prudent to conduct a detailed investigation on the age dependence of the V isotope compositions in crude oils. Although under debate because the Barbados source-rock age is less resolvable, the Venezuelan heavy crude oil (NIST RM8505) and the crude oils from Barbados have been interpreted by various investigators to be generated from similar Cretaceous source rocks under marine conditions (Hill & Schenk 2005; Galarraga *et al.* 2008). We note that the NIST RM 8505 crude oil from Venezuela has a very different average δ^{51}V value ($-0.02 \pm 0.30‰$) from the average value ($-0.61 \pm 0.07‰$) for crude oils from Barbados that have been least altered (e.g. oils with $Pr/nC_{17} < 1.0$ and $Ts/(Ts + Tm) < 0.38$; Table 1 and Fig. 3), although age-related differences are a possibility. If indeed these two sets of crude oils were actually formed at near-similar ages in the Cretaceous, the differences might indicate that the pristine V isotope composition of crude oil is controlled by factors other than the speculated δ^{51}V fluctuation of seawater. The possible factors that can affect the V isotope composition of crude oils other than the temporal variation of V isotope composition of seawater are the depositional environment-dependent speciation of V and their coordination geometries in aqueous solutions (Wu *et al.* 2015). Theoretical calculations showed that V isotope fractionation could occur between different species with the same valence state due to their different bond lengths and coordination numbers (Wu *et al.* 2015).

In aqueous solution, both V (III) and V (IV) species are cationic while V (V) species are anionic (Baes & Mesmer 1976; Zhang 2003). In aqueous solutions, both vanadate (V^{4+}) and vanadyl (V^{5+}) ions undergo a number of the self-condensation reactions. These reactions are very sensitive to the pH of the solution and the presence of potential ligands which can coordinate to the vanadyl and form a number of complexes with different coordination geometries (Baes & Mesmer 1976; Zhang 2003). Vanadium in oxic seawater should be present as V (V) and be hydrolyzed to $VO_2(OH)_3^{2-}$ at neutral pH, whilst under more acidic conditions (below pH of *c.* 3) the dioxovanadium cation VO_2^+ is the stable species due to the increased coordination number of its hydrated form (Baes & Mesmer 1976; Seewald 2003). Under moderately reducing conditions V (IV) is stable as the oxovanadium cation VO^{2+}, which should also hydrolyze at seawater pH to VO $(OH)_3^-$ (Baes & Mesmer 1976).

Theoretical calculations show that V isotopes can also be significantly fractionated due to mineral adsorption (Wu *et al.* 2015). Both V(IV) and V(V) are surface-reactive and thus subjected strongly to adsorption processes in natural waters (Wehrli & Stumm 1989). Adsorption of V to biogenic or terrigenous detritus is the most commonly considered mechanism of V removal from seawater and enrichment (Amdurer *et al.* 1983; Prange & Kremling 1985). In the pH range of most natural waters, vanadate strongly adsorbs to ferric and aluminum oxides (Honeyman 1984; Micera & Dallocchia 1988; Shieh & Duedall 1988; Wehrli & Stumm 1989) and kaolinite (Breit & Wanty 1991). The vanadyl ion adsorbs more strongly than vanadate, but its stability as an adsorbed species is limited in oxic-water environments (Wehrli & Stumm 1989). Vanadyl (VO^{2+}) ions in nature are characteristically associated with organic ligands and humic and fulvic acids (Cheshire *et al.* 1977; Wehrli & Stumm 1989). In aqueous solutions, therefore, competition for the V (IV) species exists between dissolved organic matter and solid surfaces, such as clays and iron oxides (Shieh & Duedall 1988; Trefry & Metz 1989). Adsorption of vanadyl ions to oxides is preferred over complexation by dissolved organic ligands, except at high ligand concentrations (Micera & Dallocchia 1988;

Wehrli & Stumm 1989). In crude oil, V mainly occurs as vanadyl ions (VO^{2+}) in the form of organometallic complexes with porphyrins (vanadyl porphyrins) for the substitution of the Mg of plant chlorophylls, and with other largely unknown non-porphyrins or as the cation of organic acids (Branthaver & Filby 1987; Filby & Van Berkel 1987; Filby 1994). The metalation of organometallic complexes occurs in early diagenesis where metal ion speciation in interstitial water is controlled by the Eh–pH conditions during sedimentation as the buried sediment may not be in open contact with the water column above the sediment (Lewan 1984; Filby 1994).

Hence it is reasonable to expect a first-order dependence of $\delta^{51}V$ in crude oils on the depositional environment of petroleum source rocks. During the deposition and early diagenesis, the Eh–pH conditions determine the speciation and coordination of V ions in the water column or pore water, whilst the lithology of the source rock (i.e. the relative abundance of carbonate, clay and iron oxides) defines the co-existing phases that are competing with organic molecules (e.g. porphyrins) for the available V ions in solution. The observed overall correlation between the V isotope compositions and the $V/(V + Ni)$ ratios (Fig. 3) indicate that the crude oils ordinated from strongly reducing, H_2S-rich, marine, carbonaceous sedimentary environments tend to have heavier $\delta^{51}V$ values compared with those from the less anoxic, H_2S-poor, non-marine siliciclastic environments.

Biodegradation

As shown in Figure 3, the co-genetic crude oils from Barbados have a wide variation of $\delta^{51}V$ values ranging from $-1.60‰$ to $+0.40‰$ (Fig. 3; Table 1). This clearly indicates that the pristine V isotopic compositions of crude oils can be significantly modified by secondary processes, such as migration, maturation or biodegradation in a single location. These secondary processes may result in the gain or loss of V during the petroleum system evolution through burial, catagenesis, migration and entrapment (e.g. Filby 1994).

Based on the measured ratio of Pr/nC_{17} and the shape of the chromatograms of light organic compounds in the samples – which are reliable indicators for biodegradation (Peters & Moldowan 1993) – the analysed crude oils from Barbados are subdivided into two groups: biodegraded crude oils ($Pr/nC_{17} > 1.0$) and non-biodegraded crude oils ($Pr/nC_{17} < 1.0$). Compared to the average $\delta^{51}V$ value of $-0.61 \pm 0.07‰$ for the most pristine crude oils (samples BBD-7 and BBD-9, Table 1), the V isotope composition of the biodegraded crude oils appears to be fractionated significantly to both heavier and lighter values (Fig. 4).

Bacteria in the reservoir utilize nutrients such as phosphate ions and any oxidants to oxidize hydrocarbons and non-hydrocarbons and create altered petroleum. It is generally accepted that most surface and subsurface biodegradation is caused by aerobic bacteria (e.g. Palmer 1993), with sufficient oxygen and nutrient supply provided by moving water within the reservoir (Connan 1984; Volkman et al. 1984; Larter et al. 2006). It has now been recognized, however, that anaerobic sulphate-reducing and fermenting bacteria also can degrade petroleum where oxygen is not available (Connan et al. 1995; Coates et al. 1996; Caldwell et al. 1998; Zengler et al. 1999). Under anoxic conditions, NO, SO_4^{2-}, Mn^{4+}, Fe^{3+} and CO_2 as well as vanadate can all potentially act as alternative electron acceptors (oxidants) to oxidize organic compounds during biodegradation (Rehder 1991; Crans et al. 2004; Huang 2004; Winter & Moore 2009). With increasing biodegradation, oils become more viscous, richer in resins, asphaltenes, S and metals (Connan 1984; Volkman et al. 1984; Palmer 1993; Peters & Moldowan 1993; Meredith et al. 2000). It therefore seems logical to expect a fractionation of V isotopes associated with the modification of its concentration during biodegradation. If organic V bonds are ruptured, kinetic isotope effects favouring the lighter ^{50}V would increase the $\delta^{51}V$ value of the unreacted material. An alternative explanation is that the change in $\delta^{51}V$ values is due to microbial activity-induced changes of the species of V ions (and their coordination geometries in the waters) that are ready to be added to the oils. Generally, biodegradation does not involve major V bond changes because the V complexes usually have high molecular weight and are aromatic in structure, and thus are resistant to bacterial attack (Palmer 1983; Strong & Filby 1987; Sundararaman & Hwang 1993). The pH of the solution and the presence of potential ligands, however, are very sensitive to biodegradation which can simultaneously lead to elevated contributions of organic acids that act as ligand binding sites (Hughey et al. 2007; Skaare et al. 2007).

Maturation

As shown in Figure 5, the $\delta^{51}V$ values of the non-biodegraded crude oils from Barbados are negatively correlated with the Ts/Tm maturation indices ((Ts/(Ts + Tm)) and the Tricyclic Index (Peters & Moldowan 1993). This observed correlation thermal maturity index is significant ($R^2 = 0.87$ with Ts/(Ts + Tm) and $R^2 = 0.93$ with the Tricyclic Index), though it is based only on a small dataset with a total variation on $\delta^{51}V$ of c. 1‰. With increasing burial, the temperatures in the source rocks rise and the chemically labile portions of the crude oil begin to crack. As shown in Figure 6, the V

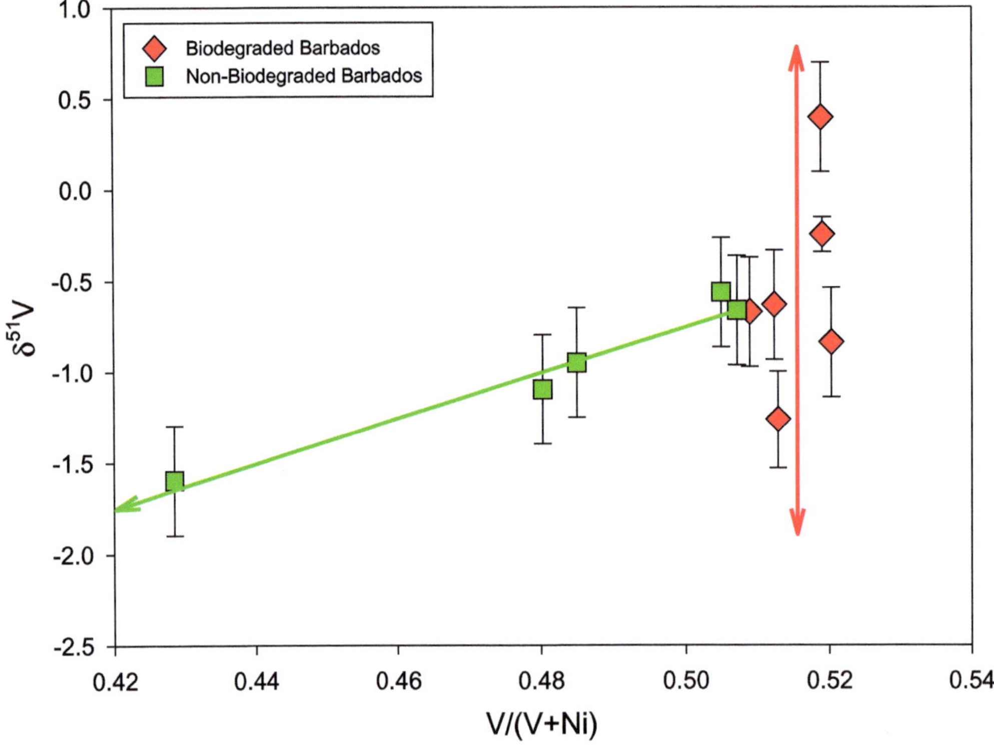

Fig. 4. V isotopic compositions of non-biodegraded vs. biodegraded crude oils from Barbados plotted against V/(V + Ni) ratios. Error bars are 2 SD as listed in Table 1. Values of $\delta^{51}V$ are reported relative to the AA V solution standard defined as 0‰ (Nielsen *et al.* 2011). Note the small change (*c.* 20%) in V/(V + Ni) in the most mature sample.

concentrations of the non-biodegraded crude oils generally decrease with the increase of maturation as measured by these indices. The V in crude oils exists as both vanadyl-porphyrin compounds and non-porphyrin compounds with various chemical structures (Branthaver & Filby 1987; Filby & Van Berkel 1987; Filby 1994; Amorim *et al.* 2007). In the porphyrin complexes, V is bound by four N donor atoms while in the non-porphyrin compounds N, O and S can all act as donor atoms in various combinations (Dickson *et al.* 1972; Sebor *et al.* 1975). The bonds between C and heteroatoms (N, S and O) are more labile and hence easier to break (Mackenzie & Quigley 1988). Bonding of V to sulphur ligands in organic matter has been reported (Yen 1975; Baker & Louda 1986). Though vanadyl porphyrins are stable up to 400°C during laboratory heating (Premović *et al.* 1996), it has been shown that the type of V porphyrins could change from deoxophylloerythroetioporphyrin (DPEP) to etioporphyrin (ETIO) at lower temperatures during thermal maturation of crude oils (Didyk *et al.* 1975; Mackenzie *et al.* 1980; Barwise & Park 1983; Barwise & Roberts 1984; Barwise 1987; Sundararaman

et al. 1988; Sundararaman & Hwang 1993). Increasing maturity has also been shown to change the distribution of VO^{2+} ETIO porphyrins towards lower carbon numbers (Gallango & Cassani 1992; Sundararaman & Moldowan 1993) indicating that for a given porphyrin type, the thermal degradation rate is higher for higher carbon- number species. It has also been speculated that during maturation of a crude oil, decomposition of the metallo-porphyrins would lead to a free cation, e.g. Ni^{2+} or VO^{2+}, which would either be complexed by an asphaltene functionality (or other polar molecules, e.g. naphthenic acids), or form a metal sulphide molecule that would be trapped in asphaltene micelles (Filby 1994). Day & Filby (1992) have shown that de-metalation–metalation of VO^{2+} porphyrins on montmorillonite is also reversible. The exchange and/or transformation of V between different species during these processes are all potentially able to cause V isotope fractionation due to the variation of the V bonding strength among different chemical structures.

Thus, the observed progressive decrease of $\delta^{51}V$ values and changes in V abundance in the co-genetic

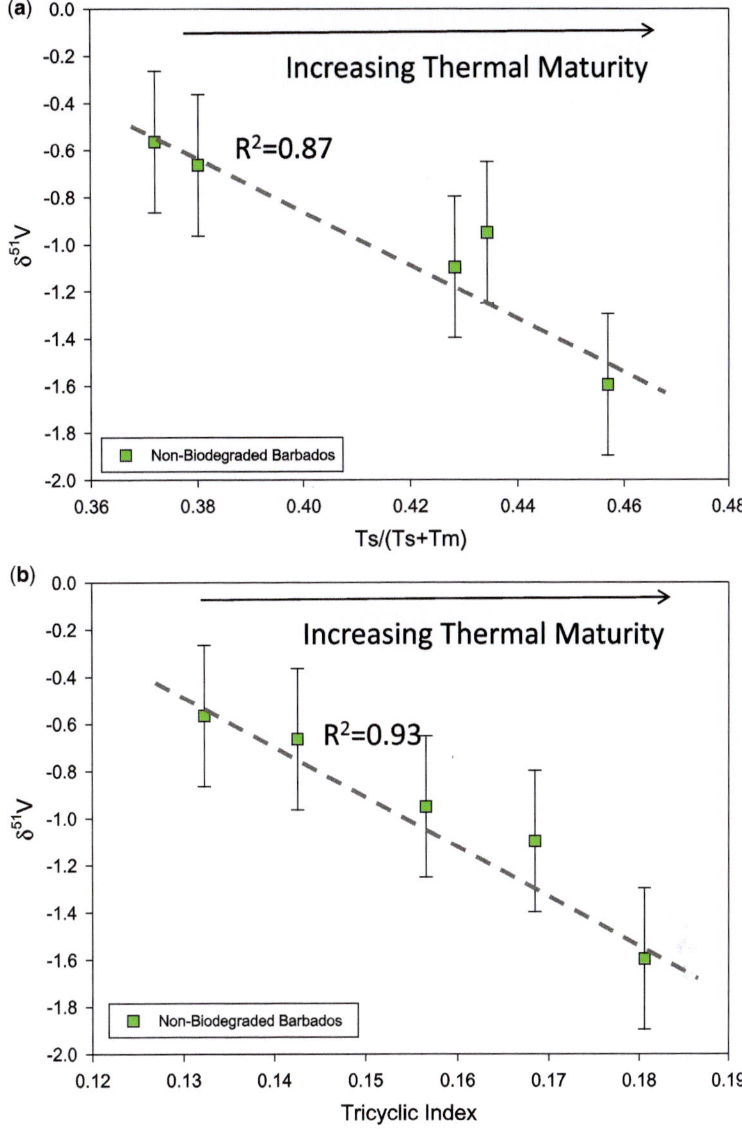

Fig. 5. Correlation of V isotope compositions of non-biodegraded crude oils from Barbados with thermal maturities defined by the Ts/(Ts + Tm) ratio (**a**) and Tricyclic Index (**b**). Error bars are 2 SD given in Table 1. Note the decreasing $\delta^{51}V$ values with increasing thermal maturation. Values of $\delta^{51}V$ are reported relative to the AA V solution standard defined as 0‰ (Nielsen *et al.* 2011).

Barbados crude oils with maturation (Fig. 5) might indicate a preferential loss of ^{51}V during de-metalation and/or a preferential incorporation of ^{50}V in the newly formed V-organometallic compounds. Alternatively, the observed variation of $\delta^{51}V$ values with maturation might be simply caused by a temperature-dependent isotope fractionation process. The typical oil window is *c.* 50–150°C, depending on how quickly the source rock is heated (Tissot & Welte 1984). Theoretical calculations have shown that the V isotope fractionation between different species varies with temperature in the oil window temperature range (Wu *et al.* 2015). However, the magnitude of temperature-induced isotope fractionation diminishes along with the increase of temperature (Wu *et al.* 2015) and is contradictory to the observed trend with increasing fractionation away from its pristine value (Fig. 5); it is therefore unlikely to be

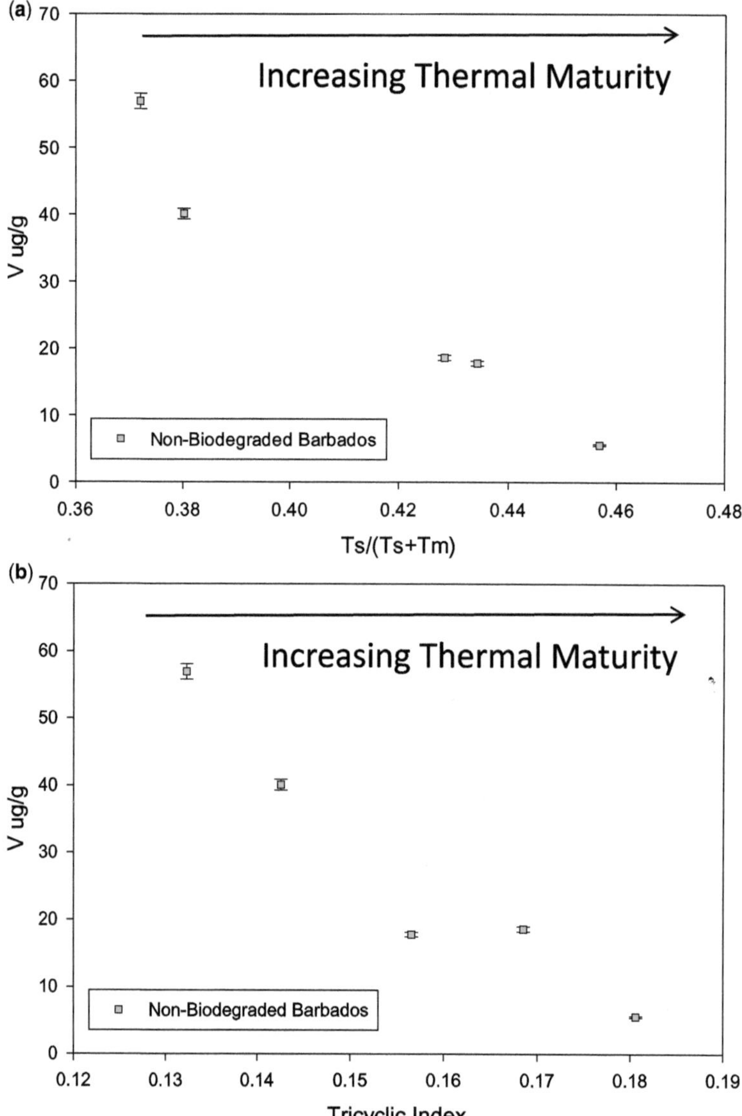

Fig. 6. Correlation of V isotope compositions of only non-biodegraded crude oils from Barbados with thermal maturity indices defined by the Ts/(Ts + Tm) ratio (**a**) and Tricyclic Index (**b**). Error bars are calculated based on 2 RSD of 2%. Note the de-metalation trend with increasing thermal maturity. Values of δ^{51}V are reported relative to the AA V solution standard defined as 0‰ (Nielsen *et al.* 2011).

the actual driver of the V isotope fractionation during maturation.

Conclusions

We have presented a detailed study on the V isotope composition of 17 crude oils spanning a wide range of V concentrations, localities, and variable formation ages using a procedure capable of yielding external precisions of better than ±0.3‰. Notable V isotope fractionation exists in these globally distributed crude oils.

The δ^{51}V isotope compositions in the unaltered crudes are primarily correlated with Ni/V ratios and are most likely determined by the depositional environment of petroleum source rocks. The Eh–pH conditions determine the speciation and coordination of V ions in the water column or pore water, whereas the lithology of the source rocks (i.e. the relative abundance of carbonate, clay and iron oxides) defines the co-existing phases that are competing with organic molecules (e.g. porphyrins) for the available V ions in solution.

Correlations between $\delta^{51}V$ composition and biomarkers indicate that V isotope compositions are subjected to significant modification by both maturation and biodegradation.

V isotope fractionation during biodegradation may depend on the conditions of biodegradation, i.e. aerobic v. anaerobic. The most likely explanation for this dependence is the change of the species of V ions and their coordination geometries induced by the microbial activity at the water/oil interface.

Significant modification of V content in crude oils occurs as a result of de-metalation–re-metalation processes of VO^{2+} porphyrins and/or non-porphyrins during the progressive thermal maturation of crude oils. The observed progressive decrease of $\delta^{51}V$ values and V content in the co-genetic Barbados crude oils with increasing maturation might suggest a preferential loss of ^{51}V during de-metalation and/or a preferential incorporation of ^{50}V in the newly formed V-organometallic compounds.

The combination of V isotope data with traditional biomarker and multi-elemental analysis data on crude oils could potentially become a powerful new geochemical marker and exploration tool. However, more complete understanding of the fundamental mechanisms that control V isotope fractionation in crude oils requires further investigation of crude oils from restricted localities and source rocks. Systematic, coupled organic/inorganic geochemical investigations of crude oils generated under specific, well-understood conditions will be essential for more complete understanding of these processes. Correlation studies of V isotope data with other elemental and biogeochemical data, as well as high precision constraints on oil formation age are essential to more completely understand V fractionation in crude oils.

We thank Sune Nielsen for kindly providing two V solution standards (AA and BDH) used in our method development. We also appreciate the support of Agilent Technology Corporation for providing technical and other support for QQQ-ICP-MS and ICP-OES instruments utilized in our method development. We acknowledge reviews by John Eiler and two anonymous reviewers that improved the manuscript.

References

AMDURER, M., ADLER, D. & SANTSCHI, P. 1983. Studies of the chemical forms of trace elements in sea water using radiotracers. *In*: WONG, C.S., BOYLE, E., BRULAND, K.W., BURTON, J.D. & GOLDBERG, E.D. (eds) *Trace Metals in Sea Water*. NATO Conference Series (IV Marine Sciences), **9**. Springer, Boston, MA, 537–562.

AMORIM, F.A.C., WELZ, B., COSTA, A., LEPRI, F.G., VALE, M.G.R. & FERREIRA, S.L. 2007. Determination of vanadium in petroleum and petroleum products using atomic spectrometric techniques. *Talanta*, **72**, 349–359.

ANBAR, A.D. & KNOLL, A. 2002. Proterozoic ocean chemistry and evolution: a bioinorganic bridge? *Science*, **297**, 1137–1142.

BABAIE, H.A., SPEED, R.C., LARUE, D.K. & CLAYPOOL, G.E. 1992. Source rock and maturation evaluation of the Barbados accretionary prism. *Marine and Petroleum Geology*, **9**, 623–632.

BAES, C.F. & MESMER, R.E. 1976. *Hydrolysis of Cations*. John Wiley and Sons, New York.

BAKER, E.W. & LOUDA, J.W. 1986. Porphyrins in the geological record. *In*: JOHNS, R.B. (ed.) *Biological Markers in the Sedimentary Record: Methods in Geochemistry and Geophysics*. Elsevier, Amsterdam, 126–225.

BARWISE, A. 1987. Mechanisms involved in altering deoxophylloerythroetioporphyrin-etioporphyrin ratios in sediments and oils. *In*: FILBY, R.H. & BRANTHAVER, J.F. (eds) *Metal Complexes in Fossil Fuels*. American Chemical Society Symposium Series, **344**, 100–109, https://doi.org/10.1021/bk-1987-0344.ch006

BARWISE, A. 1990. Role of nickel and vanadium in petroleum classification. *Energy & Fuels*, **4**, 647–652.

BARWISE, A. & PARK, P. 1983. Petroporphyrin fingerprinting as a geochemical marker. *In*: BJOROY, M. *ET AL.* (eds) *Advances in Organic Geochemistry 1981*. Wiley, Chichester, 668–674.

BARWISE, A. & ROBERTS, I. 1984. Diagenetic and catagenetic pathways for porphyrins in sediments. *Organic Geochemistry*, **6**, 167–176.

BELLENGER, J., WICHARD, T., KUSTKA, A. & KRAEPIEL, A. 2008. Uptake of molybdenum and vanadium by a nitrogen-fixing soil bacterium using siderophores. *Nature Geoscience*, **1**, 243–246.

BISSADA, K., ELROD, L., DARNELL, L., SZYMCZYK, H. & TROSTLE, J. 1993. Geochemical inversion–a modern approach to inferring source-rock identity from characteristics of accumulated oil and gas. *Energy Exploration and Exploitation*, **11**, 295–328.

BLOOMFIELD, C. & KELSO, W. 1973. The mobilization and fixation of molybdenum, vanadium, and uranium by decomposing plant matter. *Journal of Soil Science*, **24**, 368–379.

BOSTRÖM, K. & FISHER, D.E. 1971. Volcanogenic uranium, vanadium and iron in Indian Ocean sediments. *Earth and Planetary Science Letters*, **11**, 95–98.

BRANTHAVER, J.F. & FILBY, R.H. 1987. Application of metal complexes in petroleum to exploration geochemistry. *In*: FILBY, R.H. & BRANTHAVER, J.F. (eds) *Metal Complexes in Fossil Fuels*. American Chemical Society Symposium Series, **344**, 84–99, https://doi.org/10.1021/bk-1987-0344.ch005

BREIT, G.N. & WANTY, R.B. 1991. Vanadium accumulation in carbonaceous rocks: a review of geochemical controls during deposition and diagenesis. *Chemical Geology*, **91**, 83–97.

BURGGRAF, A., LUNG-CHUAN, K., WEINZAPFEL, A. & SENNESETH, O. 2002. A new assessment of Barbados onshore oil characteristics and implications for regional petroleum exploration. *16th Caribbean Geological Congress Abstracts*, Barbados, West Indies, **32**.

BUTLER, A. 1998. Acquisition and utilization of transition metal ions by marine organisms. *Science*, **281**, 207–209.

CALDWELL, M. & GARRETT, R., PRINCE, R.C. & SUFLITA, J.M. 1998. Anaerobic biodegradation of long-chain n-alkanes under sulfate-reducing conditions. *Environmental Science and Technology*, **32**, 2.

CASEY, J.F., GAO, Y. & YANG, W. 2015. Analysis of low abundance trace metals and $^{50}V/^{51}V$ isotopes in crude oils: new methods for characterization and exploration. *Goldschmidt Abstracts 2015*, **479**, Prague.

CASEY, J.F., GAO, Y., THOMAS, R. & YANG, W. 2016. New approaches in sample preparation and precise multielement analysis of crude oils and refined petroleum products using single-reaction-chamber microwave digestion and triple-quadrupole ICP-MS. *Spectroscopy*, **31**, 11–22.

CHESHIRE, M., BERROW, M., GOODMAN, B. & MUNDIE, C. 1977. Metal distribution and nature of some Cu, Mn and V complexes in humic and fulvic acid fractions of soil organic matter. *Geochimica et Cosmochimica Acta*, **41**, 1131–1138.

COATES, J.D., ANDERSON, R.T., WOODWARD, J.C., PHILLIPS, E.J. & LOVLEY, D.R. 1996. Anaerobic hydrocarbon degradation in petroleum-contaminated harbor sediments under sulfate-reducing and artificially imposed iron-reducing conditions. *Environmental Science & Technology*, **30**, 2784–2789.

COMER, J.B. 1991. *Stratigraphic Analysis of the Upper Devonian Woodford Formation, Permian Basin, West Texas and Southeastern New Mexico*. Bureau of Economic Geology, University of Texas at Austin, USA.

CONNAN, J. 1984. Biodegradation of crude oils in reservoirs. *Advances in Petroleum Geochemistry*, **1**, 299–335.

CONNAN, J., LACRAMPE-COULOUME, G. & MAGOT, M. 1995. Origin of gases in reservoirs. *In*: DOLENC, A. (ed.) *Proceedings of the 1995 International Gas Research Conference*. **I**. Government Institutes, Rockville, Madison, USA, 21–61.

CRANS, D.C., SMEE, J.J., GAIDAMAUSKAS, E. & YANG, L. 2004. The chemistry and biochemistry of vanadium and the biological activities exerted by vanadium compounds. *Chemical Reviews*, **104**, 849–902.

CURIALE, J.A. 1987. Distribution and occurrence of metals in heavy crude oils and solid bitumens – implications for petroleum exploration: Section II. Characterization, maturation, and degradation. *In*: MEYER, R.F. (ed.) *Exploration for Heavy Crude Oil and Natural Bitumen*. AAPG Studies in Geology, **25**, 207–219.

DAY, J. & FILBY, R. 1992. The role of clay mineral acidity in the evolution of copper, nickel and vanadyl geoporphyrins. *Proceedings of Catalytic Selective Oxidation*. American Chemical Society national meeting 1992. American Chemical Society, Washington DC, 1399.

DEVILLE, E., BATTANI, A. *ET AL.* 2003. Processes of mud volcanism in the Barbados–Trinidad compressional system: new structural, thermal and geochemical data. AAPG Search and Discovery Article 30017.

DICKSON, F., KUNESH, C., MCGINNIS, E. & PETRAKIS, L. 1972. Use of electron spin resonance to characterize the vanadium (IV)-sulfur species in petroleum. *Analytical Chemistry*, **44**, 978–981.

DIDYK, B.M., ALTURKI, Y.I., PILLINGER, C.T. & EGLINTON, G. 1975. Petroporphyrins as indicators of geothermal maturation. *Nature*, **256**, 563–565, https://doi.org/10.1038/256563a0

EMERSON, S.R. & HUESTED, S.S. 1991. Ocean anoxia and the concentrations of molybdenum and vanadium in seawater. *Marine Chemistry*, **34**, 177–196.

FEDONKIN, M., VICKERS-RICH, P., SWALLA, B., TRUSLER, P. & HALL, M. 2012. A new metazoan from the Vendian of the White Sea, Russia, with possible affinities to the ascidians. *Paleontological Journal*, **46**, 1–11.

FILBY, R.H. 1994. Origin and nature of trace element species in crude oils, bitumens and kerogens: implications for correlation and other geochemical studies. *In*: PARNELL, J. (ed.) *Geofluids: Origin, Migration and Evolution of Fluids in Sedimentary Basins*. Geological Society, London, Special Publications, **78**, 203–219, https://doi.org/10.1144/GSL.SP.1994.078.01.15

FILBY, R.H. & VAN BERKEL, G.J. 1987. Geochemistry of metal complexes in petroleum, source rocks and coal: an overview. *In*: FILBY, R.H. & BRANTHAVER, J.F. (eds) *Metal Complexes in Fossil Fuels*. American Chemical Society Symposium Series, **344**, 2–39, https://doi.org/10.1021/bk-1987-0344.ch001

GALARRAGA, F., REATEGUI, K., MARTINEZ, A., MARTINEZ, M., LLAMAS, J. & MÁRQUEZ, G. 2008. V/Ni ratio as a parameter in palaeoenvironmental characterisation of nonmature medium-crude oils from several Latin American basins. *Journal of Petroleum Science and Engineering*, **61**, 9–14.

GALLANGO, O. & CASSANI, F. 1992. Biological marker maturity parameters of marine crude oils and rock extracts from the Maracaibo Basin, Venezuela. *Organic Geochemistry*, **18**, 215–224.

GOLDSTEIN, T. 1983. Geocatalytic reactions in formation and maturation of petroleum. *AAPG Bulletin*, **67**, 152–159.

HILL, R.J. & SCHENK, C.J. 2005. Petroleum geochemistry of oil and gas from Barbados: implications for distribution of Cretaceous source rocks and regional petroleum prospectivity. *Marine and Petroleum Geology*, **22**, 917–943.

HILL, R.J., JARVIE, D.M., ZUMBERGE, J., HENRY, M. & POLLASTRO, R.M. 2007. Oil and gas geochemistry and petroleum systems of the Fort Worth Basin. *AAPG Bulletin*, **91**, 445–473.

HOEFS, J. 2008. *Stable Isotope Geochemistry*. Springer Science & Business Media, Berlin.

HONEYMAN, B.D. 1984. *Cation and Anion Adsorption at the Oxide/Solution Interface in Systems Containing Binary Mixtures of Adsorbents: an Investigation of the Concept of Adsorptive Additivity*. Stanford University, Stanford, Ca, USA.

HUANG, H. 2004. *Effects of biodegradation on crude oil compositions and reservoir profiles in the Liaohe basin, NE China*. PhD thesis, University of Newcastle-upon-Tyne, UK.

HUGHEY, C.A., GALASSO, S.A. & ZUMBERGE, J.E. 2007. Detailed compositional comparison of acidic NSO compounds in biodegraded reservoir and surface crude oils by negative ion electrospray Fourier transform ion cyclotron resonance mass spectrometry. *Fuel*, **86**, 758–768.

JARVIE, D., CLAXTON, B., HENK, F. & BREYER, J. 2001. Oil and shale gas from the Barnett Shale, Fort Worth Basin, Texas. *AAPG Annual Meeting Abstracts.* Denver, Colorado, USA, p. A100.

JARVIE, D.M., HILL, R.J., RUBLE, T.E. & POLLASTRO, R.M. 2007. Unconventional shale-gas systems: the Mississippian Barnett Shale of north central Texas as one model for thermogenic shale-gas assessment. *In*: HILL, R.J. & JARVIE, D.M. (eds) 'Special Issue: Barnett Shale'. *American Association of Petroleum Geologists Bulletin*, **91**, 475–499.

JEANDEL, C., CAISSO, M. & MINSTER, J. 1987. Vanadium behaviour in the global ocean and in the Mediterranean Sea. *Marine Chemistry*, **21**, 51–74.

KEELY, B., PROWSE, W. & MAXWELL, J. 1990. The Treibs hypothesis: an evaluation based on structural studies. *Energy & Fuels*, **4**, 628–634.

KELLEY, K.A., PLANK, T., LUDDEN, J. & STAUDIGEL, H. 2003. Composition of altered oceanic crust at ODP Sites 801 and 1149. *Geochemistry, Geophysics, Geosystems*, **4**, https://doi.org/10.1029/2002GC000435

KHUHAWAR, M., MIRZA, M.A. & JAHANGIR, T. 2012. *Determination of Metal Ions in Crude Oils.* INTECH Open Access Publisher, https://doi.org/10.5772/36945

KIRIYAMA, T. & KURODA, R. 1983. Anion-exchange separation and spectrophotometric determination of vanadium in silicate rocks. *Talanta*, **30**, 261–264.

KRAEPIEL, A., BELLENGER, J., WICHARD, T. & MOREL, F. 2009. Multiple roles of siderophores in free-living nitrogen-fixing bacteria. *Biometals*, **22**, 573–581.

LARTER, S., HUANG, H. *ET AL.* 2006. The controls on the composition of biodegraded oils in the deep subsurface: Part II – Geological controls on subsurface biodegradation fluxes and constraints on reservoir-fluid property prediction. *AAPG Bulletin*, **90**, 921–938.

LARUE, D., SCHOONMAKER, J., TORRINI, R., LUCAS-CLARK, J., CLARK, M. & SCHNEIDER, R. 1985. Barbados: maturation, source rock potential and burial history within a Cenozoic accretionary complex. *Marine and Petroleum Geology*, **2**, 96–110.

LAWRENCE, S., CORNFORD, C., KELLY, R., MATHEWS, A. & LEAHY, K. 2002. Kitchens on a conveyor belt-petroleum systems in accretionary prisms. *16th Caribbean Geological Congress Abstracts*, Barbados, West Indies, **41**.

LEAHY, K., LAWRENCE, S. & THRIFT, J. 2004. Caribbean source rocks may point toward buried treasure. *Oil & Gas Journal*, **102**, 35–39.

LEWAN, M.D. 1984. Factors controlling the proportionality of vanadium to nickel in crude oils. *Geochimica et Cosmochimica Acta*, **48**, 2231–2238.

LEWAN, M.D. 1997. Experiments on the role of water in petroleum formation. *Geochimica et Cosmochimica Acta*, **61**, 3691–3723.

LEWAN, M.D. & MAYNARD, J. 1982. Factors controlling enrichment of vanadium and nickel in the bitumen of organic sedimentary rocks. *Geochimica et Cosmochimica Acta*, **46**, 2547–2560.

MACKENZIE, A.S. & QUIGLEY, T.M. 1988. Principles of geochemical prospect appraisal. *AAPG Bulletin*, **72**, 399–415.

MACKENZIE, A.S., PATIENCE, R., MAXWELL, J., VANDENBROUCKE, M. & DURAND, B. 1980. Molecular parameters of maturation in the Toarcian shales, Paris Basin, France – I. Changes in the configurations of acyclic isoprenoid alkanes, steranes and triterpanes. *Geochimica et Cosmochimica Acta*, **44**, 1709–1721.

MAKISHIMA, A., ZHU, X.-K., BELSHAW, N.S. & O'NIONS, R.K. 2002. Separation of titanium from silicates for isotopic ratio determination using multiple collector ICP-MS. *Journal of Analytical Atomic Spectrometry*, **17**, 1290–1294.

MANGO, F.D. 1996. Transition metal catalysis in the generation of natural gas. *Organic Geochemistry*, **24**, 977–984.

MARÉCHAL, C.N., TÉLOUK, P. & ALBARÈDE, F. 1999. Precise analysis of copper and zinc isotopic compositions by plasma-source mass spectrometry. *Chemical Geology*, **156**, 251–273.

MAYLOTTE, D., WONG, J., PETERS, R.S., LYTLE, F. & GREEGOR, R. 1981. X-ray absorption spectroscopic investigation of trace vanadium sites in coal. *Science*, **214**, 554–556.

MCBRIDE, M.B. 1978. Transition metal bonding in humic acid: an ESR study. *Soil Science*, **126**, 200–209.

MEREDITH, W., KELLAND, S.-J. & JONES, D. 2000. Influence of biodegradation on crude oil acidity and carboxylic acid composition. *Organic Geochemistry*, **31**, 1059–1073.

MICERA, G. & DALLOCCHIA, R. 1988. Metal complex formation of the surface of amorphous aluminum hydroxide, Part IV. Interactions of oxovanadium (IV) and vanadate (V) with aluminum hydroxide in the presence of succinic, malic and 2-mercaptosuccinic acids. *Colloids Surfaces*, **34**, 185–196.

MOLDOWAN, J.M., SUNDARARAMAN, P. & SCHOELL, M. 1986. Sensitivity of biomarker properties to depositional environment and/or source input in the Lower Toarcian of SW-Germany. *Organic Geochemistry*, **10**, 915–926.

NIEDERER, F., PAPANASTASSIOU, D. & WASSERBURG, G. 1985. Absolute isotopic abundances of Ti in meteorites. *Geochimica et Cosmochimica Acta*, **49**, 835–851.

NIELSEN, S.G. 2015. *Stable Vanadium Isotopes in Crude Oils and Their Source Rocks: A New Tool to Understand the Processes Governing Petroleum Generation.* Report DN12, American Chemical Society, https://acswebcontent.acs.org/prfar/2013/Paper12260.html

NIELSEN, S.G., PRYTULAK, J. & HALLIDAY, A.N. 2011. Determination of precise and accurate V-51/V-50 isotope ratios by MC-ICP-MS, Part 1: chemical separation of vanadium and mass spectrometric protocols. *Geostandards and Geoanalytical Research*, **35**, 293–306, https://doi.org/10.1111/j.1751-908X.2011.00106.x

NIELSEN, S.G., PRYTULAK, J., WOOD, B.J. & HALLIDAY, A.N. 2014. Vanadium isotopic difference between the silicate Earth and meteorites. *Earth and Planetary Science Letters*, **389**, 167–175.

NIELSEN, S.G., OWENS, J.D. & HORNER, T.J. 2016. Analysis of high-precision vanadium isotope ratios by medium resolution MC-ICP-MS. *Journal of Analytical Atomic Spectrometry*, **31**, 531–536.

NOMURA, M., NAKAMURA, M., SOEDA, R., KIKAWADA, Y., FUKUSHIMA, M. & OI, T. 2012. Vanadium isotopic composition of the sea squirt (*Ciona savignyi*). *Isotopes in Environmental and Health Studies*, **48**, 434–438.

PALMER, M., FALKNER, K.K., TUREKIAN, K. & CALVERT, S. 1988. Sources of osmium isotopes in manganese nodules. *Geochimica et Cosmochimica Acta*, **52**, 1197–1202.

PALMER, S.E. 1983. Porphyrin distributions in degraded and nondegraded oils from colombia. *Abstracts of Papers of the American Chemical Society*. 186th ACS national convention, August. American Chemical Society, Washington, DC, 23.

PALMER, S.E. 1993. Effect of biodegradation and water washing on crude oil composition. *In*: ENGEL, M.H. & MACKO, S.A. (eds) *Organic Geochemistry*. Springer, New York, NY, 511–533.

PETERS, K.E. & MOLDOWAN, J.M. 1993. *The Biomarker Guide: Interpreting Molecular Fossils in Petroleum and Ancient Sediments*. Prentice Hall, Englewood Cliffs, NJ.

POLLASTRO, R.M., HILL, R.J., JARVIE, D.M. & HENRY, M.E. 2003. Assessing undiscovered resources of the Barnett-Paleozoic total petroleum system, Bend Arch-Fort Worth Basin Province, Texas. AAPG Search and Discovery Article #10034.

POLLASTRO, R.M., JARVIE, D.M., HILL, R.J. & ADAMS, C.W. 2007. Geologic framework of the Mississippian Barnett Shale, Barnett-Paleozoic total petroleum system, Bend archFort Worth Basin, Texas. *AAPG Bulletin*, **91**, 405–436.

PRANGE, A. & KREMLING, K. 1985. Distribution of dissolved molybdenum, uranium and vanadium in Baltic Sea waters. *Marine Chemistry*, **16**, 259–274.

PREMOVIĆ, P.I. 1978. Electron spin resonance methods for trace metals in fossil fuels. *Proceedings of the 7th Yugoslav Conference on General and Applied Spectroscopy*. Serbian Chemical Society, Belgrade, Yugoslavia, 11–16.

PREMOVIĆ, P.I. 1984. Vanadyl ions in ancient marine carbonaceous sediments. *Geochimica et Cosmochimica Acta*, **48**, 873–877.

PREMOVIĆ, P.I., PAVLOVIĆ, M.S. & PAVLOVIĆ, N.A.Z. 1986. Vanadium in ancient sedimentary rocks of marine origin. *Geochimica et Cosmochimica Acta*, **50**, 1923–1931.

PREMOVIĆ, P.I., JOVANOVIĆ, L.S. & NIKOLIĆ, G.S. 1996. Thermal stability of the asphaltene/kerogen vanadyl porphyrins. *Organic Geochemistry*, **24**, 801–814.

PREMOVIĆ, P.I., ĐORĐEVIĆ, D. & PAVLOVIĆ, M. 2002. Vanadium of petroleum asphaltenes and source kerogens (La Luna Formation, Venezuela): isotopic study and origin. *Fuel*, **81**, 2009–2016.

PRYTULAK, J., NIELSEN, S.G. & HALLIDAY, A.N. 2011. Determination of precise and accurate 51V/50V isotope ratios by multi-collector ICP-MS, Part 2: isotopic composition of six reference materials plus the Allende chondrite and verification tests. *Geostandards and Geoanalytical Research*, **35**, 307–318.

PRYTULAK, J., NIELSEN, S.G. *ET AL.* 2013. The stable vanadium isotope composition of the mantle and mafic lavas. *Earth and Planetary Science Letters*, **365**, 177–189, https://doi.org/10.1016/j.epsl.2013.01.010

REHDER, D. 1991. The bioinorganic chemistry of vanadium. *Angewandte Chemie International Edition* [in English], **30**, 148–167.

REQUEJO, A., SASSEN, R. *ET AL.* 1995. Geochemistry of oils from the northern Timan-Pechora Basin, Russia. *Organic Geochemistry*, **23**, 205–222.

SCHAUBLE, E., ROSSMAN, G. & TAYLOR, H. 2001. Theoretical estimates of equilibrium Fe-isotope fractionations from vibrational spectroscopy. *Geochimica et Cosmochimica Acta*, **65**, 2487–2497.

SEBOR, G., LANG, I., VAVREČKA, P., SYCHRA, V. & WEISSER, O. 1975. The determination of metals in petroleum samples by atomic absorption spectrometry: Part I. Determination of vanadium. *Analytica Chimica Acta*, **78**, 99–106.

SEEWALD, J.S. 1994. Evidence for metastable equilibrium between hydrocarbons under hydrothermal conditions. *Nature*, **370**, 285–287, https://doi.org/10.1038/370285a0

SEEWALD, J.S. 2003. Organic–inorganic interactions in petroleum-producing sedimentary basins. *Nature*, **426**, 327–333.

SEYFRIED, W.E. & MOTTL, M.J. 1982. Hydrothermal alteration of basalt by seawater under seawater-dominated conditions. *Geochimica et Cosmochimica Acta*, **46**, 985–1002.

SHIEH, C.-S. & DUEDALL, I.W. 1988. Role of amorphous ferric oxyhydroxide in removal of anthropogenic vanadium from seawater. *Marine Chemistry*, **25**, 121–139.

SHIELDS, W.R., MURPHY, T.J., CATANZARO, E.J. & GARNER, E.L. 1966. Absolute isotopic abundance ratios and the atomic weight of a reference sample of chromium. *Journal of Research of the National Bureau of Standards A*, **70**, 193–197.

SKAARE, B.B., WILKES, H., VIETH, A., REIN, E. & BARTH, T. 2007. Alteration of crude oils from the Troll area by biodegradation: analysis of oil and water samples. *Organic Geochemistry*, **38**, 1865–1883.

SPEED, R., BARKER, L. & PAYNE, P. 1991*a*. Geologic and hydrocarbon evolution of Barbados. *Journal of Petroleum Geology*, **14**, 323–342.

SPEED, R., RUSSO, R., WEBER, J. & ROWLEY, K. 1991*b*. Evolution of Southern Caribbean plate boundary, vicinity of Trinidad and Tobago: Discussion (1). *AAPG Bulletin*, **75**, 1789–1794.

STRELOW, F. & BOTHMA, C. 1967. Anion exchange and a selectivity scale for elements in sulfuric acid media with a strongly basic resin. *Analytical Chemistry*, **39**, 595–599.

STRELOW, F., RETHEMEYER, R. & BOTHMA, C. 1965. Ion exchange selectivity scales for cations in nitric acid and sulfuric acid media with a sulfonated polystyrene resin. *Analytical Chemistry*, **37**, 106–111.

STRONG, D. & FILBY, R.H. 1987. Vanadyl porphyrin distribution in the Alberta oil-sand bitumens. *In*: FILBY, R.H. & BRANTHAVER, J.F. (eds) *Metal Complexes in Fossil Fuels*. American Chemical Society Symposium Series, **344**, 154–172, https://doi.org/10.1021/bk-1987-0344.ch010

SUNDARARAMAN, P. & HWANG, R.J. 1993. Effect of biodegradation on vanadylporphyrin distribution. *Geochimica et Cosmochimica Acta*, **57**, 2283–2290.

SUNDARARAMAN, P. & MOLDOWAN, J.M. 1993. Comparison of maturity based on steroid and vanadyl porphyrin parameters: a new vanadyl porphyrin maturity parameter for higher maturities. *Geochimica et Cosmochimica Acta*, **57**, 1379–1386.

SUNDARARAMAN, P., BIGGS, W.R., REYNOLDS, J.G. & FETZER, J.C. 1988. Vanadylporphyrins, indicators of kerogen breakdown and generation of petroleum. *Geochimica et Cosmochimica Acta*, **52**, 2337–2341.

SUNDARARAMAN, P., SCHOELL, M., LITTKE, R., BAKER, D., LEYTHAEUSER, D. & RULLKÖTTER, J. 1993. Depositional environment of Toarcian shales from northern Germany as monitored with porphyrins. *Geochimica et Cosmochimica Acta*, **57**, 4213–4218.

SZALAY, A. & SZILÁGYI, M. 1967. The association of vanadium with humic acids. *Geochimica et Cosmochimica Acta*, **31**, 1–6, https://doi.org/10.1016/0016-7037(67)90093-2

THOMPSON, D.M. 1982. *Atoka Group (Lower to Middle Pennsylvanian), northern Fort Worth Basin, Texas: terrigenous depositional systems, diagenesis, and reservoir distribution and quality*. Report of investigations No. 125. Bureau of Economic Geology, University of Texas at Austin, USA.

TISSOT, B.P. & WELTE, D.H. 1984. *Petroleum Formation and Occurrence*. Springer Verlag, Germany.

TREFRY, J.H. & METZ, S. 1989. Role of hydrothermal precipitates in the geochemical cycling of vanadium. *Nature*, **342**, 531–533.

TRIBOVILLARD, N., ALGEO, T.J., LYONS, T. & RIBOULLEAU, A. 2006. Trace metals as paleoredox and paleoproductivity proxies: an update. *Chemical Geology*, **232**, 12–32.

UREY, H.C. 1947. The thermodynamic properties of isotopic substances. *Journal of the Chemical Society (Resumed)*, 1947, 562–581.

VENTURA, G.T., GALL, L., SIEBERT, C., PRYTULAK, J., SZATMARI, P., HÜRLIMANN, M. & HALLIDAY, A.N. 2015. The stable isotope composition of vanadium, nickel, and molybdenum in crude oils. *Applied Geochemistry*, **59**, 104–117.

VOLKMAN, J.K., ALEXANDER, R., KAGI, R.I., ROWLAND, S.J. & SHEPPARD, P.N. 1984. Biodegradation of aromatic hydrocarbons in crude oils from the Barrow Sub-basin of Western Australia. *Organic Geochemistry*, **6**, 619–632.

WANTY, R.B., GOLDHABER, M.B. & NORTHROP, H.R. 1990. Geochemistry of vanadium in an epigenetic, sandstone-hosted vanadium–uranium deposit, Henry Basin, Utah. *Economic Geology*, **85**, 270–284.

WEDEPOHL, K. 1964. Untersuchungen am Kupferschiefer in Nordwestdeutschland; ein Beitrag zur Deutung der Genese bituminöser Sedimente. *Geochimica et Cosmochimica Acta*, **28**, 305–364.

WEDEPOHL, K. 1971. Environmental influences on the chemical composition of shales and clays. *Physics and Chemistry of the Earth*, **8**, 305–333.

WEHRLI, B. & STUMM, W. 1989. Vanadyl in natural waters: adsorption and hydrolysis promote oxygenation. *Geochimica et Cosmochimica Acta*, **53**, 69–77.

WINTER, J.M. & MOORE, B.S. 2009. Exploring the chemistry and biology of vanadium-dependent haloperoxidases. *Journal of Biological Chemistry*, **284**, 18577–18581.

WOOD, L.J. 2000. Chronostratigraphy and tectonostratigraphy of the Columbus Basin, eastern offshore Trinidad. *AAPG Bulletin*, **84**, 1905–1928.

WU, F., QIN, T., LI, X., LIU, Y., HUANG, J.H., WU, Z. & HUANG, F. 2015. First-principles investigation of vanadium isotope fractionation in solution and during adsorption. *Earth and Planetary Science Letters*, **426**, 216–224.

WU, F., QI, Y., YU, H., TIAN, S., HOU, Z. & HUANG, F. 2016. Vanadium isotope measurement by MC-ICP-MS. *Chemical Geology*, **421**, 17–25.

YANG, W.H. 2014. *High Precision Determination of Trace Elements in Crude Oils by Inductively Coupled Plasma-Optical Emission Spectrometry and Inductively Coupled Plasma-Mass Spectrometry*. University of Houston, Texas, USA.

YEN, T.F. 1975. *Role of Trace Metals in Petroleum*. Ann Arbor Science, Ann Arbor, MI, USA

ZENGLER, K., RICHNOW, H.H., ROSSELLÓ-MORA, R., MICHAELIS, W. & WIDDEL, F. 1999. Methane formation from long-chain alkanes by anaerobic microorganisms. *Nature*, **401**, 266–269.

ZHANG, Y.H. 2003. *Vanadium isotope analysis and vanadium isotope effects in complex formation systems*. PhD thesis, Tokyo Institute of Technology.

ZHAO, X., SHI, Q., GRAY, M.R. & XU, C. 2014. *New Vanadium Compounds in Venezuela Heavy Crude Oil Detected by Positive-ion Electrospray Ionization Fourier Transform Ion Cyclotron Resonance Mass Spectrometry*. Scientific Reports, **4**, 5373.

ZUMBERGE, J.E. 1987. Prediction of source rock characteristics based on terpane biomarkers in crude oils: a multivariate statistical approach. *Geochimica et Cosmochimica Acta*, **51**, 1625–1637.

Carbon and hydrogen isotopic compositions of *n*-alkanes as a tool in petroleum exploration

NIKOLAI PEDENTCHOUK[1]* & COURTNEY TURICH[2]

[1]*School of Environmental Sciences, University of East Anglia, Norwich NR4 7TJ, UK*

[2]*Schlumberger, 1, Rue Henri Becquerel, Clamart, France 92140*

**Correspondence: n.pedentchouk@uea.ac.uk*

Abstract: Compound-specific isotope analysis (CSIA) of individual organic compounds is a powerful but underutilized tool in petroleum exploration. When integrated with other organic geochemical methodologies it can provide evidence of fluid histories including source, maturity, charge history and reservoir processes that can support field development planning and exploration efforts. The purpose of this chapter is to provide a review of the methodologies used for generating carbon and hydrogen isotope data for mid- and high-molecular-weight *n*-alkanes.

We discuss the factors that control stable carbon and hydrogen isotope compositions of *n*-alkanes and related compounds in sedimentary and petroleum systems and review current and future applications of this methodology for petroleum exploration. We discuss basin-specific case studies that demonstrate the usefulness of CSIA either when addressing particular aspects of petroleum exploration (e.g. charge evaluation, source rock–oil correlation, and investigation of maturity and in-reservoir processes) or when this technique is used to corroborate interpretations from integrated petroleum systems analysis, providing unique insights which may not be revealed when using other methods. CSIA of *n*-alkanes and related *n*-alkyl structures can provide independent data to strengthen petroleum systems concepts from generation and expulsion of fluids from source rock, to charge history, connectivity, and in-reservoir processes.

Gold Open Access: This article is published under the terms of the CC-BY 3.0 license.

Petroleum geoscientists use organic geochemistry as an essential tool in oil and gas exploration and field development planning. Relatively low-cost, high-throughput bulk data are commonly used to screen for source rock quality (e.g. per cent total organic carbon (%TOC), hydrogen and oxygen indices) and thermal maturity (Tmax, vitrinite reflectance equivalent). More in-depth geochemical analytical techniques are used in the context of full fluid and reservoir properties to correlate source rocks and reservoir oils, to determine fluid generation and migration history, including present-day reservoir connectivity, and to understand in-reservoir processes, such as biodegradation of in-reservoir oils. These tools are especially powerful when coupled with other measurements made during the exploration and development process, such as compositional analysis during drilling, downhole fluid analysis and other wireline measurements, and pressure, volume, temperature (PVT) and chemical analyses, integrated in the context of geological static and reservoir dynamic models.

Molecular biomarkers have been employed in petroleum exploration for several decades (Peters *et al.* 2005). The usefulness of bulk stable isotope measurements of gases and oils was well demonstrated in the petroleum industry through the decades of the 1970s and 1980s (Stahl 1977; Schoell 1984; Sofer 1984). However, the use of compound-specific isotopic composition of light hydrocarbons, alkanes and biomarkers is less common. Nonetheless, these types of data can provide valuable additional information to distinguish oil families, perform oil–oil and oil–source rock correlation, and better understand in-reservoir processes that have had an impact upon fluid properties over time and that explain current emplacement. Compound-specific isotope analysis (CSIA) provides distinctive and substantive support to a fully integrated interpretation of fluid properties in petroleum exploration and development.

The purpose of this chapter is (a) to provide a review of mid- and high-molecular-weight alkane carbon (C) and hydrogen (H) isotope analytical methodologies and the factors that control stable C and H isotopes of *n*-alkane (C_6+) compounds in sedimentary and petroleum systems, and (b) to review current and future applications of this methodology for petroleum exploration. CSIA of gas range (C_1–C_5) *n*-alkane and other molecular compositions is beyond the scope of this contribution.

From: LAWSON, M., FORMOLO, M. J. & EILER, J. M. (eds) 2018. *From Source to Seep: Geochemical Applications in Hydrocarbon Systems*. Geological Society, London, Special Publications, **468**, 105–125.
First published online December 14, 2017, https://doi.org/10.1144/SP468.1
© 2018 The Author(s). Published by The Geological Society of London.
Publishing disclaimer: www.geolsoc.org.uk/pub_ethics

CSIA of *n*-alkanes

We focus on compound-specific analysis of higher-molecular-weight *n*-alkanes and related compounds because: (a) they are the most abundant hydrocarbon groups present both in the source rock extracts and reservoir oils; (b) they are easy to extract, separate and analyse; (c) they can be analysed for both stable C and H isotope compositions using the same sample and using the same gas chromatograph isotope ratio mass spectrometer (GC-IRMS) instrument; (d) they provide a reasonable scope for in-depth review; and (e) they demonstrate the potential for growth of these underutilized techniques.

When integrated with the bulk isotope methodology, CSIA expands the usefulness of the stable isotope approach in petroleum exploration and adds several advantages. The methodology:

(1) allows investigation of multiple organic matter sources and/or processes (e.g. by comparing the isotopic composition of organic compounds of different chain length) using a single sample;

(2) enhances the ability to compare the chemical properties of individual organic compounds at different stages of their geochemical history (e.g. alkanes extracted from immature source rock are comparable to *n*-alkanes generated and expelled during thermal maturation of organic matter) so as to better understand the processes that have had an impact upon current and past reservoir fluid properties;

(3) supports information to identify potential source origin and oil families, conduct oil–oil and oil–source rock correlations, and improve our understanding of the processes that have influenced fluid properties over time;

(4) requires relatively small sample volumes.

The disadvantages are similar to other fluid evaluation techniques: the samples must be representative, and the interpretative strategy must include an integrated approach to enable the unravelling of the complex physical and chemical processes that occur over very long time periods.

Analytical methodology for compound-specific stable isotope analysis

Several previous reviews provide detailed information on analytical methods and on the use of compound-specific isotopic data on organic compounds in the natural and applied sciences (Meier-Augenstein 1999; Schmidt *et al.* 2004; Glaser 2005; Benson *et al.* 2006; Philp 2006; Sessions 2006; Evershed *et al.* 2007). The following section briefly describes the general principles of acquiring C and H stable isotope data for individual organic compounds extracted from natural samples.

Sample preparation

The initial step of sample clean-up and fraction separation depends on the matrix, i.e. whether it is a source or reservoir rock or a liquid. For solid samples, the total extractable fraction can be collected using the Soxhlet apparatus, sonication, accelerated solvent, or microwave-assisted extraction systems (Lundanes & Greibrokk 1994; Rieley 1994; Letellier & Budzinski 1999; Smith 2003; Peters *et al.* 2005; Péres *et al.* 2006). The isolation and clean-up steps to obtain specific fractions will vary according to specific project needs. Alkanes subjected to CSIA can be isolated from either the whole oil or individual fractions (e.g. saturates, aromatics). Generally, especially for δ^2H measurements, an additional step to separate the branched from straight-chain compounds is recommended using urea adduction or molecular sieves (Grice *et al.* 2008).

Analytical procedures for compound-specific stable isotope measurements

The whole oil or the saturate fraction with *n*-alkyl compounds usually must first be analysed using a gas chromatograph flame ionization detector (GC-FID) or gas chromatograph mass spectrometer (GC-MS) to quantify the amount of sample needed to achieve reproducible results on the IRMS. To achieve the most precise $\delta^{13}C$ and δ^2H measurements, *n*-alkane peaks should have baseline resolution (if the sample contains other compounds in addition to *n*-alkanes) and a sufficient signal-to-background ratio (which is system specific); compound-specific measurements are made using the GC-IRMS coupled with combustion ($\delta^{13}C$) or high-temperature conversion (δ^2H) reactors, respectively. Modern mass spectrometers equipped with a gas chromatograph and this type of 'on-line' set-up generally provide precision for $\delta^{13}C$ in the range ±0.1–0.3‰ for compounds containing 0.1–5 nmol C, and for δ^2H in the range of ±2–5‰ for compounds containing 10–50 nmol H (Sessions 2006). The user has to be aware of precision levels associated with C and H isotope measurements when designing a study and interpreting the results. Case studies discussed below demonstrate that this level of precision is sufficient for the use of the CSIA methodology in petroleum basin studies. We recommend that the end users carefully evaluate the IRMS chromatograms to ensure the separation of compounds is adequate, the baseline is clean, and the integration of individual peaks is consistent throughout the run and from sample to sample.

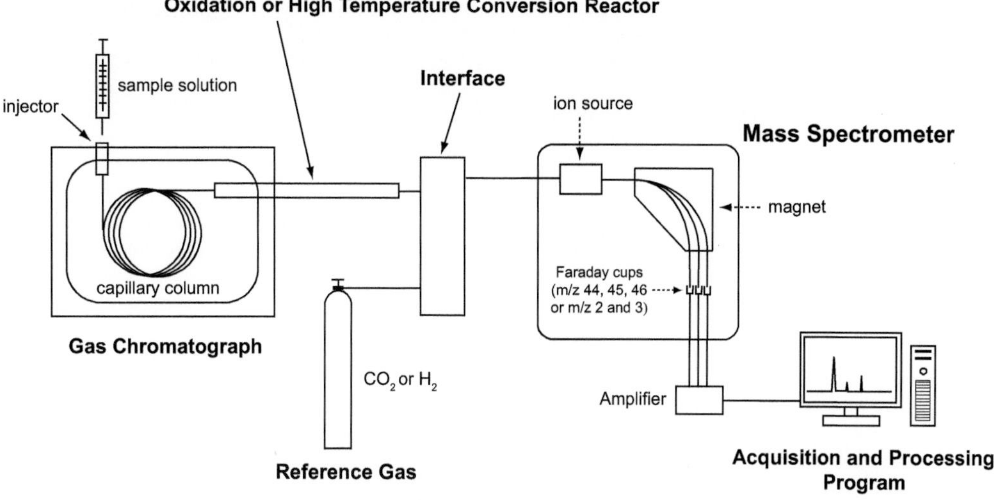

Fig. 1. A simplified schematic of a GC-IRMS system that can be configured for either $\delta^{13}C$ or $\delta^{2}H$ measurements.

Figure 1 shows a simplified schematic of a GC-IRMS system equipped with a combustion reactor for $\delta^{13}C$ measurements. The n-alkane-containing fraction is injected into the GC, where compounds are separated on a capillary column and then converted to CO_2 in the reactor, using a source of O_2 and a catalyst. The CO_2 gas is then transferred into the mass spectrometer, where it is ionized. Faraday cups for m/z 44, 45 and 46 are then used to collect ions corresponding to $^{12}C^{16}O_2$, $^{13}C^{16}O_2$ and $^{12}C^{18}O^{16}O$ isotopomers, respectively. (Other isotopomers potentially adding to m/z 45 and 46 are quantitatively insignificant.) The $\delta^{13}C$ values of individual compounds are calculated relative to either a reference gas or a co-injected compound with a known isotopic $\delta^{13}C$ value. Figure 2 shows a typical GC-IRMS chromatogram. (The trace represents m/z 44 corresponding to CO_2 gas generated from the combustion of individual organic compounds.) The chromatogram shows six peaks of reference gas and a homologous series of n-alkanes from the saturate fraction of a Nigerian oil sample. The chromatograph displays a relatively low background and, for the majority of the n-alkane peaks, an absence of co-elution, which is the key to precise $\delta^{13}C$ measurements.

The $\delta^{2}H$ measurements require a pyrolysis reactor to generate H_2 gas; organic compounds are

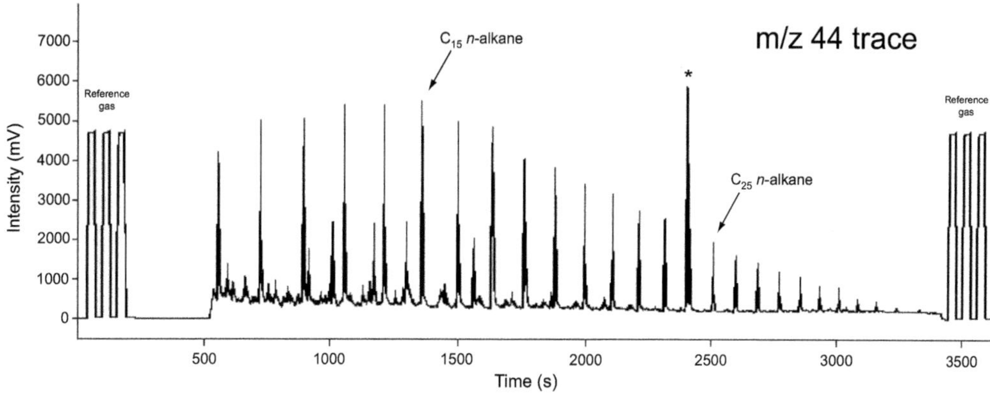

Fig. 2. A chromatogram showing an m/z 44 trace corresponding to CO_2 produced from the combustion of individual compounds (mainly a homologous series of n-alkanes) in the saturate fraction of a Nigerian oil. The symbol * designates a contaminant peak co-eluting with n-C$_{24}$ alkane. The $\delta^{13}C$ values can be obtained for each individual peak in a single run.

carried from the GC column through a high-temperature conversion reactor typically set at 1450°C. Reduced H gas is then transferred to the mass spectrometer, with Faraday cups for m/z 2 and 3 corresponding to $^1H_2$ and $^1H^2H$, respectively. Hydrogen isotope measurements require careful monitoring of the effect of protonation reactions (taking place in the ion source) on the δ^2H values (Sessions *et al.* 2001*a*, *b*). The 'H$_3$-factor' needs to be determined daily, using a series of reference gas pulses of different magnitudes. Ideally, peaks corresponding to the compounds of interest should have δ^2H values similar to that of the reference gas for a reliable H$_3$-factor correction. Because of the complications when applying H$_3$-factor correction and peak integration by the acquisition software, δ^2H measurements are particularly sensitive to chromatography issues, such as high background and peak co-elution. Therefore, a urea or other adduction step is almost always required.

Modern GC-IRMS systems are typically configured for both $\delta^{13}C$ and δ^2H measurements, with both combustion and high-temperature reactors interfaced to a mass spectrometer with a multi-collector set-up for recording multiple m/z. Therefore, the amount of time for switching between $\delta^{13}C$ and δ^2H measurements is minimized. A single GC-IRMS set-up can be used to generate $\delta^{13}C$ and/ or δ^2H compound-specific data on organic compounds of interest, depending on the needs of the organic geochemist. The user has to be aware, however, that even though $\delta^{13}C$ and/or δ^2H compound-specific data can be obtained using the same instrument, C and H isotope measurements require different acquisition modes. Therefore, separate injections are needed for these measurements.

Controls on $\delta^{13}C$ and δ^2H values of *n*-alkanes

Isotope data interpretation relies on understanding the context of the fluid samples, petroleum system and reservoir characteristics. In this review, we highlight the main controls that need to be considered when interpreting compound-specific $\delta^{13}C$ and δ^2H data. Figure 3 provides an overview of the main factors and mechanisms that affect C and H isotopes of sedimentary organic matter (OM) and petroleum fluids.

OM formation

C isotopes. The C isotope composition of extant biomass contributing to sedimentary OM is controlled by the isotopic composition of the C source and several isotope effects associated with C uptake during biosynthesis. Early reviews by Fogel & Cifuentes (1993), Hayes (1993), Farquhar *et al.* (1989) and

Hayes (2001) have identified the most significant factors that control C isotope composition of biosynthates: (a) the isotopic composition of the primary C source; (b) the isotope effect associated with C uptake; (c) the isotope effect due to organism-specific biosynthetic and metabolic pathways; and (d) cellular C budgets. The reviews by Freeman (2001) and Pancost & Pagani (2006) further refined the knowledge of these controls and discussed potential applications of the compound-specific methodology in biogeochemistry and palaeoclimate studies.

One of the most striking features of C isotope composition of *n*-alkanes is the difference between terrestrial- and marine-derived OM. The review by de Leeuw *et al.* (1995) provides a summary of C isotope fractionations characteristic of aquatic and terrestrial plants, and we briefly describe the main observations here.

The $\delta^{13}C$ values of terrestrial higher plants are controlled by C isotope composition of atmospheric CO_2 and depend significantly on the biosynthetic pathway, i.e. C3, C4 or Crassulacean acid metabolism (CAM) pathways (Deines 1980). C$_4$ terrestrial plants are typically more enriched in ^{13}C (bulk $\delta^{13}C$ = –6 to –23‰, Schidlowski 1988) in comparison with other terrestrial plants. The diversity of C$_3$ plants and their diverse ecological zones lead to a large range in $\delta^{13}C$ values. The C isotope fractionation associated with terrestrial biosynthesis has remained relatively conservative (Arthur *et al.* 1985; Popp *et al.* 1989) over geological time.

In contrast, the $\delta^{13}C$ values of aquatic plants are controlled mainly by $\delta^{13}C$ values of dissolved inorganic carbon (DIC). Aquatic plants are almost exclusively C$_3$ plants, and their $\delta^{13}C$ values vary significantly depending on the life form: cyanobacteria (–8 to –24‰); mat communities (–8 to –30‰); photosynthetic bacteria (–8 to –30‰); *Chlorobium* mat community (–23 to –25‰); sulphur-oxidizing bacteria (–30 to –32‰); and methanogenic archaeans (–18 to –38‰) (Schopf 2000). Marine algae are characterized by the following $\delta^{13}C$ values: green (–9 to –20‰); brown (–11 to –21‰); CO_2-using red algae (–30 to –35‰); and HCO$_3$-using red algae (–10 to –23‰) (Maberly *et al.* 1992). The $\delta^{13}C$ values of freshwater macrophytes are between –23 and –31‰ (Keeley & Sandquist 1992). Additionally, the $\delta^{13}C$ values of aquatic plants are strongly influenced by the level of biological productivity. During the production of marine OM over geological time, there have been major changes in C isotope fractionation (up to 10‰, Hayes *et al.* 1999).

Because of these known variations, CSIA investigations can be used to compare and contrast short n-C$_{17}$ and n-C$_{19}$ alkanes typically associated with algae, and n-C$_{29}$ and n-C$_{31}$ alkanes derived from higher plants, present in the same sample. C isotope

CONTROLS ON ISOTOPIC COMPOSITION

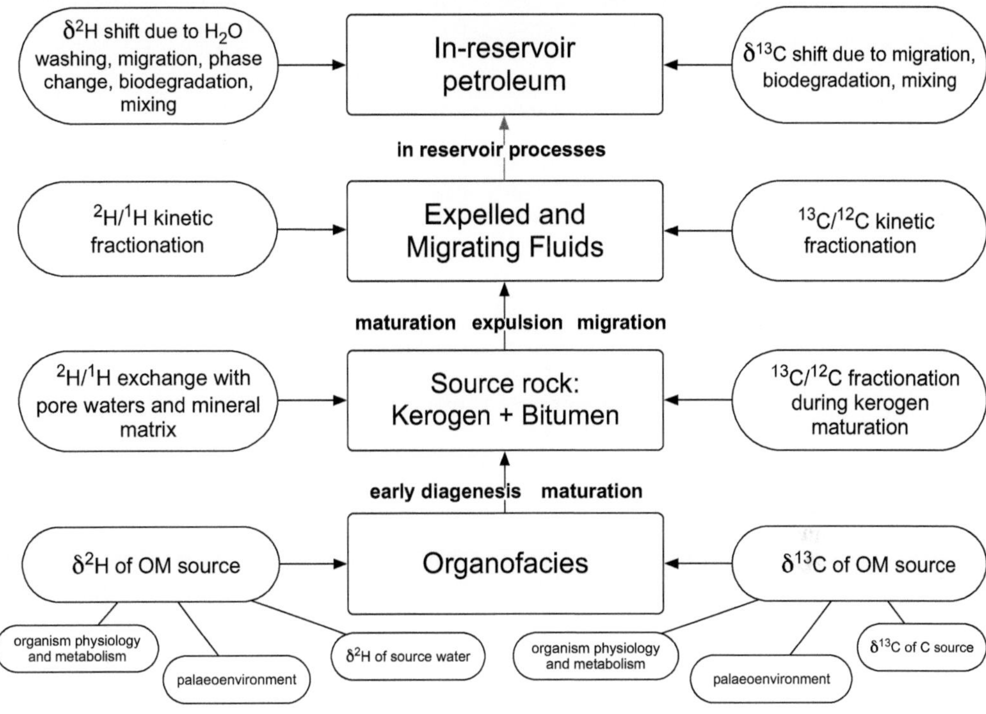

Fig. 3. The main factors that control C and H isotope compositions of *n*-alkyl lipids in the source rock and petroleum reservoir.

composition of individual organic compounds from a single sample can therefore provide information about multiple OM sources that contribute to the total sedimentary OM pool. This approach becomes particularly powerful when different groups of organisms, having specific HMW biomarkers (e.g. terpanes, steranes), contribute to the OM. The pioneering papers by Freeman *et al.* (1990), Hayes *et al.* (1990) and Rieley *et al.* (1991), and subsequent work by Summons *et al.* (1994), Grice *et al.* (1998) and Thiel *et al.* (1999), clearly demonstrated the advantage of CSIA for identifying different sources (e.g. primary producers v. bacterially mediated OM; algal v. higher plant input) of organic compounds within a single OM extract. Applying this approach in petroleum exploration can provide unique insights into different sources of hydrocarbons.

H isotopes. Two recent reviews provide comprehensive coverage of the state of knowledge on the use of compound-specific H isotope composition of organic compounds in biogeochemistry. Sachse *et al.* (2012) provide a detailed discussion of H

isotope systematics as applied to palaeohydrology, and Sessions (2016) evaluates factors that control H isotope composition of hydrocarbons in sedimentary settings. Here we focus on the main controls that have particular relevance to determining H isotope composition of organic compounds in source rock organofacies during OM synthesis. Two key factors play a role at this stage, with regard to both aquatic- and terrestrial-derived organic compounds, including *n*-alkanes: (a) the ^{2}H/^{1}H composition of source water for the organism, and (b) the physiological and biochemical processes involved in fixing water-derived H into organic compounds.

The δ^2H values of source water for plants are initially determined by the ^{2}H/^{1}H composition of meteoric precipitation. Additionally, terrestrial plants are subjected to a broad range of physiological and environmental factors that significantly influence H isotope composition of soil and leaf water used by the plant during photosynthesis. While the environmental controls on isotopic composition of precipitation and lacustrine/evaporative settings are generally well understood (Craig & Gordon 1965; Gonfiantini 1986; Rozanski *et al.* 1993), the mechanisms responsible for controlling the isotopic composition of soil and leaf water, particularly to what extent the signal is incorporated into leaf waxes, are still unclear (Sessions 2016 and references therein). The importance of the latter was also recently highlighted by Tipple *et al.* (2014), Gamarra *et al.* (2016), Oakes & Hren (2016) and the Yale University group (Dirghangi & Pagani 2013; Tipple & Pagani 2013).

Significantly, in spite of the multiple steps and processes involved in the transfer of H from environmental water to plant biochemicals to organofacies to petroleum fluids, the original meteoric water H isotope signal can still have a strong influence on the δ^2H values of oils. The broad range of the δ^2H values of terrestrially derived oils (Schimmelmann *et al.* 2004), in comparison with those from marine systems, stems from the significantly greater variability of H isotope composition of precipitation over continents than in marine basins. There is also a considerable difference between the δ^2H values of environmental water from hypersaline v. marine settings, which could be used to distinguish these depositional environments in petroleum basin studies (Santos Neto & Hayes 1999).

Fractionation of H isotopes during biosynthesis is another major factor controlling the δ^2H values of organic compounds. Early studies that demonstrated large ^2H-depletion (at the bulk level) of OM relative to that of environmental water (Schiegl & Vogel 1970; Smith & Epstein 1970; Estep & Hoering 1980) were subsequently confirmed and expanded to include compound-specific H isotope investigations of various classes of lipids (e.g. isoprenoids v. *n*-alkyl lipids) as well as different ecological (marine and lacustrine algae v. terrestrial plants) and trophic (e.g. autotrophs v. heterotrophs) groups (Sessions 2016). Compound-specific H isotope studies have revealed that biosynthetic $^2H/^1H$ fractionation can lead to ^2H-depletion relative to source water by 100–250‰, though in rare cases fractionation can lead to values from +200‰ to –450‰ (Sessions 2016). The processes and mechanisms resulting in H isotope fractionation during biosynthesis are multiple and complex. The biochemistry and stable isotope systematics of H fractionation are described in detail by Hayes (2001), Schmidt *et al.* (2003) and Sachse *et al.* (2012). A number of laboratory and field-based studies (e.g. Sessions *et al.* (1999), Chikaraishi *et al.* (2009) and Zhang *et al.* (2009)) have demonstrated the complexity of the processes responsible for H isotope fractionation in lipids and highlighted the need for further research in this rapidly developing area of stable isotope biogeochemistry. We highlight the importance of understanding the complex interplay among the environmental and organism-specific physiological and biochemical processes that control δ^2H values of organic compounds synthesized by extant biota.

As indicated above, the reader is referred to Figure 3 for a summary of those processes that control and subsequently influence H (and C) isotope composition of individual organic compounds in a petroleum system. Depending on the geological context and history, highly specific information about the source (e.g. isotopic composition of precipitation, palaeoenvironment and plant physiology) may or may not be preserved in the sedimentary record. The 'primary' isotopic signature that would characterize a specific organofacies can be subsequently 'averaged', altered or totally erased by a number of diagenetic and post-diagenetic processes in the source rock and reservoir. The reader needs to keep this in perspective and interpret CSIA data with caution.

OM during diagenesis and early maturation

Three types of processes affect OM during diagenesis: (a) selective degradation of biomolecules; (b) preservation, alteration, condensation and vulcanization of sedimentary organic compounds; and (c) generation and incorporation of new organic compounds synthesized by soil/sediment biota. The first two processes have the potential to affect *n*-alkane concentrations and their $\delta^{13}C$ and δ^2H values by either destroying or fractionating C and H isotopes during proto-kerogen formation. The third group of processes, however, results in an addition of isotopically different organic compounds, which, if synthesized in sufficient amounts, could lead to a significant alteration of the original isotope signal.

C isotopes. Lipids are among the most resistant biomolecules and are recalcitrant during diagenesis and early stages of OM maturation. Several early studies have indicated minimal to no diagenetic effects on C isotope composition of this group of biomolecules. Hayes *et al.* (1990) used theoretical considerations and compound-specific $\delta^{13}C$ data on porphyrins and isoprenoids to argue for a lack of diagenetic effect on these compounds in Cretaceous sediments. Huang *et al.* (1997) reported no significant alteration of the C isotope signature of higher-plant-derived *n*-C_{23} to *n*-C_{35} alkanes in a litter-bag experiment. Furthermore, Freeman *et al.* (1994) showed only minor (1.2‰ on average) ^{13}C-depletion of diagenetic polycyclic aromatic hydrocarbons (PAHs) extracted from Eocene sediments.

More recent work, however, has shown that the assumption about the conservative nature of lipid isotopic composition might not necessarily be correct, particularly when considering terrestrial OM sources. Terrestrially derived lipids could potentially undergo significant alteration in the soil and during transport before being deposited in lacustrine or marine depositional settings. On the basis of data from a litter-bag experiment, Nguyen Tu *et al.* (2004) reported a *c.* 3‰ ^{13}C-enrichment of *n*-alkanes in comparison with those from fresh leaves. An intriguing aspect of this observation is that diagenetic alteration of the primary C isotope signal observed in *n*-alkyl lipids can result from the generation of new biomass by soil microbes, which could also explain a *c.* 4‰ ^{13}C-enrichment of soil *n*-alkanes and other *n*-alkyl compounds reported by Chikaraishi & Naraoka (2006). Clearly, there is a need for further investigation of the potential effect of diagenesis on δ^{13}C values of sedimentary *n*-alkyl compounds under different depositional conditions.

H isotopes. As with C isotope systematics, there is uncertainty about the effect of diagenesis on the δ^2H values of *n*-alkyl lipids in soil and immature sediments. Yang & Huang (2003) argued for a lack of H isotope effect on *n*-alkyl lipids recovered from fossil leaves of Miocene lacustrine deposits. On the other hand, based on the results of a litter-bag experiment, Zech *et al.* (2011) argued for a significant effect of leaf degradation and seasonality on the δ^2H values of *n*-alkane biomarkers. The authors used the argument invoked by Nguyen Tu *et al.* (2011) – i.e. microbial contribution of *n*-alkanes with different H isotope composition. Further work is required to better constrain the effect of early diagenesis and accompanying soil microbiological processes on the H isotope record of sedimentary *n*-alkyl lipids.

The integrated environmental and biochemical H isotope signal acquired by organic compounds during OM formation can also be influenced by the exchange of C-bound H with H atoms in pore/formation H_2O and with H in clay minerals. Early H isotope studies by Yeh & Epstein (1981) and Schoell (1984) conducted on bulk oil have shown minimal exchange of C-bound H over geological timescales. More recent studies (Sessions *et al.* 2004; Wang *et al.* 2009*a, b*, 2013) involving H isotope investigation at the compound-specific level, however, do show that H exchange can take place. A broad range of δ^2H values typical of organic compounds from different compound classes (H in *n*-alkyl and aromatic structures, and H bound to and adjacent to heteroatoms), and thus a different degree of susceptibility of these compounds to H exchange reactions (both the extent and rates), provide ample opportunity for investigating the extent of H isotope exchange over geological timescales. Sessions *et al.* (2004), Schimmelmann *et al.* (2006) and Sessions (2016) gave detailed accounts of the processes involved in H exchange reactions in geological settings, for example, (a) mechanisms and rates of ^{2}H/^{1}H exchange, and (b) equilibrium ^{2}H/^{1}H fractionation.

Previous work clearly showed that the CSIA approach is well suited for investigating the occurrence and the magnitude of H isotope exchange during OM sedimentation and diagenesis. Of particular interest are integrated δ^2H analyses of *n*-alkanes and isoprenoids, which can provide information about the extent of diagenetic alteration and the level of OM transformation at the early stages and more advanced stages of OM thermal maturation (Radke *et al.* 2005; Pedentchouk *et al.* 2006; Dawson *et al.* 2007; Kikuchi *et al.* 2010). These authors provided empirical data (supported by theory and experimental kinetic data) that suggest a more conservative nature of *n*-alkanes, in comparison with isoprenoids (pristane and phytane), with regard to H isotope exchange during early OM diagenesis.

OM maturation and petroleum generation

The primary environmental/biological C and H isotope signature of *n*-alkyl lipids is further modified by thermal maturation and hydrocarbon expulsion during petroleum generation. At least two outcomes are possible: first, the original environmental/source information can partially be preserved in petroleum hydrocarbons, and second, new additional information about the extent of stable C and H isotope overprinting can be acquired. Depending on the balance between the two, both types of information can be used in petroleum basin analysis, particularly in OM source and/or OM maturity investigations.

C isotopes. Carbon isotope composition of petroleum hydrocarbons is largely determined by the isotopic composition of the kerogen type (i.e. OM source material) and the depositional environment. Petroleum generation and maturation, however, affects the C isotope properties of kerogen and expelled products. These processes involve C—C bond cleavage, with kinetic effects resulting in the preferential breaking of ^{12}C—^{12}C bonds relative to ^{13}C—^{12}C bonds (Peters *et al.* 1981; Lewan 1983). Expelled gas and oil tend to be ^{13}C-depleted in comparison with the residual kerogen by *c.* 2‰. Even though the knowledge of these effects and of the extent of their control on bulk δ^{13}C values of petroleum has been used in petroleum exploration since the 1970s (Stahl 1977; Schoell 1984; Sofer 1991), the use of CSIA (Chung *et al.* 1994; Rooney *et al.* 1998; Whiticar & Snowdon 1999; Odden *et al.* 2002) would benefit from further knowledge of the effects of thermal maturation on individual hydrocarbons.

A number of laboratory and field-based studies have demonstrated the effect of thermal maturation on the C isotope composition of various petroleum hydrocarbons (Clayton 1991; Clayton & Bjorøy 1994; Cramer *et al.* 1998; Lorant *et al.* 1998). Theoretical, experimental and field information, however, are rarely integrated to derive a mechanistic understanding of the processes that control C isotope composition of individual organic compounds. Tang *et al.* (2005) were among the first to not only link theoretical and empirical laboratory-based investigation of C isotope systematics but also integrate it with the CSIA of H (see the discussion below). On the basis of the quantitative kinetic model and controlled closed-system pyrolysis experiment, Tang *et al.* (2005) identified a *c.* 4‰ increase in the $\delta^{13}C$ values from pyrolysate extracts of samples with Ro = 1.5% v. immature samples (Fig. 4). In contrast to the observations with regard to H isotopes (see the following *H isotopes* section), there was no clear link between ^{13}C-enrichment and *n*-alkane chain length.

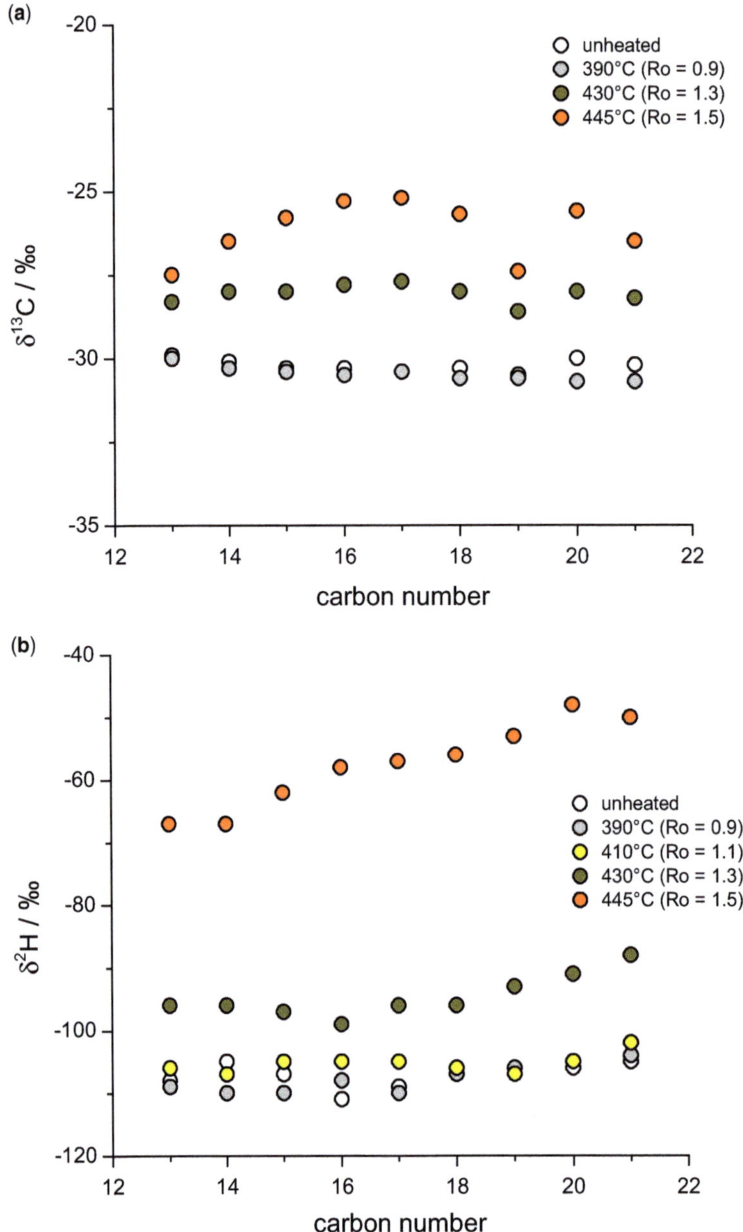

Fig. 4. (a) C and (b) H isotope compositions of *n*-C₁₃ to *n*-C₂₁ alkanes from North Sea oil used for pyrolysis experiment. Redrawn from Tang *et al.* (2005).

Further studies are needed to fully understand these processes and effects.

H isotopes. Depending on the structure and the amount of time available for equilibrium $^2H/^1H$ fractionation during diagenesis, an organic molecule can provide an isotopic record of OM source/environment and the extent of diagenetic alteration during OM deposition. From a petroleum geochemist's perspective, however, the key processes and reactions start at the level of OM maturation when kerogen cracking, bitumen formation and petroleum fluid expulsion occur. During these processes, a C-bound H will undergo additional $^2H/^1H$ fractionation as a result of kinetic isotope effects.

Several field-based studies have shown that higher-maturity oils are typically characterized by higher δ^2H values (Li *et al.* 2001; Schimmelmann *et al.* 2004; Dawson *et al.* 2005). This observation would imply that processes leading to ^2H-enrichment of the residual fraction (kerogen, remaining oil) might be similar to those that lead to ^{13}C-enrichment of the remaining products during cracking. The effects of cracking on the δ^2H values of the products and remaining fraction, however, are difficult to separate from those that could result from equilibrium H exchange with formation H_2O. Therefore, tightly controlled laboratory investigations provide the best source of information with regard to the effect of cracking during oil generation. To our knowledge, the study by Tang *et al.* (2005) is unique in providing a thorough investigation of the kinetic isotope effects on CSIA δ^2H values, using the combined approach of theoretical calculations and heating experiments. The study used a simple kinetic model of oil cracking for qualitative prediction of ^2H-enrichment of *n*-alkanes of different chain lengths at different thermal maturities (Fig. 4). There was an increase in δ^2H values of up to 60‰ at 445°C. The effect was more noticeable for *n*-alkanes with longer chain lengths. The main outcome of this study is that the kinetic model can be used for qualitative prediction of $^2H/^1H$ fractionation during kerogen/oil cracking in natural settings. The kinetic isotope effects are likely to be significant at thermal maturities of Ro > 1.5.

Petroleum fluid migration

Isotope effects of fluid migration are different for (a) primary migration, i.e. the process of expulsion of generated petroleum from source rock, and (b) secondary migration, i.e. the process of fluid movement following expulsion. Partition coefficients govern the rate at which compound classes are released, but the expelled fluid composition will approach the composition of the generated petroleum (under steady-state conditions). However, it appears that solution/dissolution processes do not discriminate isotopically for mid- and long-chain *n*-alkanes, and therefore the isotopic composition of expelled fluids should represent the generated fluid and source rock (Liao & Geng 2009).

There are many published accounts using petroleum geochemistry and stable isotope composition of gases to help assess migration issues (e.g. Seifert & Moldowan 1986; Curiale & Bromley 1996; Zhang *et al.* 2013) but few published contributions using CSIA of high-molecular-weight (HMW) alkanes to assess migration. Li *et al.* (2001) compared two oils from the same genetic source in Western Canada, with one proximal to the source rock (Pembina field) and one that had migrated 150 km updip (Joarcam field) (Creaney *et al.* 1994; Larter *et al.* 1996). Despite differences in migration, the δ^2H composition of the individual alkanes does not appear to have been affected, varying by only 4–8‰. The *n*-hexane/benzene ratio may have increased only about six-fold in the Joarcum field oils, which correlates to a relatively small volume of exchange with water along the migration route and may provide an explanation for the lack of δ^2H variation.

The use of gasoline-range hydrocarbons along with other markers of secondary migration, such as quinolines (Larter & Aplin 1995; Li *et al.* 2001), can create a more detailed picture of migration (e.g. distance and extent) and can help develop an understanding of the utility of the δ^2H of alkanes for correlations or as a sensitive tool for assessing the impact of migration. However, additional published case studies are needed.

Other in-reservoir processes

In-reservoir processes, such as evaporative fractionation (including gas washing), deasphalting, gravity separation, water washing, biodegradation and thermochemical sulphate reduction lead to secondary alteration of petroleum during and after emplacement (Wenger *et al.* 2002). To varying degrees, compound-specific C and H isotope composition of hydrocarbons, including straight and branched moieties in the C_5–C_{30} range, may be influenced by, and therefore useful in understanding, in-reservoir processes.

Evaporative fractionation. Thompson (1987, 1988) originally defined evaporative fractionation as a multistep process, involving the addition of gas to an oil accumulation followed by a phase separation as the gas escapes, carrying additional components based on vapour–liquid partition coefficients (also referred to as gas washing – e.g. van Graas *et al.* 2000). The remaining liquid is enriched typically with higher-molecular-weight compounds. Evaporative fractionation may also occur from loss of solution gas or gas cap (Masterson *et al.* 2001). Variations in the composition of *n*-alkanes and C$_7$ components are used to compare oils and determine and possibly quantify the impact of evaporative fractionation on fluid properties, which can also be used to determine the history of the petroleum system (Masterson *et al.* 2001; Losh *et al.* 2002*a, b*; Thompson 2010; Murillo *et al.* 2016). Experimental separation of liquid and gas phases shows the δ^{13}C of C$_6$, C$_7$ and C$_8$ *n*-alkanes are identical in both the gas phase and the original oil. However, *n*-C$_9$ through *n*-C$_{14}$ as well as 1-methylcyclopentane show a 1‰ ^{13}C-depletion in the gas phase, suggesting that the isotopic effect of in-reservoir phase partitioning is very minor (Carpentier *et al.* 1996). Therefore, in systems where evaporative fractionation has occurred, δ^{13}C values are conserved, and can therefore still be used in evaluating correlations, charge, etc.

Biodegradation. Subsurface biodegradation leads to a well-characterized sequence of compound class losses as microbial groups with anaerobic hydrocarbon-degrading enzymes (Head *et al.* 2003; Aitken *et al.* 2004; Bian *et al.* 2015) metabolize hydrocarbons, leading to generation of acidic compounds and loss of hydrocarbons in a characteristic sequence (*n*-alkanes > monocyclic alkanes > alkyl benzenes > isoprenoid alkanes > alkyl naphthalenes > bicyclic alkanes > steranes > hopanes) (Peters *et al.* 2005). Biodegradation preferentially removes ^{12}C and ^{1}H, leaving ^{13}C- and ^{2}H-enriched organic compounds (Stahl 1980; Clayton 1991; Odden *et al.* 2002; Jones *et al.* 2008). The impact of biodegradation on whole oil δ^{13}C values is minor, but, with increasing levels of biodegradation – as evidenced by the disappearance of *n*-alkanes (e.g. Marcano *et al.* 2013) – the δ^{13}C values of organic compounds in the saturate fraction will increase.

Analysis of a suite of seven Liaohe basin oils, from pristine to heavily biodegraded, showed that the δ^{13}C values are relatively conservative even at severe levels of biodegradation for HMW (C$_{19+}$) compounds (Sun *et al.* 2005). However, low-molecular-weight (LMW) compounds can show up to 4‰ ^{13}C-enrichment relative to unaltered oil from the same system. Only the LMW *n*-alkane compounds are significantly affected isotopically during progressive biodegradation. This also means that the HMW *n*-alkanes should still be well correlated with source rock δ^{13}C even in highly biodegraded reservoirs, assuming the compounds are still present and that no other significant processes have influenced the original values.

Experiments have also been conducted (Vieth & Wilkes 2006) on the CSIA of gasoline-range hydrocarbons in the Gullfaks field (North Sea) to assess how biodegradation changes the δ^{13}C composition of LMW hydrocarbons. Butane through nonane from biodegraded oil were 3–7‰ ^{13}C-enriched, but toluene or cyclohexanes were not. (Surprisingly, toluene was not degraded at all in the Gullfaks oils, which led the authors to suggest that the microbial community in this particular field were not capable of degrading toluene.) The experimentally derived isotopic fractionation factor for *n*-hexane was used to apply the Rayleigh equation to data from the Gullfaks field to quantify hydrocarbons that had been lost to biodegradation (Vieth & Wilkes 2006; Wilkes *et al.* 2008).

Hydrogen isotope compositions appear to be less conservative as a function of biodegradation; variations of up to 35‰ have been observed (Sun *et al.* 2005), and therefore stable H isotopes are potentially useful in understanding and possibly quantifying the impact of biodegradation.

Thermochemical sulphate reduction. As thermochemical sulphate reduction (TSR) destroys organic compounds, the remaining organic compounds become ^{13}C-enriched (Rooney 1995; Whiticar & Snowdon 1999). There are also compositional changes as TSR oxidizes petroleum constituents to CO$_2$ through a range of polar, volatile and non-volatile intermediates (Walters *et al.* 2015). Therefore, the combination of compositional changes and the CSIA of gasoline-range hydrocarbons is a sensitive method to discriminate fluids influenced by TSR. Rooney (1995), as described in Peters & Fowler (2002), showed a 22‰ increase in the δ^{13}C of the *n*-alkane and branched hydrocarbons in TSR-affected oils, compared to a 2–3‰ increase in oils influenced only by increased thermal maturity. Other compounds, such as toluene, showed much smaller shifts in δ^{13}C. Routine analysis of the C (and sulphur) isotope composition of the TSR-intermediates could potentially be used to create additional correlation and classification tools.

Applications of compound-specific stable isotopes of *n*-alkanes

The application of compound-specific stable isotopes of *n*-alkanes and related compounds can contribute to an understanding of the various aspects of fluid migration history, in-reservoir processes, and provide insights into oil–source and oil–oil correlations. Caveats to the applications are the same as for any reservoir fluid study. The user needs to (a) ensure that the samples are representative, (b) acknowledge any analytical uncertainties, and (c) understand that many complex processes – that have occurred over geological timescales – can change fluid properties.

Integrating CSIA with other geochemical methodologies for reservoir studies has several advantages. CSIA provides more resolution than bulk methods. It separates HMW and LMW compounds, which are influenced differently by different processes, and facilitates a direct comparison of individual compounds (n-alkanes and other biomarkers) from different sources (e.g. kerogen pyrolysates, oils, source rock extracts, etc.). CSIA may also give many additional components for resolving similarities and differences for correlation purposes. Ultimately, CSIA contributes to a more complete understanding of petroleum systems, within the context of other geological and fluid properties.

Case studies

Here we review a number of regionally organized case studies that show applications of the C and H isotopic compositions of n-alkanes in support of petroleum exploration and development activities. Figure 5 shows specific applications of CSIA as demonstrated by these case studies. A review of the full history of each basin is well beyond the scope of this contribution; our goal is to highlight cases in which the CSIA of n-alkanes have improved our understanding of fluid histories and properties.

Europe

Austria (application in OM source, OM maturity and charge/migration investigations). In the Alpine foreland basin, a variety of Oligocene source rocks in the Schöneck, Dynow and Eggerding formations, with shaly to marly lithologies, varying laterally and vertically from west to east, generated and expelled petroleum to two main reservoirs – sandstones from the Cretaceous and Eocene. In a study of the C and H isotopic compositions of the n-alkanes from both source rock and reservoir fluids, Bechtel *et al.* (2013) found ^{13}C-depletion varied from west to east by *c.* 2–3‰ for δ^{13}C and *c.* 30‰ for δ^{2}H. The differences in fluid CSIA reflect the changing contribution of source rocks, with greater contribution from the ^{13}C-depleted unit 'C' of the Schöneck Formation towards the east, where that formation also thickens (Gratzer *et al.* 2011; Bechtel *et al.* 2012). The pattern of δ^{13}C values in the n-alkanes is also consistent with changing source rock properties, with δ^{13}C values decreasing to n-C$_{21}$ and then increasing from n-C$_{21}$ to n-C$_{31}$, which we illustrate in a cross plot of the C$_{19}$ and C$_{26}$ average δ^{13}C values for fluids and source rock (Fig. 6). The variations in δ^{2}H of n-alkanes also reflect source variations and are attributed to the ^{2}H-depletion in the more brackish, less marine depositional environment in unit 'C'. The δ^{2}H values increase with n-alkane chain length and also increase with increasing maturity, providing additional evidence for maturity variations. The study also used benzocarbozole ratios as migration parameters, reflecting useful integration of additional molecular properties to further resolve charge history. This and related studies show that the Alpine foreland

SED. BASIN APPLICATION	Tarim	Niger	Barents	North Sea	WCSB	Sirte	West Sak	Austria	Potwar	Perth	Liaohe
OM Source	■	■	■	■	■	■		■	■		
OM Maturity	■	■	■	■	■			■		■	
Charge/Migration	■	■	■	■	■		■	■			
Oil-Oil Correlation	■	■	■	■	■	■			■		
Oil-Source Rock Correlation	■	■	■	■	■	■			■		
Biodegradation	■	■	■						■		■

Fig. 5. Simplified table showing the specific applications of CSIA of n-alkanes from a number of case studies. Studies by Bjorøy *et al.* (1994) and Odden *et al.* (2002) in the North Sea were among the first to use CSIA data for correlation and then for additional follow-up investigations. Studies in other basins (Tarim, Niger, Barents and Western Canada Sedimentary) are further excellent examples of integrated geochemical basin analysis (Li *et al.* 2001, 2010; Samuel *et al.* 2009; Jia *et al.* 2010, 2013; He *et al.* 2012; Murillo *et al.* 2016). Other basin studies are examples of investigations that have focused on particular applications, such as charge evaluation (West Sak, Masterson *et al.* 2001), source rock correlation (Austria, Bechtel *et al.* 2012; Sirte Basin, Aboglia *et al.* 2010), and the interplay of maturity and in-reservoir processes (Perth Basin, Dawson *et al.* 2005; Potwar Basin, Asif *et al.* 2009; Liaohe Basin, Sun *et al.* 2005).

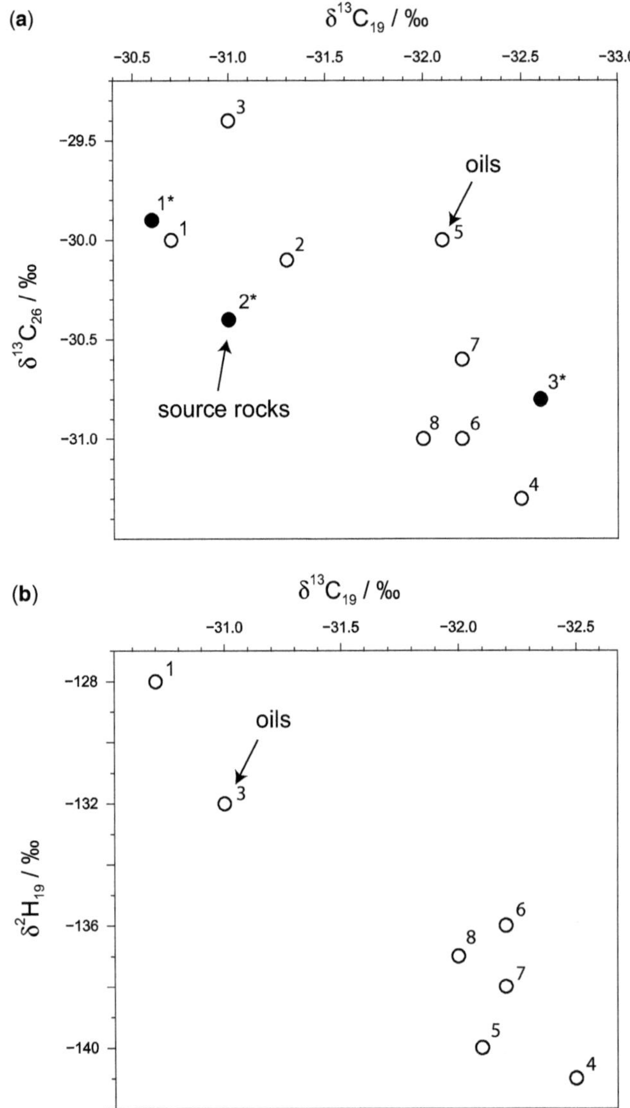

Fig. 6. Bechtel *et al.* (2013) provide average CSIA values from *n*-alkanes and isoprenoids in oils from the Molasse Basin, reflecting the west–east trend. We replotted these averages in cross plots of $\delta^{13}C$ of C_{19} alkane and (a) $\delta^{13}C$ of the C_{26} alkane, including three regional source rocks, and (b) $\delta^{2}H$ of C_{19} alkane. Labels on the graphs correspond to those used in Bechtel *et al.* (2013). Oil fields: 1 – K, Ktg, MS, R, St; 2 – Li, Sch, W, P; 3 – Trat; 4 – Gruenau (Alpine subthrust); 5 – Mdf, Sat, Eb, Ob, Ra, Sths, Wels; 6 – V. Hier; 7 – BH, Ke, En, Pi; 8 – Sier, Wir. Source rocks 1* – Obhf Schöneck Fm; 2* – Mlrt (Rupelian); 3 – Molln Schöneck Fm. This study illustrates the identification of patterns and correlations possible when comparing source rock and oils, and utilizing dual ($\delta^{13}C$ and $\delta^{2}H$) isotope systems.

basin provides an excellent natural laboratory to study the impact of source and maturity variations as well as migration on compound-specific isotope compositions and to further develop these tools in petroleum exploration. Applications of CSIA should also extend to other basins where lateral and vertical variations in source rock are important (e.g. the Bakken, Eagle Ford).

Barents Sea (integrated basin analysis). Murillo *et al.* (2016) used a full suite of geochemical analyses on 16 fluid samples (from 15 wells) and ten source rock extracts to investigate questions on the Barents Sea Hammerfest Basin petroleum system, including source–oil correlation, oil–oil correlation, assessment of maturation and in-reservoir processes. In this case, $\delta^{13}C$ of *n*-alkanes ($>C_{15}$) were valuable for oil–source rock correlation and even quantification of the contribution from different source rocks. This proved especially useful for differentiating Triassic and Jurassic source contributions. For example, oil families III and IV are not distinguishable on the basis of $\delta^{13}C$ alone. However, Family II is obviously different from Families I, III and IV. Additionally, Family I oils have unique pristane and phytane $\delta^{13}C$ values, in comparison with *n*-alkane $\delta^{13}C$ values in the same oils. The oil families have the following characteristics:

Family I: C_{10}–C_{14}, −28 to −30‰; $C_{15}+$, −30 to −34‰; pristane and phytane, −31 to −32‰; branched alkanes, slight ^{12}C-enrichment with increasing molecular weight;
Family II: C_{10}–C_{14}, −31 to −33‰; $C_{15}+$, −33 to −36‰; pristane and phytane, −32.5 to −33.5‰;
Family III: C_{10}–C_{14}, −29 to −31‰; pristane and phytane, −30 to −31‰; branched alkanes, −28 to −29‰; cyclic alkanes, −25 to −29‰;
Family IV: C_{10}–C_{14}, −28 to −31‰; pristane and phytane, −32 to −33‰; branched alkanes, −28 to −30‰.

The source rock extract $\delta^{13}C$ values are differentiating as well. The Triassic Kobbe Formation values range from −32 to −35‰. The Upper Jurassic Hekkingen Formation ($>C_{20}$) *n*-alkanes have $\delta^{13}C$ values ranging from −28 to −31‰ – a 4–6‰ difference compared with the Triassic Kobbe Formation. These variations enable source assessment for the different oil families; the Triassic is the likely source for Families I, II and IV, while the Jurassic contribution is higher in Family III oils. A mixing model was applied to calculate more precisely the contributions of the different sources to the oils. This type of analysis forms a useful baseline to be used in future exploration, and also when assessing changes in production over time, depending on the field and reservoir properties.

Hydrogen and carbon CSIA of *n*-alkanes also provide information about the extent of thermal maturation and other physical processes. Oils show increasing $\delta^{2}H$ values with increasing C number, reflecting thermal maturation. Condensates also show the same trend, but with even higher $\delta^{2}H$ values, showing the impact of evaporative fractionation revealed as greater ^{2}H-enrichment with increasing C number (Murillo *et al.* 2016). As one of the most complete and comprehensive recently published studies utilizing *n*-alkane CSIA in the context of other fluid properties, Murillo *et al.* (2016) provide an excellent case study on how to use CSIA to improve the understanding of the source origin of oil families, and how this methodology can contribute to long-term field exploration, development planning, and potentially production monitoring.

He *et al.* (2012) performed another regional study of the Barents region, focusing on the Timan–Pechora Basin. The study consisted of 32 oil samples from 25 fields and also included two surface samples from the island of Spitsbergen. The samples were allocated into six families (using a chemometric approach with 20 biomarker parameters and two isotopic parameters) and were inferred to correlate with the respective source rocks (Devonian marl, Devonian carbonate, Triassic/Devonian carbonate, Triassic, Lower/Middle Jurassic and Upper Jurassic). CSIA of *n*-alkanes was used on a subset of presumed end-member and mixed samples, to better identify source rocks for the mixed oil families. The Upper Jurassic family (V), the Triassic family (I) and the two Devonian families (marl, II and carbonate, III) tend to have values that bracket those of the mixed Triassic/Devonian carbonate (IV, *c.* −29 to −32‰). The Upper Jurassic family (V) has the most ^{13}C-enriched values (*c.* −27 to −29‰), and the Devonian families (II, III) are generally the most ^{13}C-depleted (−33 to −34‰). Therefore, again the additional information provided uniquely by CSIA of *n*-alkanes supports and refines the oil–source rock correlations and strengthens the understanding of the regional petroleum system.

Asia and Australasia
Potwar Basin (applications in source rock depositional environment, source rock–oil relationships, oil–oil correlation, and biodegradation investigations). The Potwar Basin contains sedimentologically diverse Precambrian through Tertiary units and is structurally complex because of the intense tectonism associated with the Tertiary Himalayan Orogeny. There are multiple reservoir targets, including Cambrian, Jurassic and Eocene formations, with fluids ranging from 16° to 49° API gravity. Organic-rich potential source rocks include Precambrian evaporite/carbonate/clastic facies and Permian shale and carbonate units (Asif *et al.* 2011).

The petroleum geochemistry of the Potwar Basin was examined using 18 crude oil samples, biomarkers and $\delta^{13}C$ and $\delta^{2}H$ of whole oil, saturates and aromatics. On the basis of OM source, Asif *et al.* (2011) indentified three distinct oil groups: a terrigenous-origin oil family, and two marine oil families, differentiated by suboxic and oxic depositional conditions.

A more in-depth study, including the $\delta^{2}H$ CSIA on *n*-alkanes, pristane and phytane, was used to assess the level of biodegradation in eight crude oils that ranged in gravity from 16° to 41° API (Asif *et al.* 2009). The $\Delta\delta^{2}H$ ($\delta^{2}H$ isoprenoids −$\delta^{2}H$ *n*-alkanes) has a positive correlation with

API gravity ($\Delta\delta^2H$ decreasing with decreasing API gravity), showing preferential ^2H-enrichment in *n*-alkanes. The $\Delta\delta^2H$, therefore, provides a useful tool for classifying and differentiating oil families and also for possibly assessing low levels of biodegradation.

Taken together, these studies contribute to a deeper understanding of potential source rocks, oil classification and in-reservoir processes in the Potwar petroleum system.

Perth Basin (application in OM maturity investigation). This study used $^2H/^1H$ measurements: source rock and oil molecular properties were compared to show the relationship between thermal maturity δ^2H values of sedimentary hydrocarbons, and the δ^2H values of *n*-alkanes and acyclic isoprenoids. Dawson *et al.* (2005) studied nine samples, representing immature to mature OM, collected from the Triassic Hovea Formation, onshore in the northern Perth Basin. They compared the δ^2H of *n*-alkanes and isoprenoids with data from two crude oils, which are thought to be sourced from the organic-rich sapropels in the same formation. Generally, with increasing maturity, *n*-alkane δ^2H values remained consistent except in the highest maturity samples, where a 42‰ increase in δ^2H was observed. Pristane and phytane, however, became ^2H-enriched even at lower levels of maturity. The authors suggest that the H isotopic exchange mechanism (exchange at chiral C) is more rapid in isoprenoids (e.g. compounds containing tertiary C centres). Therefore, δ^2H values of pristane and phytane are well correlated with vitrinite reflectance equivalent values and maturity values derived from the terpane biomarker C_{27} 18α(H)-22,29,30-trinorneohopane to C_{27} 17α(H)-22,29,30-trinorhopane (Ts/Tm) ratio. The magnitude of the offset between the δ^2H values of *n*-alkanes and the δ^2H values of isoprenoids suggests that the $^2H/^1H$ content of these compounds could be used in determining the thermal maturity of a source rock. Relatively simple targeted studies of δ^2H of pristine and phytane from a wide variety of basins would expand on the observations from this study and be a useful addition to standard molecular geochemistry studies.

Tarim Basin (integrated basin analysis). The Tarim Basin is one of the most important petroleum basins in China and has been extensively reported in the literature. A full review is well beyond the scope of this contribution, and we focus on a few examples in which CSIA has provided important insights and utility to petroleum exploration in the Tarim Basin, especially source correlation and charge history, connectivity and in-reservoir alterations.

Briefly, the Tarim Basin is large and geologically complex; fluid types range from light to heavy, normal to waxy, early mature to secondarily cracked to gas. A range of in-reservoir processes related to multiple charge events and biodegradation also influence fluid properties. Large variations in stable C isotopic composition of *n*-alkanes in fluid compositions reflect mixing from many distinct source rocks (Jia *et al.* 2010; Li *et al.* 2010). More recently, using both C and H CSIA of *n*-alkanes, Jia *et al.* (2013) untangled complex source contributions in the Tabei and Tazhong uplift areas and ultimately distinguished oils from the same source rock at different maturity levels as well as contributions from different sources also at different maturity levels (Jia *et al.* 2013).

On the basis of biomarker and isotopic properties, the Tabei and Tazhong basin oils fall into two groups. Most oils fall into Group I, which have relatively low $\delta^{13}C$ values (−31.0 to −34.5‰) and *n*-alkane δ^2H values from −75 to −110‰. (Group II includes only two samples of heavy oil, with $\delta^{13}C$ values of −29 and −30‰, and δ^2H values of −142 to −145‰.) The $\delta^{13}C$ values of *n*-alkanes generally correlate with the biomarker maturity indices, and this positive correlation shows maturity is controlling, at least in part, the isotopic composition of the *n*-alkanes. Because the impact on δ^2H is stronger during thermal maturation (e.g. Tang *et al.* 2005), the relative ^2H-enrichment of *n*-alkanes is seen as the result of kinetic fractionation during oil maturation.

The fluids of the Tazhong Basin show much greater variation than those from Tabei, with light and waxy oils displaying a larger $\delta^{13}C$ range, from −31 to −36‰. In addition, there are greater variations with respect to molecular weight – the shorter-chain compounds (C_{10}–C_{16}) are ^{13}C-depleted, while the HMW compounds (C_{16+}) are slightly ^{13}C-enriched. These patterns, along with distinctive biomarker correlations, led the authors to propose several scenarios for source and charge history. One suite of oils also contains 25 nor-hopanes and features similar, relatively low $\delta^{13}C$ values (<−33.5‰) in the saturate fraction *n*-alkanes and in the asphaltene-released *n*-alkanes. Moreover, the asphaltenes do not contain biodegraded residues (Jia *et al.* 2008). This led to the conclusion that a biodegraded fluid was charged later by a non-biodegraded fluid. Thus $\delta^{13}C$ values of the *n*-alkanes helped reveal both charge history and in-reservoir biodegradation of the earlier charge.

Another suite of oils shows mixing of oil from one source but at different maturities. In this example, cross plots of maturity parameters and average $\delta^{13}C$ of *n*-alkanes were used to propose a likely mixing scenario in which fluids generated from a Middle Ordovician source filled reservoirs during Cretaceous through Tertiary times. To further explain the more complex fluid properties in the Tazhong Basin, the authors used both biomarkers and $\delta^2H/\delta^{13}C$ of *n*-alkanes to show possible mixing of oils from different sources, namely the Cambro-Ordovician and the Middle–Upper Ordovician.

As Murillo *et al.* (2016) have demonstrated in the Barents area, the Tarim Basin studies also show that the CSIA of *n*-alkanes provides evidence of variable fluid histories in source, maturity, charge history and reservoir processes, which can be integrated in petroleum systems concepts and models to aid field development planning and future exploration efforts.

Africa

North Africa (applications in OM source and oil–source rock correlation investigations). Source age and family correlations were also made in the Sirte Basin petroleum system using biomarkers and δ^{13}C and δ^2H of *n*-alkanes. Aboglia *et al.* (2010) analysed 24 wellhead-sampled oils in seven areas of the Sirte Basin, across a roughly 120-km north–south transect, representing six different reservoir formations, from Precambrian to Eocene age. Two oil families were defined based on maturity, as derived from several parameters: Family A is of higher maturity, with source rock possibly showing a higher percentage of terrigenous component; Family B is less mature, with dominantly marine source markers. The authors found that the *n*-alkanes in Family A have higher δ^{13}C values, which is consistent with a terrigenous source rock origin and a higher level of thermal maturity. Additional evidence for the more extensive thermal maturity of Family A oils came from higher δ^2H values of pristane and phytane in these oils in comparison with those from Family B oils. In another study in North Africa, Peters & Creaney (2004) found substantial variations in the δ^{13}C in *n*-C$_{17}$, *n*-C$_{18}$, and pristane and phytane, that enabled them to distinguish between Silurian- and Devonian-sourced oils in Algeria. These observations are therefore age-diagnostic, and can also be useful markers for long-term production monitoring as well as in exploration.

Niger Delta (integrated basin analysis). A study by Samuel *et al.* (2009) reassessed the evidence for petroleum charge from multiple source rocks in the Niger Delta petroleum system. They studied 58 crude oil samples from both shallow and deepwater fields and core samples from the Late Cretaceous Araromi Formation in SW Nigeria. The CSIA of *n*-alkanes proved useful in distinguishing the influence of different source facies on present-day reservoir fluid distributions. From alkane and biomarker distributions, three oil families were deduced on the basis of source (terrigenous, mixed marine–terrigenous and dominant marine). Biomarker distributions including C$_{30}$ tetracyclic polyprenoids (TPPs) were used to calculate the TPP proxy (Holba *et al.* 2003), oleanane indices (Eneogwe & Ekundayo

2003; Matava *et al.* 2003), and the tricyclic terpane index (TTTI) to discriminate between the marine (mostly western deepwater oils) and terrigenous source rocks, and also to suggest mixing of lacustrine and terrestrial-derived oils.

Generally, however, the *n*-alkanes in the western deepwater and some western shallow-water oils show a nearly flat trend of *n*-alkane δ^{13}C values, ranging from –24‰ to –30‰, with little variation with increasing C number. Other western shallow-water oils also have a relatively flat trend of δ^{13}C values with increasing C number but are also more ^{13}C-enriched than the western deepwater oils, especially in the *n*-C$_{12}$–C$_{16}$ range. Samuel *et al.* (2009) attributed these values to a marine source with uniform C isotopic composition during deposition (e.g. a well-mixed, buffered carbonate system). In contrast, the rest of the shallow-water samples (central, east and west) all demonstrate decreasing δ^{13}C values with increasing C number; values range from –23 to –28‰ at *n*-C$_{12}$, and decrease to values between –28 and –34‰ at *n*-C$_{30}$. (*n*-C$_{27}$ is also systematically ^{13}C-depleted by *c.* 0.5–1.5‰). ^{13}C-depletion with increasing C number is observed in many other basins around the world, notably the Tertiary systems of Australasia (Murray *et al.* 1994; Wilhelms *et al.* 1994) and south Texas (Bjorøy *et al.* 1991).

The authors found no clear correlation between maturity and the δ^{13}C values. Therefore, CSIA of *n*-alkanes appears to reflect mainly source kerogen isotopic properties. However, available source rock samples were limited in this study and only the Araromi formation samples from the Dahomey Basin were extracted and compared with the oils in this study. The Araromi formation is the youngest unit in the Late Cretaceous Abeokuta Group, with TOC ranging from *c.* 0.5% to 5% and hydrogen index (HI) values of 0 to *c.* 400 mg HC g^{-1} TOC. The biomarker and *n*-alkane distributions, as well as the C isotopic composition of the *n*-alkanes, proved to be a poor match to the studied oils.

The geochemical complexities of the Niger Basin petroleum system are beyond the scope of this contribution and/or are also largely unpublished, but the inclusion of CSIA-alkane analysis provides an additional tool for comparing and separating oil families and excluding specific source rock contributions.

North America

West Sak–Kuparuk–Prudhoe (application in charge evaluation investigation). C isotopes of gasoline-range hydrocarbons were used to understand the source of primary and secondary charge, as well as in-reservoir processes and their impacts on fluid properties, and ultimately they aided field development planning in the West Sak reservoirs

(Masterson *et al.* 2001). The similarity in biomarker ratios was used to show that West Sak reservoirs shared primary charge with the Prudhoe reservoirs to the east. However, the $\delta^{13}C$ values of an apparent secondary charge, rich in gasoline-range hydrocarbons, matched the underlying Kuparuk reservoirs. Faulting in the thick shale units separating Kuparuk and West Sak reservoirs is hypothesized to provide the conduit for this charge. The $\delta^{13}C$ values and distribution of the gasoline-range components in the upper West Sak fluids were shown to be further affected by biodegradation. The biodegradation results in more viscous fluid, and thus reduces producibility of the West Sak oils. Therefore, the light hydrocarbon secondary charge is the key to reducing viscosity of the reservoired fluid, and to the economic viability of West Sak production. Understanding the source of the secondary charge, using CSIA of gasoline-range hydrocarbons, provided a tool to identify the origin of the secondary charge and to therefore de-risk future development targets. The $\delta^{13}C$ values of organic compounds have provided an important piece of information for understanding charge history, basin development and economics.

Western Canada Sedimentary Basin (integrated basin analysis). Large variations in the difference between H isotopic compositions of *n*-alkanes and acyclic isoprenoids (pristane and phytane) was used effectively for source correlation in the Western Canada Sedimentary Basin (WCSB), including the Alberta and the Williston basins (Li *et al.* 2001). From Alberta, 13 oils from 12 fields were analysed, representing at least eight source units including Devonian, Triassic, Jurassic and Upper Cretaceous organofacies. From the Williston Basin, 13 oils from seven fields were studied, representing putative Cambrian, Ordovician, Devonian and Mississippian facies.

Within this rich dataset, there are examples that show how the H isotopic composition of the source material is preserved in the compound-specific H isotopic composition of the *n*-alkanes and acyclic isoprenoids, which then enables correlation of oils to source rocks. For example, the δ^2H of the *n*-alkanes from both the Cambrian and the Upper Cretaceous range from −160 to −190‰, and are more ^{2}H-depleted that the Mississippian–Devonian source oils in both the Alberta and Williston basins (−80 to −160‰). Also, the δ^2H values of the *n*-alkanes from the Devonian–Mississippian oils of the Williston Basin are intermediate between a Lodgepole and a Bakken end-member signature – oils from different vertical depths in the same well – which gives a strong indication of mixing of these two sources. Furthermore, known lacustrine-, marine- and evaporite-derived oils show different

δ^2H values, which demonstrate preservation of source-water H isotope compositions, e.g. oils derived from evaporites show more ^{2}H-enriched values.

As also shown in the Perth Basin (Dawson *et al.* 2005), maturity too has an impact on δ^2H, which potentially allows δ^2H values to be used as evidence of thermal maturity level, especially as other biomarker ratios of thermal maturity either reach equilibrium or disappear altogether in the highest maturity liquids. In this case, the ^{2}H-enrichment leads to a 40‰ increase in the weighted-average δ^2H values.

This study illustrates the interpretive power and resolution of H isotopic compositions on *n*-alkanes for oil–oil and oil–source rock correlation, as well as how they can provide evidence for mixing and thermal maturity. Although such studies may be undertaken internally within exploration and production groups in national and international petroleum companies, there are few examples in the literature or published datasets. Yet, such datasets can benefit both palaeoclimate and petroleum geochemists.

Summary

CSIA of *n*-alkanes provides evidence of fluid histories including source, maturity, charge history and reservoir processes that – when integrated in petroleum systems concepts and models – can support field development planning and exploration efforts. For source rock determination, the C and H isotope compositions of *n*-alkanes and related compounds can provide differentiation between marine, lacustrine and evaporitic palaeoenvironments. These observations can add confidence to correlation among oils, and between oils and source rocks.

In some cases, the use of CSIA is targeted to determine particular processes. Single-basin studies are examples of particular applications, such as charge evaluation (West Sak), source rock correlation (Austria, Sirte, Potwar), and interplay of maturity and in-reservoir processes (Perth, Potwar, Western Canada Sedimentary Basin).

In other cases, CSIA of *n*-alkanes is used to corroborate interpretations from integrated petroleum systems analysis, providing unique insights, which may not be revealed when using other methods. Studies in the Tarim, Niger, Barents and Western Canada Sedimentary basins are good examples of integrated geochemical basin analysis.

Overall, CSIA of *n*-alkanes and related *n*-alkyl structures can provide independent data to strengthen petroleum systems concepts from generation and expulsion of fluids from source rock, to charge history, connectivity and in-reservoir processes (Fig. 5). The studies should be fully integrated in basin analysis, with CSIA being used to reduce uncertainty and increase confidence in basin evaluation.

As highlighted by Curiale (2008), whether directly comparing organic geochemical data from putative source rocks and oils, or inferring possible source rock properties from oil data, correlation and in-reservoir process determination can be continually updated and improved as data are incorporated into conceptual, basin, geological and reservoir models. Likewise, the integration of the full range of fluid and rock properties, the creation and/or consolidation of large datasets, and the use of the next level of analytical tools such as compound-specific isotopic compositions, can significantly refine and improve basin and reservoir understanding, and reduce the risk of field development planning.

We thank Michael Lawson and Michael Formolo for the invitation to contribute this review to the volume and for editorial handling of the manuscript. We are thankful to Schlumberger Ltd for permission to publish and for editing the manuscript during the revision of the manuscript. Many thanks to the anonymous reviewers whose comments significantly improved the manuscript. The School of Environmental Sciences, University of East Anglia, UK provided funding for Gold Open Access.

References

ABOGLIA, S., GRICE, K., TRINAJSTIC, K., DAWSON, D. & WILLIFORD, K.H. 2010. Use of biomarker distributions and compound specific isotopes of carbon and hydrogen to delineate hydrocarbon characteristics in the East Sirte Basin (Libya). *Organic Geochemistry*, **41**, 1249–1258.

AITKEN, C., JONES, D.M. & LARTER, S.R. 2004. Anaerobic hydrocarbon biodegradation in deep subsurface oil reservoirs. *Nature*, **431**, 291–294.

ARTHUR, M.A., DEAN, W.E. & CLAYPOOL, G.E. 1985. Anomalous ^{13}C enrichment in modern marine organic carbon. *Nature*, **315**, 216–218.

ASIF, M., GRICE, K. & FAZEELAT, T. 2009. Assessment of petroleum biodegradation using stable hydrogen isotopes of individual saturated hydrocarbon and polycyclic aromatic hydrocarbon distributions in oils from the Upper Indus Basin, Pakistan. *Organic Geochemistry*, **40**, 301–311.

ASIF, M., FAZEELAT, T. & GRICE, K. 2011. Petroleum geochemistry of the Potwar Basin, Pakistan: 1. Oil–oil correlation using biomarkers, δ^{13}C and δD. *Organic Geochemistry*, **42**, 1226–1240.

BECHTEL, A., HAMOR-VISO, M., GRATZER, R., SACHSENHOFER, R. & PUTTMANN, W. 2012. Facies evolution and stratigraphic correlation in the early Oligocene Tard Clay of Hungary as revealed by maceral, biomarker and stable isotope composition. *Marine and Petroleum Geochemistry*, **35**, 55–74.

BECHTEL, A., GRATZER, R., LINZER, H.-G. & SACHSENHOFER, R. 2013. Influence of migration distance, maturity and facies on the stable isotopic composition of alkanes and on carbazole distributions in oils and source rocks of the Alpine Foreland Basin of Austria. *Organic Geochemistry*, **62**, 74–85.

BENSON, S., LENNARD, C., MAYNARD, P. & ROUX, C. 2006. Forensic applications of isotope ratio mass spectrometry – a review. *Forensic Science International*, **157**, 1–22.

BIAN, X.-Y., MBADINGA, S.M. *ET AL.* 2015. Insights into the anaerobic biodegradation pathway of *n*-alkanes in oil reservoirs by detection of signature metabolites. *Scientific Reports*, **5**, 9801.

BJORØY, M., HALL, K., GILLYON, P. & JUMEAU, J. 1991. Carbon isotope variations in *n*-alkanes and isoprenoids of whole oils. *Chemical Geology*, **93**, 13–20.

BJORØY, M., HALL, P. & MOE, R. 1994. Variation in the isotopic composition of single components in the C_4–C_{20} fraction of oils and condensates. *Organic Geochemistry*, **21**, 761–776.

CARPENTIER, B., UNGERER, P., KOWALEWSKI, I., MAGNIER, C., COURCY, J.P. & HUC, A.Y. 1996. Molecular and isotopic fractionation of light hydrocarbons between oil and gas phases. *Organic Geochemistry*, **24**, 1115–1139.

CHIKARAISHI, Y. & NARAOKA, H. 2006. Carbon and hydrogen isotope variation of plant biomarkers in a plant–soil system. *Chemical Geology*, **231**, 190–202.

CHIKARAISHI, Y., TANAKA, R., TANAKA, A. & OHKOUCHI, N. 2009. Fractionation of hydrogen isotopes during phytol biosynthesis. *Organic Geochemistry*, **40**, 569–573.

CHUNG, H.M., CLAYPOOL, G.E., ROONEY, M.A. & SQUIRES, R.M. 1994. Source characteristics of marine oils as indicated by carbon isotopic ratios of volatile hydrocarbons. *AAPG Bulletin*, **78**, 396–408.

CLAYTON, C. 1991. Effect of maturity on carbon isotope ratios of oils and condensates. *Organic Geochemistry*, **17**, 887–899.

CLAYTON, C.J. & BJORØY, M. 1994. Effect of maturity on ratios of individual compounds in North Sea oils. *Organic Geochemistry*, **21**, 737–750.

CRAIG, H. & GORDON, L.I. 1965. Deuterium and oxygen 18 variations in the ocean and the marine atmosphere. *In*: TONGIORGI, E. (ed.) *Stable Isotopes in Oceanographic Studies and Paleotemperatures*. Consiglio Nazionale Delle Richerche, Laboratorio de Geologia Nucleare, Spoleto, Italy, 9–131.

CRAMER, B., KROOSS, B.M. & LITTKE, R. 1998. Modelling isotope fractionation during primary cracking of natural gas: a reaction kinetic approach. *Chemical Geology*, **149**, 235–250.

CREANEY, S., ALLAN, J. *ET AL.* 1994. Petroleum generation and migration in the Western Canada Sedimentary Basin. *In*: MOSSOP, G. & SHETSON, I. (eds) *Geological Atlas of the Western Canada Sedimentary Basin*. Canadian Society of Petroleum Geologists and Alberta Research Council, Canada, 455–468.

CURIALE, J.A. 2008. Oil-source rock correlations – limitations and recommendations. *Organic Geochemistry*, **39**, 1150–1161.

CURIALE, J.A. & BROMLEY, B.W. 1996. Migration induced compositional changes in oils and condensates of a single field. *Organic Geochemistry*, **24**, 1097–1113.

DAWSON, D., GRICE, K. & ALEXANDER, R. 2005. Effect of maturation on the indigenous δD signatures of individual hydrocarbons in sediments and crude oils from the

Perth Basin (Western Australia). *Organic Geochemistry*, **36**, 95–104.

DAWSON, D., GRICE, K., ALEXANDER, R. & EDWARDS, D. 2007. The effect of source and maturity on the stable isotopic compositions of individual hydrocarbons in sediments and crude oils from the Vulcan Sub-basin, Timor Sea, Northern Australia. *Organic Geochemistry*, **38**, 1015–1038.

DEINES, P. 1980. The isotopic composition of reduced organic carbon. *In*: FRITZ, P. & FONTES, J.C. (eds) *Handbook of Environmental Geochemistry*. Elsevier, New York, **1**, 239–406.

DE LEEUW, J.W., FREWIN, N.L., VAN BERGEN, P.F., SINNINGHE DAMSTÉ, J.S. & COLLINSON, M.E. 1995. Organic carbon as a palaeoenvironmental indicator in the marine realm. *In*: BOSENCE, D.W.J. & ALLISON, P.A. (eds) *Marine Palaeoenvironmental Analysis from Fossils*. Geological Society, London, Special Publications, **83**, 43–71, https://doi.org/10.1144/GSL.SP.1995.083.01.04

DIRGHANGI, S.S. & PAGANI, M. 2013. Hydrogen isotope fractionation during lipid biosynthesis by *Tetrahymena thermophila*. *Organic Geochemistry*, **64**, 105–111.

ENEOGWE, C. & EKUNDAYO, O. 2003. Geochemical correlation of crude oils in the NW Niger Delta, Nigeria. *Journal of Petroleum Geology*, **26**, 95–103.

ESTEP, M.F. & HOERING, T.C. 1980. Biogeochemistry of the stable hydrogen isotopes. *Geochimica et Cosmochimica Acta*, **44**, 1197–1206.

EVERSHED, R.P., BULL, I.D. *ET AL.* 2007. Compound-specific stable isotope analysis in ecology and paleoecology. *In*: MICHENER, R. & LAJTHA, K. (eds) *Stable Isotopes in Ecology and Environmental Science*. Blackwell, Oxford, 480–540.

FARQUHAR, G.D., EHLERINGER, J.R. & HUBICK, K.T. 1989. Carbon isotope discrimination and photosyntheis. *Annual Review of Plant Physiology and Plant Molecular Biology*, **40**, 503–537.

FOGEL, M.L. & CIFUENTES, L.A. 1993. Isotope fractionation during primary production. *In*: ENGEL, M.H. & MACKO, S.A. (eds) *Organic Geochemistry*. Plenum Press, New York, 73–98.

FREEMAN, K.H. 2001. Isotopic biogeochemistry of marine organic carbon. *In*: VALLEY, J.W. & COLE, D.R. (eds) *Stable Isotope Geochemistry*. Reviews in Mineralogy & Geochemistry, **43**. The Mineralogical Society of America, Washington, DC, 579–605.

FREEMAN, K.H., HAYES, J.M., TRENDEL, J.-M. & ALBRECHT, P. 1990. Evidence from carbon isotope measurements for diverse origins of sedimentary hydrocarbons. *Nature*, **343**, 254–256.

FREEMAN, K.H., BOREHAM, C.J., SUMMONS, R.E. & HAYES, J.M. 1994. The effect of aromatization on the isotopic compositions of hydrocarbons during early diagenesis. *Organic Geochemistry*, **21**, 1037–1049.

GAMARRA, B., SACHSE, D. & KAHMEN, A. 2016. Effects of leaf water evaporative ^{2}H-enrichment and biosynthetic fractionation on leaf wax *n*-alkane δ^2H values in C_3 and C_4 grasses. *Plant, Cell & Environment*, **39**, 2390–2403.

GLASER, B. 2005. Compound-specific stable-isotope (δ^{13}C) analysis in soil science. *Journal of Plant Nutrition and Soil Science*, **168**, 633–648.

GONFIANTINI, R. 1986. Environmental isotopes in lake studies. *In*: FRITZ, P. & FONTES, J.CH. (eds) *Handbook of Environmental Isotope Geochemistry, The Terrestrial Environment B*. Elsevier, Amsterdam, **II**, 113–168.

GRATZER, R., BECHTEL, A., SACHSENHOFER, R.F., LINZER, H.-G., REISCHENBACHER, D. & SCHULZ, H.-M. 2011. Oil–oil and oil–source rock correlations in the Alpine Foreland basin of Austria: insights from biomarker and stable carbon isotope studies. *Marine and Petroleum Geology*, **28**, 1171–1186.

GRICE, K., SCHOUTEN, S., NISSENBAUM, A., CHARRACH, J. & SINNINGHE DAMSTÉ, J.S. 1998. Isotopically heavy carbon in the C_{21} to C_{25} regular isoprenoids in halite-rich deposits from the Sdom Formation, Dead Sea Basin, Israel. *Organic Geochemistry*, **28**, 349–359.

GRICE, K., MESMAY, R., GLUCINA, A. & WANG, S. 2008. An improved and rapid 5Å molecular sieve method for gas chromatography isotope ratio mass spectrometry of *n*-alkanes (C_8–C_{30}+). *Organic Geochemistry*, **39**, 284–288.

HAYES, J.M. 1993. Factors controlling ^{13}C contents of sedimentary organic compounds: principles and evidence. *Marine Geology*, **113**, 111–125.

HAYES, J.M. 2001. Fractionation of carbon and hydrogen isotopes in biosynthetic processes. *In*: VALLEY, J.W. & COLE, D.R. (eds) *Stable Isotope Geochemistry*. Reviews in Mineralogy & Geochemistry, **43**. The Mineralogical Society of America, Washington, DC, 225–277.

HAYES, J.M., FREEMAN, K.H. & POPP, B.N. 1990. Compound-specific isotopic analyses: a novel tool for reconstruction of ancient biogeochemical processes. *Organic Geochemistry*, **16**, 1115–1128.

HAYES, J.M., STRAUSS, H. & KAUFMAN, A.J. 1999. The abundance of ^{13}C in marine organic matter and isotopic fractionation in the global biogeochemical cycle of carbon during the past 800 Ma. *Chemical Geology*, **161**, 103–125.

HE, M., MOLDOWAN, J.M., NEMCHENKO-ROVENSKAYA, A. & PETERS, K.E. 2012. Oil families and their inferred source rocks in the Barents Sea and northern Timan–Pechora basin, Russia. *AAPG Bulletin*, **96**, 1121–1146.

HEAD, I.M., JONES, D.M. & LARTER, S.R. 2003. Biological activity in the deep subsurface and the origin of heavy oil. *Nature*, **426**, 344–352.

HOLBA, A.G., DZOU, L.I. *ET AL.* 2003. Application of tetracyclic polyprenoids as indicators of input from fresh-brackish water environments. *Organic Geochemistry*, **34**, 441–469.

HUANG, Y., EGLINTON, G., INESON, P., LATTER, P.M., BOL, R. & HARKNESS, D.D. 1997. Absence of carbon isotope fractionation of individual *n*-alkanes in a 23-year field decomposition experiment with *Calluna vulgaris*. *Organic Geochemistry*, **26**, 497–501.

JIA, W.L., PENG, P.A. & XIAO, Z.Y. 2008. Carbon isotopic compositions of 1,2,3,4-tetramethylbenzene in marine oil asphaltenes from the Tarim Basin: evidence for the source formed in a strongly reducing environment. *Science in China, Series D: Earth Sciences*, **51**, 509–514.

JIA, W.L., PENG, P.A., YU, C.L. & XIAO, Z.Y. 2010. Molecular and isotopic compositions of bitumens in Silurian tar sands from the Tarim Basin, NW China: characterizing biodegradation and hydrocarbon charging in an old composite basin. *Marine and Petroleum Geology*, **27**, 13–25.

JIA, W., WANG, Q., PENG, P., XIAO, Z. & LI, B. 2013. Isotopic compositions and biomarkers in crude oils from the Tarim basin: oil maturity and oil mixing. *Organic Geochemistry*, **57**, 95–106.

JONES, D.M., HEAD, I.M. *ET AL.* 2008. Crude-oil biodegradation via methanogenesis in subsurface petroleum reservoirs. *Nature*, **451**, 176–180.

KEELEY, J.E. & SANDQUIST, D.R. 1992. Carbon: freshwater plants. *Plant, Cell & Environment*, **15**, 1021–1035.

KIKUCHI, T., SUZUKI, N. & SAITO, H. 2010. Change in hydrogen isotope composition of n-alkanes, pristane, phytane, and aromatic hydrocarbons in Miocene siliceous mudstones with increasing maturity. *Organic Geochemistry*, **41**, 940–946.

LARTER, S.R. & APLIN, A.C. 1995. Reservoir geochemistry: methods, applications and opportunities. *In*: CUBITT, J. M. & ENGLAND, W.A. (eds) *The Geochemistry of Reservoirs*. Geological Society, London, Special Publications, **86**, 5–32, https://doi.org/10.1144/GSL.SP.1995.086.01.02

LARTER, S.R., BOWLER, B.F.J. *ET AL.* 1996. Molecular indicators of secondary oil migration distances. *Nature*, **383**, 593–597.

LETELLIER, M. & BUDZINSKI, H. 1999. Microwave assisted extraction of organic compounds. *Analusis*, **27**, 259–270.

LEWAN, M.D. 1983. Effects of thermal maturation on stable organic carbon isotopes as determined by hydrous pyrolysis of Woodford Shale. *Geochimica et Cosmochimica Acta*, **47**, 1471–1479.

LI, M., HUANG, Y., OBERMAJER, M., JIANG, C., SNOWDON, L. & FOWLER, M. 2001. Hydrogen isotopic compositions of individual alkanes as a new approach to petroleum correlation: case studies from the Western Canada Sedimentary Basin. *Organic Geochemistry*, **32**, 1387–1399.

LI, S.M., PANG, X.Q., JIN, Z.J., YANG, H.J., XIAO, Z.Y., GU, Q.Y. & ZHANG, B.S. 2010. Petroleum source in the Tazhong Uplift, Tarim Basin: new insights from geochemical and fluid inclusion data. *Organic Geochemistry*, **41**, 531–553.

LIAO, Y. & GENG, A. 2009. Stable carbon isotopic fractionation of individual n-alkanes accompanying primary migration: evidence from hydrocarbon generation-expulsion simulations of selected terrestrial source rocks. *Applied Geochemistry*, **24**, 2123–2132.

LORANT, F., PRINZHOFER, A., BEHAR, F. & HUC, A.Y. 1998. Carbon isotopic and molecular constraints on the formation and the expulsion of thermogenic hydrocarbon gases. *Chemical Geology*, **147**, 249–264.

LOSH, S., CATHLES, L. & MEULBROEK, P. 2002a. Gas washing of oil along a regional transect, offshore Louisiana. *Organic Geochemistry*, **33**, 655–663.

LOSH, S., WALTER, L., MEULBROEK, P., MARTINI, A., CATHLES, L. & WHELAN, J. 2002b. Reservoir fluids and their migration into the south Eugene Island block 330 reservoirs, offshore Louisiana. *AAPG Bulletin*, **86**, 1463–1488.

LUNDANES, E. & GREIBROKK, T. 1994. Separation of fuels, heavy fractions, and crude oils into compound classes: a review. *Journal of High Resolution Chromatography*, **17**, 197–202.

MABERLY, S.C., RAVEN, J.A. & JOHNSTON, A.M. 1992. Discrimination between ^{12}C and ^{13}C by marine plants. *Oecologia*, **91**, 481–492.

MARCANO, N., LARTER, S. & MAYER, B. 2013. The impact of severe biodegradation on the molecular and stable (C, H, N, S) isotopic compositions of oils in the Alberta Basin, Canada. *Organic Geochemistry*, **59**, 114–132.

MASTERSON, W.D., DZOU, L.I.P., HOLBA, A.G., FINCANNON, A.L. & ELLIS, L. 2001. Evidence for biodegradation and evaporative fractionation in West Sak, Kuparuk and Prudhoe Bay field areas, North Slope, Alaska. *Organic Geochemistry*, **32**, 411–441.

MATAVA, T., ROONEY, M.A., CHUNG, H.M., NWANKWO, B.C. & UNOMAH, G.I. 2003. Migration effects on the composition of hydrocarbon accumulations in the OML 67–70 areas of the Niger Delta. *AAPG Bulletin*, **87**, 1193–1206.

MEIER-AUGENSTEIN, W. 1999. Applied gas chromatography coupled to isotope ratio mass spectrometry. *Journal of Chromatography A*, **842**, 351–371.

MURILLO, W., VIETH-HILLEBRAND, A., HORSFIELD, B. & WILKES, H. 2016. Petroleum source, maturity, alteration and mixing in the southwestern Barents Sea: new insights from geochemical and isotope data. *Marine and Petroleum Geology*, **70**, 119–143.

MURRAY, A., SUMMONS, R., BOREHAM, C. & DOWLING, L. 1994. Biomarker and n-alkane isotope profiles for Tertiary oils: relationship to source rock depositional setting. *Organic Geochemistry*, **22**, 521–542.

NGUYEN TU, T.T., DERENNE, S., LARGEAU, C., BARDOUX, G. & MARIOTTI, A. 2004. Diagenesis effects on specific carbon isotope composition of plant n-alkanes. *Organic Geochemistry*, **35**, 317–329.

NGUYEN TU, T.T., EGASSE, C. *ET AL.* 2011. Early degradation of plant alkanes in soils: a litterbag experiment using ^{13}C-labelled leaves. *Soil Biology and Biochemistry*, **43**, 2222–2228.

OAKES, A.M. & HREN, M.T. 2016. Temporal variations in the δD of leaf n-alkanes from four riparian plant species. *Organic Geochemistry*, **97**, 122–130.

ODDEN, W., BARTH, T. & TALBOT, M.R. 2002. Compound-specific carbon isotope analysis of natural and artificially generated hydrocarbons in source rocks and petroleum fluids from offshore Mid-Norway. *Organic Geochemistry*, **33**, 47–65.

PANCOST, R. & PAGANI, M. 2006. Controls on the carbon isotopic composition of lipids in marine environments. *In*: VOLKMAN, J. (ed.) *Marine Organic Matter: Biomarkers, Isotopes and DNA*. Springer-Verlag, Berlin, 209–249.

PEDENTCHOUK, N., FREEMAN, K.H. & HARRIS, N.B. 2006. Different response of δD values of n-alkanes, isoprenoids, and kerogen during thermal maturation. *Geochimica et Cosmochimica Acta*, **70**, 2063–2072.

PÉRES, V.F., SAFFI, J. *ET AL.* 2006. Comparison of soxhlet, ultrasound-assisted and pressurized liquid extraction of terpenes, fatty acids and Vitamin E from *Piper gaudichaudianum* Kunth. *Journal of Chromatography A*, **1105**, 115–118.

PETERS, K.E. & CREANEY, S. 2004. Geochemical differentiation of Silurian from Devonian crude oils in eastern Algeria. *In*: HILL, R., LEVENTHAL, J. *ET AL.* (eds) *Geochemical Investigations in Earth and Space Science*. The Geochemical Society Special Publications, **9**, Elsevier Science, 287–301, https://doi.org/10.1016/S1873-9881(04)80021-X

PETERS, K.E. & FOWLER, M. 2002. Applications of petroleum geochemistry to exploration and reservoir management. *Organic Geochemistry*, **33**, 5–36.

PETERS, K.E., ROHRBACK, B.G. & KAPLAN, I.R. 1981. Carbon and hydrogen stable isotope variations in kerogen during laboratory-simulated thermal maturation. *AAPG Bulletin*, **65**, 501–508.

PETERS, K.E., WALTERS, C.C. & MOLDOWAN, J.M. 2005. *The Biomarker Guide: Volume 1, Biomarkers and Isotopes in Environment and Human History*. Cambridge University Press, Cambridge.

PHILP, R.P. 2006. The emergence of stable isotopes in environmental and forensic geochemistry studies: a review. *Environmental Chemistry Letters*, **5**, 57–66.

POPP, B.N., TAKIGIKU, R., HAYES, J.M., LOUDA, J.M. & BAKER, E.W. 1989. The post-Paleozoic chronology and mechanism of ^{13}C depletion in primary marine organic matter. *American Journal of Science*, **289**, 436–454.

RADKE, J., BECHTEL, A., GAUPP, R., PÜTTMANN, W., SCHWARK, L., SACHSE, D. & GLEIXNER, G. 2005. Correlation between hydrogen isotope ratios of lipid biomarkers and sediment maturity. *Geochimica et Cosmochimica Acta*, **69**, 5517–5530.

RIELEY, G. 1994. Derivatization of organic compounds prior to gas chromatographic-combustion-isotope ratio mass spectrometric analysis: identification of isotope fractionation processes. *The Analyst*, **119**, 915–919.

RIELEY, G., COLLIER, R., JONES, D., EGLINTON, G., EAKIN, P. & FALLICK, A. 1991. Sources of sedimentary lipids deduced from stable carbon-isotope analyses of individual compounds. *Nature*, **352**, 425–427.

ROONEY, M.A. 1995. Carbon isotopic ratios of light hydrocarbons as indicators of thermochemical sulfate reduction. *In*: GRIMALT, J.O. (ed.) *Organic Geochemistry: Applications to Energy, Climate, Environment and Human History*. AIGOA, San Sebastian, 523–525.

ROONEY, M., VULETICH, A. & GRIFFITH, C. 1998. Compound-specific isotope analysis as a tool for characterizing mixed oils: an example from the West of Shetlands area. *Organic Geochemistry*, **29**, 241–254.

ROZANSKI, K., ARAGUÁS-ARAGUÁS, L. & GONFIANTINI, R. 1993. Isotopic patterns in modern global precipitation. *In*: SWART, P.K., LOHMANN, K.C., McKENZIE, J. & SAVIN, S. (eds) *Climate Change in Continental Isotopic Records*. Geophysical Monograph **78**, American Geophysical Union, Washington, DC, 1–36.

SACHSE, D., BILLAULT, I. *ET AL.* 2012. Molecular palaeohydrology: interpreting the hydrogen-isotopic composition of lipid biomarkers from photosynthesizing organisms. *Annual Reviews of Earth and Planetary Science*, **40**, 221–249.

SAMUEL, O.J., CORNFORD, C., JONES, M., ADEKEYE, O.A. & AKANDE, S.O. 2009. Improved understanding of the petroleum systems of the Niger delta basin, Nigeria. *Organic Geochemistry*, **40**, 461–483.

SANTOS NETO DOS, E.V. & HAYES, J.M. 1999. Use of hydrogen and carbon stable isotopes characterizing oils from the Potiguar Basin (Onshore), northeastern Brazil. *AAPG Bulletin*, **83**, 496–518.

SCHIDLOWSKI, M. 1988. A 3800-million-year isotopic record of life from carbon in sedimentary rocks. *Nature*, **333**, 313–318.

SCHIEGL, W.E. & VOGEL, J.C. 1970. Deuterium content of organic matter. *Earth and Planetary Science Letters*, **7**, 307–313.

SCHIMMELMANN, A., SESSIONS, A., BOREHAM, C., EDWARDS, D., LOGAN, G. & SUMMONS, R. 2004. D/H ratios in terrestrially sourced petroleum systems. *Organic Geochemistry*, **35**, 1169–1195.

SCHIMMELMANN, A., SESSIONS, A.L. & MASTALERZ, M. 2006. Hydrogen isotopic (D/H) composition of organic matter during diagenesis and thermal maturation. *Annual Review of Earth and Planetary Sciences*, **34**, 501–533.

SCHMIDT, H.-L., WERNER, R.A. & EISENREICH, W. 2003. Systematics of ^{2}H patterns in natural compounds and its importance for the elucidation of biosynthetic pathways. *Phytochemistry Reviews*, **2**, 61–85.

SCHMIDT, T., ZWANK, L., ELSNER, M., BERG, M., MECKENSTOCK, R. & HADERLEIN, S. 2004. Compound-specific stable isotope analysis of organic contaminants in natural environments: a critical review of the state of the art, prospects, and future challenges. *Analytical and Bioanalytical Chemistry*, **378**, 283–300.

SCHOELL, M. 1984. Recent advances in petroleum isotope geochemistry. *Organic Geochemistry*, **6**, 645–663.

SCHOPF, J.W. 2000. The fossil record: tracing the roots of the cyanobacterial lineage. *In*: WHITTON, B.A. & POTTS, M. (eds) *The Ecology of Cyanobacteria*. Kluwer, Dordrecht, 13–35.

SEIFERT, W.K. & MOLDOWAN, J.M. 1986. Use of biological markers in petroleum exploration. *In*: JOHNS, R.B. (ed.) *Biological Markers in the Sedimentary Record*. Elsevier, Amsterdam, 261–290.

SESSIONS, A. 2006. Isotope-ratio detection for gas chromatography. *Journal of Separation Science*, **29**, 1946–1961.

SESSIONS, A. 2016. Factors controlling the deuterium contents of sedimentary hydrocarbons. *Organic Geochemistry*, **96**, 43–64.

SESSIONS, A.L., BURGOYNE, T.W. & SCHIMMELMANN, A. 1999. Fractionation of hydrogen isotopes in lipid biosynthesis. *Organic Geochemistry*, **30**, 1193–1200.

SESSIONS, A., BURGOYNE, T. & HAYES, J. 2001*a*. Correction of H_3^+ contributions in hydrogen isotope ratio monitoring mass spectrometry. *Analytical Chemistry*, **73**, 192–199.

SESSIONS, A., BURGOYNE, T. & HAYES, J. 2001*b*. Determination of the H_3 factor in hydrogen isotope ratio monitoring mass spectrometry. *Analytical Chemistry*, **73**, 200–207.

SESSIONS, A.L., SYLVA, S.P., SUMMONS, R.E. & HAYES, J.M. 2004. Isotopic exchange of carbon-bound hydrogen over geologic timescales. *Geochimica et Cosmochimica Acta*, **68**, 1545–1559.

SMITH, B.N. & EPSTEIN, S. 1970. Biogeochemistry of the stable isotopes of hydrogen and carbon in salt marsh biota. *Plant Physiology*, **46**, 738–742.

SMITH, R. 2003. Before the injection – modern methods of sample preparation for separation techniques. *Journal of Chromatography A*, **1000**, 3–27.

SOFER, Z. 1984. Stable carbon isotope compositions of crude oils: application to source depositional environments and petroleum alteration. *AAPG Bulletin*, **68**, 31–49.

SOFER, Z. 1991. Stable isotopes in petroleum exploration. *In*: MERRILL, R.K. (ed.) *Source and Migration*

Processes and Evaluation Techniques. Amoco Production, Tulsa, 103–106.

STAHL, W. 1977. Carbon and nitrogen isotopes in hydrocarbon research and exploration. *Chemical Geology*, **20**, 121–149.

STAHL, W. 1980. Compositional changes and $^{13}C/^{12}C$ fractionations during the degradation of hydrocarbons by bacteria. *Geochimica et Cosmochimica Acta*, **44**, 1903–1907.

SUMMONS, R.E., JAHNKE, L.L. & ROKSANDIC, Z. 1994. Carbon isotopic fractionation in lipids from methanotrophic bacteria: Relevance for interpretation of the geochemical record of biomarkers. *Geochimica et Cosmochimica Acta*, **58**, 2853–2863.

SUN, Y., CHEN, Z., XU, S. & CAI, P. 2005. Stable carbon and hydrogen isotopic fractionation of individual *n*-alkanes accompanying biodegradation: evidence from a group of progressively biodegraded oils. *Organic Geochemistry*, **36**, 225–238.

TANG, Y., HUANG, Y. *ET AL.* 2005. A kinetic model for thermally induced hydrogen and carbon isotope fractionation of individual *n*-alkanes in crude oil. *Geochimica et Cosmochimica Acta*, **69**, 4505–4520.

THIEL, V., PECKMANN, J., SEIFERT, R., WEHRUNG, P., REITNER, J. & MICHAELIS, W. 1999. Highly isotopically depleted isoprenoids: molecular markers for ancient methane venting. *Geochimica et Cosmochimica Acta*, **63**, 3959–3966.

THOMPSON, K.F.M. 1987. Fractionated aromatic petroleum and the generation of gas-condensates. *Organic Geochemistry*, **11**, 573–590.

THOMPSON, K.F.M. 1988. Gas-condensate migration and oil fractionation in deltaic systems. *Marine and Petroleum Geology*, **5**, 237–246.

THOMPSON, K.F.M. 2010. Aspects of petroleum basin evolution due to gas advection and evaporative fractionation. *Organic Geochemistry*, **41**, 370–385.

TIPPLE, B.J. & PAGANI, M. 2013. Environmental control on eastern broadleaf forest species' leaf wax distributions and D/H ratios. *Geochimica et Cosmochimica Acta*, **111**, 64–77.

TIPPLE, B.J., BERKE, M.A., HAMBACH, B., RODEN, J.S. & EHLREINGER, J.R. 2014. Predicting leaf wax *n*-alkane $^{2}H/^{1}H$ ratios: controlled water source and humidity experiments with hydroponically grown trees confirm predictions of Craig-Gordon model. *Plant, Cell & Environment*, **38**, 1035–1047.

VAN GRAAS, G., ELIN GILJE, A., ISOM, T.P. & AASE TAU, L. 2000. The effects of phase fractionation on the composition of oils, condensates and gases. *Organic Geochemistry*, **31**, 1419–1439.

VIETH, A. & WILKES, H. 2006. Deciphering biodegradation effects on light hydrocarbons in crude oils using their stable carbon isotopic composition: a case study from the Gullfaks oil field, offshore Norway. *Geochimica et Cosmochimica Acta*, **70**, 651–665.

WALTERS, C.C., WANG, F.C., QIAN, K., WU, C., MENNITO, A.S. & WEI, Z. 2015. Petroleum alteration by thermochemical sulfate reduction – a comprehensive molecular study of aromatic hydrocarbons and polar compounds. *Geochimica et Cosmochimica Acta*, **153**, 37–71.

WANG, Y., SESSIONS, A.L., NIELSEN, R.J. & GODDARD, W.A., III 2009*a*. Equilibrium $^{2}H/^{1}H$ fractionations in organic molecules: I. Experimental calibration of ab initio calculations. *Geochimica et Cosmochimica Acta*, **73**, 7060–7075.

WANG, Y., SESSIONS, A.L., NIELSEN, R.J. & GODDARD, W.A., III 2009*b*. Equilibrium $^{2}H/^{1}H$ fractionations in organic molecules: II. Linear alkanes, alkenes, ketones, carboxylic acids, esters, alcohols and ethers. *Geochimica et Cosmochimica Acta*, **73**, 7076–7086.

WANG, Y., SESSIONS, A.L., NIELSEN, R.J. & GODDARD, W.A., III 2013. Equilibrium $^{2}H/^{1}H$ fractionations in organic molecules: III. Cyclic ketones and hydrocarbons. *Geochimica et Cosmochimica Acta*, **107**, 82–95.

WENGER, L., DAVIS, C.L. & ISAKSEN, G.H. 2002. Multiple controls on petroleum biodegradation and impact on oil quality. *SPE Reservoir Evaluation & Engineering*, **5**, 375–383.

WHITICAR, M.J. & SNOWDON, L.R. 1999. Geochemical characterization of selected Western Canada oils by C_5–C_8 Compound Specific Isotope Correlation (CSIC). *Organic Geochemistry*, **30**, 1127–1161.

WILHELMS, A., LARTER, S.R. & HALL, K. 1994. A comparative study of the stable carbon isotopic composition of crude oil alkanes and associated crude oil asphaltene pyrolysate alkanes. *Organic Geochemistry*, **21**, 751–759.

WILKES, H., VIETH, A. & ELIAS, R. 2008. Constraints on the quantitative assessment of in-reservoir biodegradation using compound-specific stable carbon isotopes. *Organic Geochemistry*, **39**, 1215–1221.

YANG, H. & HUANG, Y. 2003. Preservation of lipid hydrogen isotope ratios in Miocene lacustrine sediments and plant fossils at Clarkia, northern Idaho, USA. *Organic Geochemistry*, **34**, 413–423.

YEH, H.-W. & EPSTEIN, S. 1981. Hydrogen and carbon isotopes of petroleum and related organic matter. *Geochimica et Cosmochimica Acta*, **45**, 753–762.

ZECH, M., PEDENTCHOUK, N., BUGGLE, B., LEIBER, K., KALBITZ, K., MARKOVIĆ, S.B. & GLASER, B. 2011. Effect of leaf litter degradation and seasonality on D/H isotope ratios of n-alkane biomarkers. *Geochimica et Cosmochimica Acta*, **75**, 4917–4928.

ZHANG, L., LI, M., WANG, Y., YIN, Z. & ZHANG, W. 2013. A novel molecular index for secondary oil migration distance. *Scientific Reports*, **3**, 2487.

ZHANG, X., GILLESPIE, A.L. & SESSIONS, A.L. 2009. Large D/H variations in bacterial lipids reflect central metabolic pathways. *Proceedings of the National Academy of Sciences*, **106**, 12580–12586.

Noble gases in conventional and unconventional petroleum systems

DAVID J. BYRNE[1]*, P. H. BARRY[1], M. LAWSON[2] & C. J. BALLENTINE[1]

[1]*Department of Earth Sciences, University of Oxford, South Parks Road, Oxford OX1 3AN, UK*

[2]*ExxonMobil Upstream Research Company, 22777 Springwoods Village Pkwy, Spring, TX 77389, USA*

**Correspondence: david.byrne@earth.ox.ac.uk*

Abstract: Petroleum systems represent complex multiphase subsurface environments. The properties of the noble gases as conservative physical tracers allow them to be used to gain insight into the physical behaviour occurring within hydrocarbon systems. This can be used to better understand the mechanisms of hydrocarbon migration, residence time of fluids, and measurement of the scale of the subsurface fluid system involved in the transport and trapping of the hydrocarbon phase. The noble gases in the subsurface derive from different sources with distinct isotopic compositions, allowing them to be resolved in any crustal fluid. We discuss the processes within petroleum systems that incorporate the noble gases from each of these sources into hydrocarbon accumulations. The dominant mechanism controlling the introduction of air-derived noble gases into petroleum systems is via subsurface groundwater, and this records key information about the interaction of the petroleum system with the hydrogeological regime. Radiogenic noble gases accumulate over time, recording information about the age and relative timing of processes within the petroleum system. We review the conceptual framework and quantitative models describing these processes using examples from previous studies, and discuss both their current limitations and the potential for their application to unconventional hydrocarbon systems.

Overview

The study of the origin and post-generation history of hydrocarbons in sedimentary basins has been the subject of thousands of investigations worldwide, both by the academic and industrial communities. Many studies have tried to better understand or characterize the elements or processes that occur within a given hydrocarbon system (e.g. Waples 1994; Hindle 1997). In a frontier basin, initial studies may focus on predicting whether a particular package of the Earth's crust has the necessary elements of sufficient quality and distribution to generate and store hydrocarbons over geological timescales (e.g. Blanc & Connan 1994; Whiticar 1994). In better studied basins, efforts are likely to focus on defining the robustness of critical hydrocarbon play elements. The application of geochemical techniques has been at the forefront of many of these studies by: (i) providing time and temperature constraints on basin evolution (Sweeney & Burnham 1990; Crowhurst *et al.* 2002; Stolper *et al.* 2014); (ii) determining the depositional environment, age, thermal maturity and type of organic matter of key source-rock intervals (Hughes *et al.* 1995); and (iii) linking hydrocarbons in reservoirs to the source rock responsible for their generation (Dow 1974; Stahl 1978; Philp 1993). However, the physical histories of accumulated hydrocarbons have proven to be more difficult to understand.

The noble gases have found strong application as physical tracers in a wide range of geochemical fields (Porcelli *et al.* 2002; Ozima & Podosek 2002; Burnard *et al.* 2013). Thanks to their chemical inertness, they are unaffected by biological activity, chemical alteration or redox reactions that complicate many other tracer systems. This means that only physical processes, such as mixing, dissolution, phase partitioning and diffusion, are recorded by the noble gases. This, coupled with the fact that the different components of the Earth's noble gas inventory (atmosphere, crust and mantle) are isotopically distinct and well constrained, means that the inputs into and the processes occurring within any given system can be accurately deduced. Given these properties, noble gas geochemistry has the theoretical potential to help constrain physical processes and timescales associated with hydrocarbon fluid generation, migration and storage in petroleum systems.

In a typical noble gas investigation of a petroleum system, samples are taken from the produced hydrocarbon and measured for their noble gas composition. From here, a range of different models can be applied to investigate different characteristics of the system, from gas, oil and water interactions to the timing of migration and accumulation. The

From: LAWSON, M., FORMOLO, M. J. & EILER, J. M. (eds) 2018. *From Source to Seep: Geochemical Applications in Hydrocarbon Systems*. Geological Society, London, Special Publications, **468**, 127–149.
First published online December 14, 2017, https://doi.org/10.1144/SP468.5
© 2018 The Author(s). Published by The Geological Society of London. All rights reserved.
For permissions: http://www.geolsoc.org.uk/permissions. Publishing disclaimer: www.geolsoc.org.uk/pub_ethics

development of these quantitative noble gas models has happened alongside improvements in analytical technology, allowing measurement of more noble gas isotopes with greater accuracy and precision, and from this an increased understanding of the processes affecting the noble gases within the subsurface. Consequently, noble gases now provide a powerful tool that can reveal information about the subsurface inaccessible by other analytical techniques. For complementary information, we refer readers to comprehensive reviews by Ballentine & Burnard (2002) and Holland & Gilfillan (2013) for detailed discussion.

Definitions

We use the term 'petroleum system' in the sense defined by Magoon & Dow (1994), and provide a schematic representation in Figure 1. It is considered to encompass a unit of active source rock, all related hydrocarbon accumulations, and the geological features and processes that are necessary for the hydrocarbon accumulations to exist. This comprises a series of elements and processes: elements include the source rock, migration pathway, reservoir and trap; whereas processes include hydrocarbon generation, migration and the geological processes affecting the system. As such, the petroleum system can cover a large geographical, stratigraphic and temporal range, and significant differences may exist between different systems.

We discuss both conventional and unconventional hydrocarbon systems. Conventional hydrocarbon systems comprise: (i) a source rock of sufficient quality and distribution that is buried to temperatures and pressures necessary to result in the generation and expulsion of hydrocarbons from the source rock; (ii) secondary migration of hydrocarbons along carrier beds to a reservoir rock in a configuration that traps hydrocarbons either stratigraphically or structurally; and (iii) an overlying low-permeability seal rock that results in the accumulation of

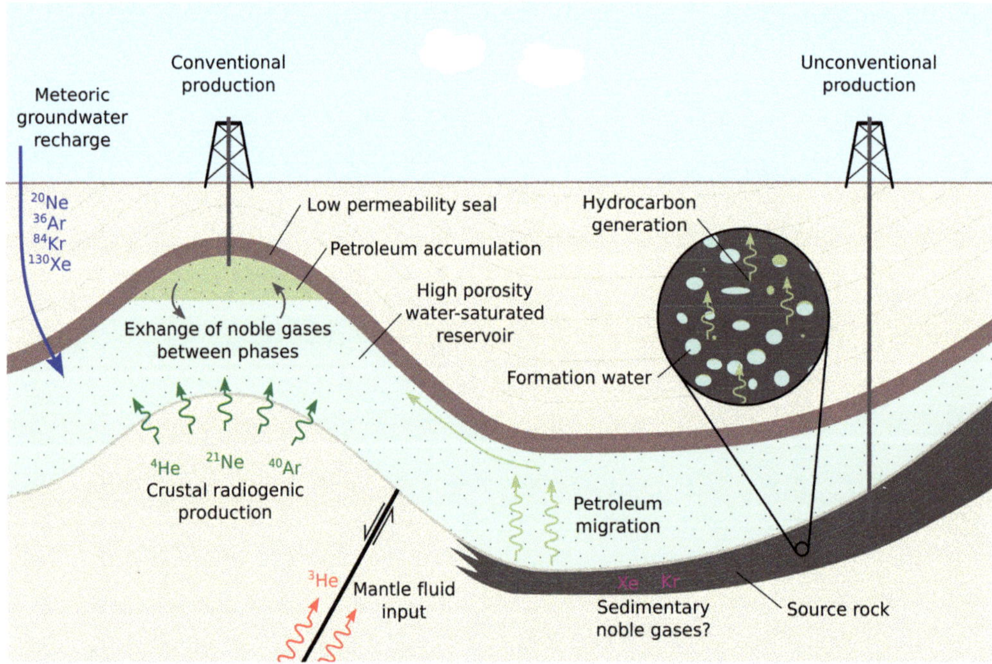

Fig. 1. A schematic conceptual model of the sources of noble gases in petroleum systems, modified from Ballentine & O'Nions (1992). For each input, the major characteristic isotopes are shown, although, in reality, each source will have varying contributions from a range of isotopes. Groundwater is usually considered to be ubiquitous in the subsurface, although the provenance of the water (meteoric recharge or formation water) and the connectivity of different sources can vary. Meteoric groundwater recharge has air-saturated water (ASW) composition and air-like isotopic ratios; the isotopes shown have no significant mantle or radiogenic sources, and so are considered characteristic of the subsurface noble gas component. Crustal radiogenic production occurs throughout the crust, although the produced isotopes can be transported via groundwater flow. Phase equilibrium between groundwater and any hydrocarbon accumulation is rapid on a geological timescale, and is the primary conduit for introducing noble gases into the hydrocarbon phase.

Table 1. *The volume mixing ratios for the noble gases in dry air*

Element	Volume mixing ratio
He	$5.24\ (\pm0.05) \times 10^{-6}$
Ne	$1.818\ (\pm0.004) \times 10^{-5}$
Ar	$9.34\ (\pm0.01) \times 10^{-3}$
Kr	$1.14\ (\pm0.01) \times 10^{-6}$
Xe	$8.7\ (\pm0.1) \times 10^{-8}$

Data from Porcelli *et al.* (2002).

hydrocarbons within the reservoir rock. Whilst the term 'unconventional' has been applied to petroleum exploitation environments ranging from tar sands to ultra-deep-water production, we use the term here in a stricter sense to apply to source-rock reservoirs such as gas shales (Curiale & Curtis 2016). These are traditionally unconventional in that they produce directly from the source rocks that generate the petroleum, with very little secondary migration. The low porosity and permeability of these rocks means that they typically require advanced production techniques, such as hydraulic fracturing or directional drilling, to produce economically. A schematic representation of conventional and unconventional production is shown in Figure 1.

Table 2. *The isotopic ratios of noble gases in air*

Isotope	Relative abundance
^{3}He	$1.399\ (\pm0.013) \times 10^{-6}$
^{4}He	$\equiv 1$
^{20}Ne	9.80 ± 0.08
^{21}Ne	0.0290 ± 0.0003
^{22}Ne	$\equiv 1$
^{36}Ar	$\equiv 1$
^{38}Ar	0.1885 ± 0.0003
^{40}Ar	298.56 ± 0.31
^{78}Kr	0.6087 ± 0.0020
^{80}Kr	3.9599 ± 0.002
^{82}Kr	20.217 ± 0.004
^{83}Kr	20.136 ± 0.021
^{84}Kr	$\equiv 100$
^{86}Kr	30.524 ± 0.025
^{124}Xe	2.337 ± 0.008
^{126}Xe	2.180 ± 0.011
^{128}Xe	47.15 ± 0.07
^{129}Xe	649.6 ± 0.9
^{130}Xe	$\equiv 100$
^{131}Xe	521.3 ± 0.8
^{132}Xe	660.7 ± 0.5
^{134}Xe	256.3 ± 0.4
^{136}Xe	217.6 ± 0.3

Data from Porcelli *et al.* (2002) and Lee *et al.* (2006).

Noble gas isotope ratio measurements are typically reported simply as raw ratios, unlike most other geochemical isotope ratios which use per mille (‰) notation (Hoefs 1997). Using the raw ratios is more convenient when relatively large isotopic variations are common (Porcelli *et al.* 2002). Although using raw ratios eliminates the requirement for a universal standard, samples in natural systems are often compared to air, which is well mixed and defined (Tables 1 & 2). In subsurface fluid systems, it is also common to use air-saturated water (ASW) as a standard, which is well defined over a range of temperatures and salinities (Table 3). As dissolution does not isotopically fractionate, isotope ratios in ASW are equal to those in air (Table 2). Helium isotope ratios (^{3}He/^{4}He) are usually reported relative to the air ratio (R_a). Quantities of noble gases are typically reported in cm^3 standard temperature and pressure (cm^3 STP or cc STP), due to the small amounts of noble gases usually present in geochemical systems. The cm^3 STP unit can be directly converted to moles using the molar gas volume ($22\,400\ cm^3$ at STP) (Burnard *et al.* 2013).

Origin of terrestrial noble gases

Primordial noble gases

Primordial elements are those indigenous to Earth, incorporated during planetary formation. The noble gases are highly volatile, and thus difficult to entrain during planetary accretion (Pepin 1991). This results in a bulk Earth that is depleted in noble gases by several orders of magnitude compared to the protosolar nebula from which the planet was formed (Wieler 2002; Grimberg *et al.* 2006). Some of the noble gas isotopes (e.g. ^{20}Ne, ^{36}Ar, ^{84}Kr, ^{130}Xe) are not produced on Earth in significant quantities,

Table 3. *The equilibrium concentrations of the noble gases in air-saturated water at atmospheric pressure for selected salinities and temperatures, calculated from Henry's constant equations in Fernández-Prini* et al. *(2003) and Setschenow coefficient equations in Smith & Kennedy (1983)*

Element	Concentration (cm^3 STP g^{-1})		
	Freshwater (10°C)	Freshwater (20°C)	Seawater (10°C)
He	4.73×10^{-8}	4.65×10^{-8}	4.04×10^{-8}
Ne	2.06×10^{-7}	1.93×10^{-7}	1.70×10^{-7}
Ar	3.91×10^{-4}	3.23×10^{-4}	3.11×10^{-4}
Kr	9.21×10^{-8}	7.27×10^{-8}	7.26×10^{-8}
Xe	1.34×10^{-8}	9.29×10^{-9}	1.04×10^{-8}

and thus their provenance in terrestrial reservoirs is assumed to be 100% primordial (Burnard *et al.* 1997; Moreira *et al.* 1998). The isotopic composition of possible accretionary precursors, such as the protosolar nebula and carbonaceous chondrites, have been compared with the terrestrial inventory of the primordial gases to reconstruct the origin and processes responsible for delivery of volatiles to Earth on its formation (e.g. Ballentine *et al.* 2005; Ballentine & Holland 2008; Holland *et al.* 2009; Marty 2012; Halliday 2013). These models of planetary noble gas acquisition successfully explain the modern terrestrial inventory for all of the primordial noble gases except Xe, whose apparent depletion in the Earth is still the subject of much debate (Sanloup *et al.* 2005; Lee & Steinle-Neumann 2006; Pepin & Porcelli 2006; Pujol *et al.* 2011).

Production of noble gases

Noble gases are produced both directly and indirectly from radioactive processes occurring within the Earth. The partitioning of different radioactive elements into chemically and physically distinct parts of the planet is the cause of much of the heterogeneity in the terrestrial noble gas reservoirs. The most productive radioactive elements in terms of abundance and decay rate (U, Th, K) are highly incompatible in the Earth's mantle, meaning that the crust is typically associated with much higher levels of noble gas production (Ballentine & Burnard 2002).

Noble gas production proceeds by three mechanisms: radiogenic, nucleogenic and fissiogenic. The laws governing the pathways and rates of radioactive decay are well understood, allowing theoretical production rates to be accurately calculated in any system if the parent isotope concentration and half-life (decay constant) is known (Rutherford 1906; Pierce *et al.* 1964). This is the basis of a number of successful dating tools such as Ar/Ar dating and ^{4}He dating (Merrihue & Turner 1966; Farley 2002; Renne *et al.* 2010).

Radiogenic production. Radiogenically produced isotopes are those produced as direct daughter products of radioactive decay. For the modern Earth, the most significant radioactive decay pathway is the U–Th series. The initial decay of ^{235}U, ^{238}U and ^{232}Th is by α-decay, significant in terms of noble gas production because an α-particle is equivalent to a ^{4}He nucleus. The vast majority of α-particles ionize the surrounding material upon ejection, collecting electrons to form a ^{4}He atom. Elevated U and Th concentrations are therefore associated with helium that is isotopically enriched in ^{4}He. The incompatibility of both U and Th in the mantle means that they, and consequently ^{4}He, are highly enriched in continental crust.

Similarly, high crustal concentrations of ^{40}K lead to elevated ^{40}Ar when ^{40}K decays by electron capture. The subsequent release of radiogenic ^{40}Ar into the atmosphere has drastically altered the isotopic composition of atmospheric argon over Earth history. Whilst the majority of primordial argon is ^{36}Ar (the proto-solar nebula has a ^{40}Ar/^{36}Ar value of *c.* 3.0×10^{-4}), the ^{40}Ar/^{36}Ar ratio in the modern atmosphere is 298.6 because of radiogenic ^{40}Ar enrichment (Göbel *et al.* 1978; Anders & Grevesse 1989; Lee *et al.* 2006).

Nucleogenic production. In addition to α-particles, the emission of neutrons is also a direct result of radioactivity. Nucleogenic production is the interaction of these particles with the nuclei of nearby atoms, leading to the formation of new elements in Wetherill reactions (Wetherill 1954).

Most commonly, the impact of an α-particle into a nucleus is followed by the emission of a neutron in an (α, n) reaction, effectively adding two protons and one neutron to the nucleus of the affected atom. Noble gases formed in measurable quantities in this fashion include ^{21}Ne, which is derived from the (α, n) reaction of ^{18}O (^{18}O$(\alpha, n)^{21}$Ne) as well as ^{22}Ne, which is formed indirectly via the β^+ decay of the short-lived ^{22}Na produced by the (α, n) reaction with ^{19}F (^{19}F$(\alpha, n)^{22}$Na$(\beta^+)^{22}$Ne).

Alternatively, an incident neutron can cause a nucleus to emit an α-particle, effectively removing two protons and one neutron in a (n, α) reaction. This pathway provides a further source of ^{21}Ne, produced by the ^{24}Mg$(\alpha, n)^{21}$Ne reaction, as well as a small contribution to ^{3}He, produced by the ^{6}Li $(\alpha, n)^3$He reaction.

As nucleogenic reactions require a source of α-particles and neutrons, their production rates are tightly coupled with nearby radioactivity, primarily U and Th. Both α-particles and neutrons have a limited range in typical crustal rock after being ejected from the parent nucleus, with typical ranges of 15–45 µm and 10–100 cm, respectively (Ziegler 1977; Martel *et al.* 1990). As such, heterogeneities on the mineral-scale can have measurable effects on the relative production of different isotopes. Notably, observed radiogenic ^{21}Ne/^{22}Ne values from the modern crust are incompatible with those predicted from average crust ^{18}O and ^{19}F concentrations. This has led to the inference that the O/F concentration in the modern crust is systematically lower in phases associated with high U and Th concentrations (Kennedy *et al.* 1990; Hiyagon & Kennedy 1992).

Fissiogenic production. A further source of noble gas production is as a result of the fission of heavy, unstable nuclei. The systematics of this process are more complex than simple radioactive decay, but the production rates of different isotopes can still be

Table 4. *Present-day production rates for the radiogenic noble gases in typical continental crust*

Isotope	Present-day production rate (cm^3 STP kg^{-1} a^{-1})
^{3}He	2.49×10^{-18}
^{4}He	3.31×10^{-10}
^{20}Ne	1.47×10^{-18}
^{21}Ne	1.51×10^{-17}
^{22}Ne	3.03×10^{-17}
^{36}Ar	2.38×10^{-18}
^{38}Ar	1.09×10^{-18}
^{40}Ar	6.05×10^{-11}
^{83}Kr	5.86×10^{-21}
^{84}Kr	2.12×10^{-20}
^{86}Kr	1.39×10^{-19}
^{131}Xe	7.89×10^{-20}
^{132}Xe	5.24×10^{-19}
^{134}Xe	7.50×10^{-19}
^{136}Xe	9.09×10^{-19}

Data from Ballentine & Burnard (2002).

relatively well constrained (Wieler & Eikenberg 1999). The dominant fission process in noble gas production is the spontaneous fission of ^{238}U, which produces ^{129,131,132,134,136}Xe, as well as smaller amounts of ^{83,84,86}Kr. The spontaneous fission of ^{232}Th, and the neutron-induced fission of ^{238}U, ^{235}U and ^{232}Th, also make minor contributions to the production of these fissiogenic Xe and Kr isotopes (Ballentine & Burnard 2002).

The production rates of fissiogenic isotopes are much lower than those of radiogenic or nucleogenic isotopes (Table 4). As such, fissiogenic Xe excesses are only observed in samples that have been isolated for significant periods of time (Reynolds 1963; Holland *et al.* 2013).

Physical chemistry of noble gases in fluids

Henry's law

The dissolution of noble gases into fluids (water, oil and gas) has been reviewed by Ballentine *et al.* (2002). Dissolution follows Henry's law, which states that the concentration of a gas in solution is directly proportional to the partial pressure of that gas in the gas phase. This can be formulated for any noble gas i, assuming ideal gas behaviour in both the gas and fluid phases:

$$p_i = K_i x_i \qquad (1)$$

where p_i is the partial pressure, x_i is the mole fraction in solution and K_i is Henry's constant. Henry's constants are specific to the solute, and are temperature and salinity dependent; their units depend on the units used to measure the partial pressure and concentration in the fluid phase.

Henry's constants of noble gases in water have been determined empirically over a temperature range from the freezing point to the critical point of water (Crovetto *et al.* 1982; Smith 1985), and equations have been formulated using a compilation of empirical data that allow the calculation of Henry's constants at any temperature (Fernández-Prini *et al.* 2003). Determination of Henry's constant for oil faces the further complexity of the natural variation in oil composition. This work is limited to just one study of two oils with API gravities of 25° and 31° (Kharaka & Specht 1988). The solubility of the noble gases was shown to be affected by the API gravity, but it is likely that the solubility is also controlled by properties other than bulk density (e.g. the concentration of polar compounds, trace element concentrations). Hence, Henry's constant for oils remains a key uncertainty in models that apply these constraints.

Non-ideality

Henry's law can be modified to account for non-ideal behaviour in the gas and fluid phases by taking into account both the gas-phase fugacity coefficient and the liquid-phase activity coefficient:

$$\Phi_i p_i = \gamma_i K_i x_i \qquad (2)$$

where Φ_i is the fugacity coefficient and γ_i is the activity coefficient; a deviation of either of these from unity represents non-ideal behaviour. Fugacity is pressure and temperature dependent, and can be calculated from empirical measurements of real molar volume (Dymond & Smith 1980). Activity is dependent on temperature and salinity, with the relationship formulated in the Setschenow equation:

$$\gamma_i = e^{C k_i(T)} \qquad (3)$$

where C is the concentration of salt in the solution and k_i is the temperature-dependent Setschenow coefficient. The response of Ar to changes in salinity is independent of the electrolyte species, and this relationship is assumed to be true for the other noble gases (Ben-Naim & Egel-Thal 1965). The Setschenow coefficients have been determined empirically for the noble gases over a range of temperatures from 270 to 340 K (Smith & Kennedy 1983). Unfortunately, the temperatures encountered in subsurface environments often greatly exceed this range, with the onset of oil formation occurring at temperatures of approximately 350 K, and secondary cracking to gas being above 420 K (Waples

1980). The curves fitted to the empirical data must therefore be extrapolated into the temperature range of interest in order to estimate the Setschenow coefficient associated with saline water in most hydrocarbon systems. This represents a potential source of uncertainty in any derivative product. Recent work on noble gas partitioning in CO_2–water systems show measurable deviations from predicted behaviours in high-pressure environments (Warr *et al.* 2015). In most petroleum system investigations, these deviations from ideality are unlikely to affect the general outcomes of any models, but care must be taken when considering the precision with which they can be applied. Further work involving a combination of empirical and model data will be crucial in refining these models in the future.

Terrestrial noble gas inventories

The atmosphere

Although small in size compared to the solid Earth, the majority of the Earth's primordial noble gases are held in the atmosphere (Porcelli & Ballentine 2002). The creation and evolution of the atmosphere is complex, and not yet fully understood. Whilst it is likely that terrestrial bodies, such as the Earth, are able to gravitationally capture volatiles from the proto-solar nebula during formation (Pepin 2006), the atmosphere is clearly distinct from a solar composition (Brown 1949; Porcelli & Ballentine 2002). Alternative proposed methods of acquiring noble gases include adsorption onto accumulating dust (Marrocchi *et al.* 2005) or the implantation by solar winds (Podosek *et al.* 2000). A contribution of volatile-rich cometary material brought in during the late heavy bombardment has also been suggested as a source of atmospheric noble gases (Owen *et al.* 1992; Dauphas 2003; Holland *et al.* 2009).

The reasons for the atmosphere's distinct noble gas composition are likely to be due to the atmospheric losses experienced by the Earth during its early history. Whether this occurred by aggregate catastrophic loss, or by the gradual escape of volatiles, is still debated (Pepin 2006; Pepin & Porcelli 2006; Tucker & Mukhopadhyay 2014). However, the atmospheric composition of noble gases has been remarkably constant over geological timescales. Whilst degassing of the solid Earth at volcanoes and mid-ocean ridges releases mantle noble gases into the atmosphere, it is in such small quantities as to be insignificant even over billions of years. However, exceptions to this include isotopes with high radiogenic production, such as ^{40}Ar, which are produced within the Earth to such an extent that solid Earth degassing can influence atmospheric concentrations (Allègre *et al.* 1996; Ballentine & Burnard 2002). Helium is gravitationally unbound in the atmosphere. However, the loss into space is balanced by a volcanic degassing flux from mid-ocean ridge, ocean-island and subduction-related volcanism, resulting in an assumed steady-state helium concentration in the atmosphere (Torgersen 1989; Lupton & Evans 2013). When considering most fluids in the crust, and groundwater in particular, it is reasonable to assume that the atmospheric noble gas composition has been constant over the timescale of their introduction into the subsurface.

Water

When in direct contact with the atmosphere, water will rapidly equilibrate with air according to the processes detailed in the previous subsection. The solubilities of the noble gases vary by element; meaning that at equilibrium, the elemental composition of the water is not air-like (Kipfer *et al.* 2002). However, isotopic ratios of the individual noble gases are not significantly affected by equilibrium partitioning. As the rules governing this partitioning and the variation with respect to temperature, pressure and salinity are well known, it is possible to calculate the air-saturated water composition in any given environment. This composition is crucial for noble gas analysis of hydrocarbon reservoirs, as a common assumption in modelling and analysis is that all atmospherically derived noble gases in the subsurface will be delivered in known quantities dissolved as air-saturated water (ASW) through hydrocarbon–water interactions and noble gas partitioning.

One complication to this arises from empirical measurements of atmosphere-derived noble gas concentrations in meteoric groundwater that have often been found to be higher than the values predicted by Henry's law equilibration. This phenomenon is commonly referred to as 'excess air' (Herzberg & Mazor 1979; Heaton & Vogel 1981). The behaviour of this excess air component is generally consistent, with neon being the most strongly affected, and the heavier noble gases proportionally less so (Heaton & Vogel 1981). Theoretical models describing processes by which this can occur have been developed and refined over time to more closely match observations (Stute *et al.* 1995; Ballentine & Hall 1999; Aeschbach-Hertig *et al.* 2000). Whilst these models differ slightly, they are based on the principle that excess air is introduced into groundwater by the entrapment of air bubbles during fluctuations of the water table. A close fit to experimental observations is found by modelling a closed-system equilibration of these bubbles (Aeschbach-Hertig *et al.* 2008). Within petroleum systems deposited under marine conditions, the groundwater is likely to have a marine noble gas composition (Kipfer *et al.* 2002) and excess air within these systems is not expected.

The mantle

The Earth's mantle contains a high proportion of primordial noble gases. It is introduced to the surface in CO_2-rich volcanic gases that can permeate through the crust or get released at mid-ocean ridges and volcanoes. The measured compositions of these gases have distinctive high, but variable, $^3He/^4He$ ratios. For example, the well-mixed asthenospheric upper mantle, as sampled by depleted mid-ocean ridge basalt (MORB) mantle, ranges from 7 to 9 R_a (Graham 2002), where the subcontinental lithospheric mantle is defined as 6.1 ± 2.1 (Day et al. 2015). In contrast, many mantle plume regions extend as high as c. 50 R_a (e.g. Stuart et al. 2003). Likewise, the mantle is marked by high $^{20}Ne/^{22}Ne$ and $^{40}Ar/^{36}Ar$ ratios (Sarda et al. 1988; Staudacher et al. 1989; Holland et al. 2009). The addition of mantle noble gases into subsurface hydrocarbon accumulations, while not useful in the context of origin and history of hydrocarbon gases, may be useful for constraining the origin of non-hydrocarbon gases that in some cases accumulate concurrently in the subsurface – particularly CO_2 and N_2 (Ballentine & Sherwood Lollar 2002; Ballentine et al. 2005; Gilfillan et al. 2009).

The crust

Due to their incompatibility in the mantle, the major radioactive elements involved in noble gas production were partitioned heavily into the Earth's crust during planetary formation. These elements, which include U, Th and K, initiate the decay chains that produce the radiogenic noble gas isotopes, principally ^4He, ^{40}Ar, ^{21}Ne, ^{22}Ne, and, to a lesser extent, some Xe and Kr isotopes. Whilst the crust is a complex and heterogeneous system with inputs from many sources, crustal fluids are often characterized by high concentrations of these radiogenic isotopes.

These isotopes are produced within the minerals that make up the crust, but can be released into the surrounding fluid systems (Bach et al. 1999). The extent of this release is controlled by a number of factors including grain size, temperature and mineral alteration (Honda et al. 1982; Brooker et al. 1998; Baxter et al. 2002). Lighter isotopes are more readily released at lower temperatures and, as such, the ratios of radiogenic isotopes (e.g. $^4He/^{21}Ne^*$, $^4He/^{40}Ar^*$) can be used to determine the temperatures of release (Torgersen et al. 1992; Baxter et al. 2002).

In general, higher concentrations of radiogenically produced noble gas isotopes reflect older systems that have been isolated for longer periods of time. Data from fracture fluids in mines in Precambrian shields show concentrations that would have taken billions of years to accumulate (Holland et al. 2013).

Introduction of noble gases into the petroleum system

Source-rock formation

The genesis of the petroleum system begins with the deposition of an organic-carbon-rich sedimentary rock. Typically, >1% total organic carbon (TOC) is considered sufficient for a source rock (Gluyas & Swarbrick 2013), although this can vary. As organic carbon comes in a variety of forms, each of which has a distinct chemical behaviour in a petroleum system, these source rocks are typically classified according to their kerogen composition, as outlined in Table 5. The type of organic matter that makes up source rocks is of particular importance because it is this that ultimately determines how oil or gas prone a source rock is likely to be, as well as the volume of hydrocarbons that can be generated from the source rock on a gram for gram basis (van Krevelen 1961; Dow 1977; Peters & Cassa 1994). Laboratory simulation of this naturally occurring process has been extensively investigated in hydrous pyrolysis experiments, in which immature source rock is heated under controlled conditions to simulate the naturally occurring process of petroleum formation (Peters 1986; Peters et al. 2015). These and other experimental studies suggest that the presence of water may be required during catagenesis and hydrocarbon generation to produce the compositions that we observe in nature (Lewan 1993). Noble gases are more soluble in oil than in water, and thus the formation of a liquid petroleum phase within the

Table 5. *The classification of source rocks*

Kerogen type	Primary organic matter source	Depositional environment	Principal generated hydrocarbons
Type I (Sapropelic)	Algae	Lacustrine	Oil
Type II (Planktonic)	Plankton	Marine	Oil + Gas
Type III (Humic)	Plants	Terrestrial	Gas (+ Coal)

After Gluyas & Swarbrick (2013).

source rock will cause the partitioning of noble gases between it and the surrounding water phase that remains in the rock (Ballentine *et al.* 1991). The water in this context is likely to not represent a significant volume given the prior compaction and lithification of the source rock. Given sufficient pressure generated during the generation of hydrocarbons, oil may escape from the source rock and, as such, noble gases in the escaping fluid will evolve separately (described below) to the noble gas signature of any retained oil. The extent of kerogen decomposition can further affect the partitioning of noble gases, as the solubility of noble gases has been shown to be positively correlated with the API of the oil (Kharaka & Specht 1988). Given continued burial and increase in temperature, the retained oil itself will begin to crack to gaseous hydrocarbons. The onset of gas generation will further complicate this phase partitioning, as the noble gases have a strong affinity for the gas phase when it is present. Furthermore, it is unclear how the alteration of the source-rock composition will affect the noble gases stored within; it is possible that any component adsorbed onto the organic carbon could be released during thermal degradation.

Most noble gas studies to date have focused on characterizing or modelling the inheritance and evolution of noble gases during migration and accumulation in conventional reservoirs. However, key gaps exist in our understanding of hydrocarbon generation and expulsion from source rocks, and there have been very few studies to date that have studied the evolution of noble gases during hydrocarbon generation. Questions such as the role of water during hydrocarbon generation, the relative volumes of expelled v. retained oil and gas, and the mechanisms of storage within the source rocks (adsorption or as free gas within porosity or fractures) remain elusive.

Primary migration

Primary migration is the initial expulsion of the generated hydrocarbons out of the source rock and into the surrounding crust. This occurs by thermally activated diffusion of hydrocarbons through the residual organic-matter network (Stainforth & Reinders 1990). The rate of primary migration is an important control on the total hydrocarbons produced from the source rock, as well as the extent of secondary cracking. The effects of primary migration on the noble gas composition of the hydrocarbon phase are unclear. Although a thermal diffusive release mechanism may impart a mass-dependent kinetic fractionation signature onto the noble gases, it has not been identified in real-world samples. It is possible that the effect is small compared to that of other subsurface processes. Such processes include the interaction of noble gases co-transported in or with hydrocarbons with other fluids or the mechanisms and timescales associated with the movement of hydrocarbons within the crust.

Secondary migration

The principal mechanism driving secondary migration is the buoyancy force due to the density difference between petroleum and the surrounding water. This driving force is balanced by the resistance provided from capillary entry pressure, a function of the surrounding pore size and permeability (Schowalter 1979). Empirical investigations into this process suggest that secondary migration is likely to occur along restricted pathways, often along structural boundaries, and proceeds rapidly on geological timescales (Dembicki & Anderson 1989). Figure 1 shows the migration of hydrocarbons through a groundwater-saturated reservoir, and the potential inputs of noble gas sources into the same groundwater. Once in contact, the noble gases are expected to equilibrate between the two phases quickly relative to geological timescales (Ballentine & Burnard 2002).

Secondary migration can occur laterally for up to hundreds of kilometres and vertically through kilometres of groundwater-saturated strata (Demaison & Huizinga 1991), and have profound effects on the noble gases in the system. The migrating hydrocarbons can undergo phase partitioning with this surrounding groundwater and, as the noble gases are more soluble in hydrocarbons than in water, this can strip the groundwater of much of its noble gases (Ballentine *et al.* 1996). This occurs via well-defined Henry's law solubility-dependent partitioning, and forms the basis of a series of models that investigate hydrocarbon migration using the varying solubilities of the noble gases (Zartman *et al.* 1961; Bosch & Mazor 1988; Ballentine *et al.* 1991, 1996). While this has yet to be discussed or tested in studies performed to date, the evolution of noble gas signatures in hydrocarbons in conventional systems is therefore likely to be dependent on the mechanism and length scales of migration. In systems where migration occurs short over distances, it is possible that hydrocarbons may encounter relatively limited volumes of water with which to interact and partition noble gases (e.g. Ballentine *et al.* 1996). In contrast, migration within hydrocarbon systems, such as those of the foreland basins of the Western Canadian Sedimentary Basin, may occur over tens or hundreds of kilometres and, as such, will encounter significantly greater volumes of water. However, to date, there has been no thorough investigation into the effects of secondary migration distance upon noble gas composition – if any such pattern exists, it could prove an invaluable

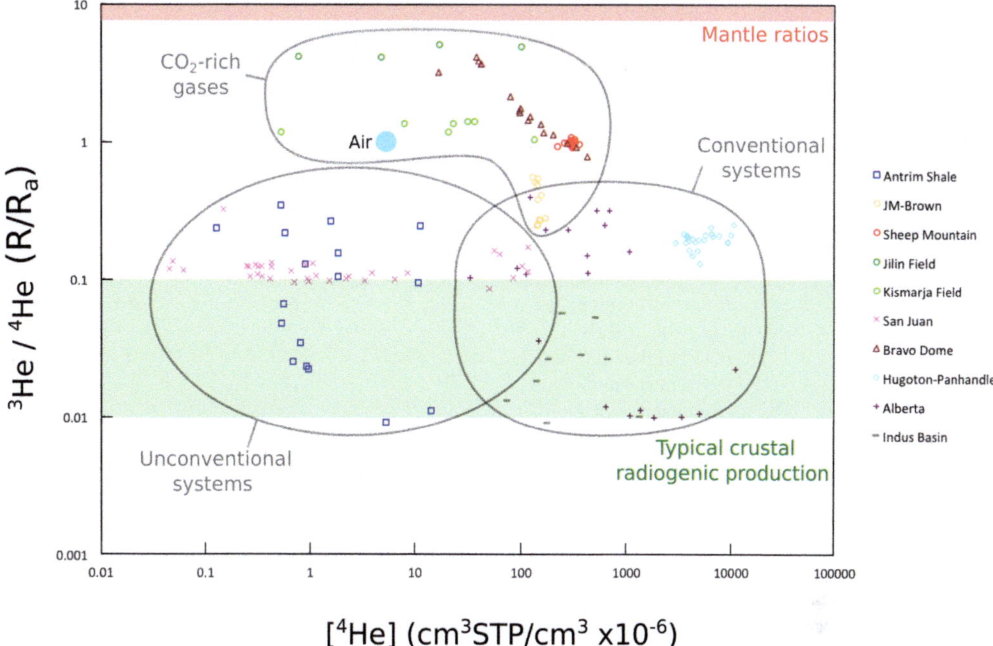

Fig. 2. Helium isotope ratio and concentration measurements from selected noble gas studies of petroleum systems. ^{3}He/^{4}He ratios are reported relative to the atmospheric ratio R_a. The observed range of ^{3}He/^{4}He ratios and ^{4}He concentrations is controlled by the regional geology and tectonic regime of the petroleum systems. The ^{3}He/^{4}He ratio of a system is usually thought to comprise 'two end-member' mixing between crustal fluids with a ratio of 0.01–0.02, and mantle fluids, which have a ratio of >8. The presence of even small contributions of mantle fluids therefore increases ^{3}He/^{4}He ratios above normal crustal values, and is a feature of extensional tectonic regimes (Marty *et al.* 1993). Concentrations of ^{4}He are affected by several factors: radiogenic production of ^{4}He occurs over time, meaning that older systems will have accumulated more ^{4}He; the regional hydrogeology can also transport ^{4}He dissolved in groundwater into the petroleum accumulations; finally, concentrations of all noble gases will be affected by the extent of secondary cracking of hydrocarbons, as this effectively increases the relative volume of hydrocarbon gases in the sample and dilutes the helium concentration. As is evident from the data, different systems show distinct ratios and concentrations, and different amounts of scatter show variations in homogeneity. Mixing lines are apparent in some systems, such as the mantle CO_2-rich Bravo Dome; others show little or no pattern. Data are broadly clustered according to the type of the system, conventional or unconventional, and are CO_2-rich. Data sources: Antrim shale (Wen *et al.* 2015); San Juan coalbed methane (Zhou *et al.* 2005); JM-Brown, Sheep Mountain, Jilin and Kismarja fields (Gilfillan *et al.* 2009); Alberta Field (Hiyagon & Kennedy 1992); Indus Basin (Battani *et al.* 2000); Hugoton–Panhandle Field (Ballentine & Sherwood Lollar 2002); Bravo Dome (Ballentine *et al.* 2005).

tool in retroactively assessing migration distances and pathways.

The general effects of secondary migration on noble gases can be considered by comparing the conventional systems with unconventional systems where no secondary migration takes place. Figure 2 shows an overlapping but distinct grouping of conventional and unconventional systems with respect to [^{4}He] concentrations. Conventional systems tend to have higher [^{4}He] concentrations, which we speculate could be due to interaction with large-scale aquifer systems during secondary migration. Previous studies of conventional systems have shown some to contain concentrations of radiogenic noble gases far higher than that possible by local production, requiring input from regional systems (Ballentine *et al.* 1991, 1996).

Accumulation

The final stage in a conventional petroleum system is the migration of the hydrocarbons into a suitable structural or stratigraphic trap, with a low-porosity seal allowing an accumulation to form (Downey 1984). Once in an accumulation, there are still several processes that can allow further alteration of the noble gas composition. The structure and fluid-flow pathways that focused the hydrocarbon migration into the trap are likely to allow fluids from other sources to also arrive at the same location. It is therefore not unusual to see mixing and interaction with CO_2-rich mantle-derived fluids, especially in tectonically active areas (Hooker *et al.* 1985; Gilfillan *et al.* 2008, 2009). These mantle fluids have a distinct elevated $^3He/^4He$ ratio, and so their presence in the hydrocarbon accumulation is often, although not always, easily identified. Continued subsurface flow of and interaction with groundwater can also transport noble gases from other parts of the crust. In many cases, this can lead to high concentrations of crustal radiogenic noble gases that require contributions from a large fetch area or catchment (Ballentine *et al.* 1991, 1996; Ballentine & Sherwood Lollar 2002).

Tertiary migration

Once in an accumulation, it is possible for the hydrocarbons to leak gradually through the seal or through mechanical structures such as faults (Downey 1984; Wiprut & Zoback 2000). Evidence for this is apparent in the many natural hydrocarbon seeps found both onshore and on the seafloor (Bojesen-Koefoed *et al.* 1999; Hornafius *et al.* 1999; Holzner *et al.* 2008). It is possible to identify this loss of hydrocarbons within the noble gas signature using a mass-balance approach if the input of noble gases into the accumulation is well constrained. This is particularly notable in systems that are filled to structural spill. A recent study by Barry *et al.* (2016) demonstrated the sensitivity of noble gases to constraining the relative hydrocarbon volumes lost from the Sleipner Vest reservoirs of the Norwegian North Sea through such a process.

Unconventional systems

As the widespread commercial exploitation of unconventional systems is a relatively recent phenomenon, there is a comparatively small compilation of literature describing the geochemical behaviour of these systems (Curtis 2002). As source-rock reservoirs typically have low porosity and hydrocarbons produced from these systems have not undergone any prior migration, we might expect the noble gas signature to be indigenous to the source rock. However, it is clear that these are not perfectly closed systems, as many source-rock reservoirs have generated hydrocarbons that have migrated to form conventional systems (Curiale & Curtis 2016).

The behaviour of noble gases within unconventional systems is therefore likely to be similar to that of noble gases within source rocks, barring any drilling or production-related effects. The lack of secondary migration should simplify the factors affecting noble gases composition, as unknowns about the timing and rate of migration can be eliminated.

Analysis of the petroleum system using noble gases

Early noble gas investigations

The presence of significant amounts of helium in natural gases was first observed over 100 years ago (Cady & McFarland 1906). The first isotopic measurements of noble gases were made nearly 50 years later and established, using $^4He/^{40}Ar$ ratios, that the presence of high concentrations of radiogenic ^4He in natural gases was not due to increased local ^4He production, but due to fluid migration transporting He from off-structure (Zartman *et al.* 1961). Wasserburg *et al.* (1963) also used measurements of natural gases to make early estimates of the terrestrial ^4He degassing flux through the crust and, hence, the residence time in the atmosphere. These pioneering studies highlighted the potential utility of noble gases to study the origin and history of hydrocarbon and non-hydrocarbon fluids in sedimentary basins, and formed the foundations for future studies targeting multiple elements and processes of numerous hydrocarbon systems.

The preparation of samples for noble gas analysis requires specialized equipment and techniques compared to sample collection for traditional hydrocarbon geochemical analyses, mostly to avoid air contamination. Concentrations of noble gas species in air are significantly higher than those of typical crustal samples (especially Ar and Ne), and so even very small air admixtures can have a large impact on measured values even when present only at the ppm level (Barry *et al.* 2016). Furthermore, the small size of the noble gas atoms means that they have a much higher propensity to slowly leak across seals that are sufficient to hold hydrocarbon compounds. Valved high-pressure cylinders are not considered to be effective at maintaining the sample integrity required for noble gas analysis over long periods of time. Typically, noble gas hydrocarbon samples are collected at low pressure in refrigeration-grade copper tubes that are either crimped or clamped to make a helium-tight seal effective during transport and storage.

End-member mixing models

Noble gases are introduced into the petroleum system from several distinct geochemical reservoirs, each with a distinct elemental and isotopic composition. Deconvolving the contribution of different end-member compositions is a key step in beginning to understand the subsurface environment of any petroleum system, and the composition of any accumulations. The detection of mantle-rich fluids can be used to gain insight into the geological history and connectivity of any accumulation. Figure 2 shows that He isotope data reveal a grouping of CO_2-rich petroleum accumulations that are associated with elevated $^3He/^4He$ isotope ratios, in which this CO_2 is postulated to have been derived from magmatic fluids passing through the petroleum system (Gilfillan *et al.* 2009; Prinzhofer 2013).

For a robust analysis of noble gas provenance by end-member mixing decomposition, it is first crucial that the end members themselves are well defined. For some systems, such as the atmosphere, this is certainly the case; however, others can be more complicated. The elemental composition of air-saturated groundwater is dependent on the temperature and salinity of recharge conditions, and these are often difficult to constrain. However, the isotopic composition of each element is unaffected by solubility partitioning, and, for He and Ne, their solubilities are similar enough that they can be considered equivalent in certain situations. Radiogenic production ratios are well known for average crust, although local parent-isotope concentrations can cause significant deviations in some isotope ratios (Ballentine & Burnard 2002). For the mantle, isotope ratios (e.g. $^3He/^4He$) are reasonably well constrained for different mantle compositions (e.g. MORB, ocean island basalt (OIB)) and the fact that these are orders of magnitude different to crustal end members makes the uncertainty in them less significant.

Processes that can isotopically or elementally fractionate noble gases, such as those detailed in the following subsections, must be identified and taken into account. For this reason, it is preferable to use isotopic ratios of the same element rather than elemental ratios when calculating mixing fractions, as they are less affected by solubility or mass-dependent fractionation. As such, mantle contributions to a petroleum system are usually deduced from $^3He/^4He$ ratios (equation 4), and radiogenic contributions from $^{20}Ne/^{22}Ne$ and $^{21}Ne/^{22}Ne$ ratios:

$$R_{sample} = f_{mantle}R_{mantle} + f_{radiogenic}R_{radiogenic} \quad (4)$$

This is an example equation for deconvolving mantle helium contributions, where R denotes a $^3He/^4He$ ratio, and f is the fractional contribution to the measured sample.

A further important use for calculations is the identification of any air contamination in a sample, as even ppm levels of air within a sample can impact measurements. Due to the varying concentrations of the noble gases in typical crustal samples and air, certain isotopes are more sensitive to air contamination. For example, the $^4He/^{20}Ne$ ratio is often used to identify atmospheric He contamination, as all of the ^{20}Ne is assumed to be originally atmospheric in origin. A $^4He/^{20}Ne$ close to the air value of 0.288 suggests atmospheric contamination, as crustal

samples typically have values many orders of magnitude higher (Ballentine *et al.* 1996). However, as Ar is much more abundant in air than He, it takes a smaller amount of air contamination to affect the $^{40}Ar/^{36}Ar$ ratios.

Hydrocarbon–water interaction models

Water is a key component in all petroleum systems. It is present in the sediments during burial (known as 'connate' or 'formation water'); however, groundwater can also migrate through the system depending on the permeability and structural configuration of reservoir lithologies or carrier beds. Understanding the interactions between hydrocarbon and water phases can help give an insight into the history of the formation and migration of the hydrocarbons themselves, as well as their present-day water contact. The present-day water contact may be particularly important in providing pressure support for the production of hydrocarbons at wells, and may have implications for water breakthrough (Toth 1980).

The noble gases have different solubilities in fluids, with the heavier noble gases typically more soluble. This phenomenon has been demonstrated empirically in noble gas investigations of groundwaters, where Xe and Kr are enriched relative to their atmospheric abundances (Mazor 1972). Theoretically, it should be possible to predict the partitioning behaviour for the noble gases between any fluid phases in the subsurface, such as water and oil or gas. This has led to the development of models describing the partitioning of the noble gases between water and oil/gas phases in the subsurface.

These models make a series of assumptions. First, they assume that the hydrocarbon phase is initially devoid of noble gases and only inherits an atmospheric noble gas inventory through interaction with air-saturated water (ASW) in the subsurface post-expulsion from the source rock (i.e. hydrocarbons in source rocks do not contain any noble gases). These models also typically assume that an initial composition of atmosphere-derived noble gases is present in groundwater, which is well constrained over a range of temperatures and salinities (Kipfer *et al.* 2002). Isotopes for each gas are chosen which have no significant radiogenic production in the subsurface (^{20}Ne, ^{36}Ar, ^{84}Kr, ^{130}Xe), so that the total amount of each isotope present in the system is constant. These isotopes have previously been used as tracers in the investigation of groundwater circulation patterns (Castro *et al.* 1998*a, b*). Helium is typically not considered in these models as ^4He is produced radiogenically in large amounts, and atmosphere-derived ^3He in the subsurface is negligible. Furthermore, the solubilities of He and Ne are often indistinguishable at crustal pressures and temperatures, meaning that little new information would be gained regardless. In systems with negligible fissiogenic Xe, ^{132}Xe is often used instead of ^{130}Xe, as it is present in higher quantities at atmospheric ratios. The result from these assumptions is that measurement of the atmosphere-derived noble gas isotopes in one phase of the subsurface system allows the volume ratios and partitioning behaviour of the entire system to be reconstructed.

The fundamentals of this approach were originally laid out by Bosch & Mazor (1988), who used the isotope ratios of atmosphere-derived Ne/Ar, Kr/Ar and Xe/Ar to predict partitioning patterns in water–oil and water–gas systems. Ballentine *et al.* (1996) used atmosphere-derived ^{20}Ne/^{36}Ar ratios measured in the North Sea Magnus oil field to quantitatively estimate oil/water volume ratios (V_o/V_w). The technique is similarly applicable to predict gas/water volume ratios (V_g/V_w) in gas-dominated petroleum systems, as shown in the following equation:

$$\frac{V_g}{V_w} = \frac{\left(\dfrac{^{20}\text{Ne}}{^{36}\text{Ar}}\right)_{\text{ASW}}}{\left(\dfrac{^{20}\text{Ne}}{^{36}\text{Ar}}\right)_g} - \frac{K^{\text{Ar}}}{K^{\text{Ne}}} \qquad (5)$$

where K^i is Henry's coefficient of noble gas i in water at reservoir conditions, and the subscript ASW refers to the air-saturated water composition of groundwater recharge. For oil–water systems, V_o/V_w is calculated analogously to the above equation, but using the ratio of Henry's coefficients of oil and water, instead of simply water.

This concept was modified by Zaikowski & Spangler (1990), who used Ne/Ar ratios combined with absolute ^{36}Ar concentrations to predict the evolution of groundwater in contact with varying gas volumes. Using absolute concentrations has the added complexity of uncertainties that arise from the conversion of concentration at STP to reservoir temperature and pressure than using ratios, but can provide an additional constraint on water/gas ratio volumes. Furthermore, at gas/water ratios of over *c.* 0.01, effectively 100% of the noble gases are partitioned into the gas phase, making ratios insensitive to changes in V_g/V_w. Concentrations, however, will still be diluted by the addition of more hydrocarbons, making them more effective at determining V_g/V_w in these scenarios. The formulation for calculating V_g/V_w using concentrations is as follows:

$$\frac{V_g}{V_w} = \frac{C^i_{\text{ASW}}}{C^i_g} - \frac{1}{K_i} \qquad (6)$$

where C is the concentration of the noble gas i in a particular phase, and K_i is Henry's coefficient

in water at reservoir conditions. Units for Henry's coefficient must be chosen to complement the units used for concentrations in the water and gas phase. For oil–water systems, again the ratio of Henry's coefficients is used.

The models and equations described thus far are applicable to a simple two-phase static closed system. In reality, petroleum systems are often more complex, exhibiting open-system behaviour in the form of reservoir leakage, gradual noble gas stripping by previous migrating hydrocarbons and dual hydrocarbon-phase accumulations (e.g. reservoired oil with a gas cap). These more complex scenarios can be accounted for by using a mass balance for hydrocarbon loss and/or partitioning, or Rayleigh fractionation to describe gradual processes (Zhou *et al.* 2005). The context of the specific system will dictate which model is most appropriate. For example, Barry *et al.* (2016) used compound models to describe volume ratio interactions in both open and closed systems from the Sleipner Field, North Sea.

A graphical representation of this approach is shown in Figure 3, illustrating the different distributions of atmosphere-derived noble gases for both gas–water and oil–water systems. Examples of data from previous studies are shown in Figure 4, which shows the observed ranges of ^{20}Ne/^{36}Ar from real-world gas and oil systems. It is clear that many systems have ^{84}Kr/^{36}Ar and ^{130}Xe/^{36}Ar in excess of what would be predicted from their ^{20}Ne/^{36}Ar ratios. This 'excess' Kr and Xe is commonly observed in petroleum phases. The origin of this excess is not well understood, although it is thought that it may originate from a sorbed component within source-rocks (Zhou *et al.* 2005). As a result, ^{84}Kr/^{36}Ar and ^{130}Xe/^{36}Ar are not typically used for quantitative volume ratio calculations.

The practical applications of this approach include measuring the extent of groundwater interaction with a known hydrocarbon phase, which can provide insight into migration patterns and regional subsurface fluid-flow regimes (Bosch & Mazor 1988; Ballentine *et al.* 1996). The extent of groundwater interaction also has implications for the quality of the hydrocarbons present. The dissolution and removal of soluble hydrocarbons by persistent groundwater flow (also known as 'water-washing') can have a negative impact on the quality of the accumulated oil (Lafargue & Barker 1988). Furthermore, the movement of groundwater through the system can introduce and maintain microbial communities, which can biodegrade oil or produce microbial methane (Leahy & Colwell 1990; Horstad *et al.* 1992).

The accuracy and precision that arise from the application of this technique is somewhat limited by its input parameters. As detailed in previous

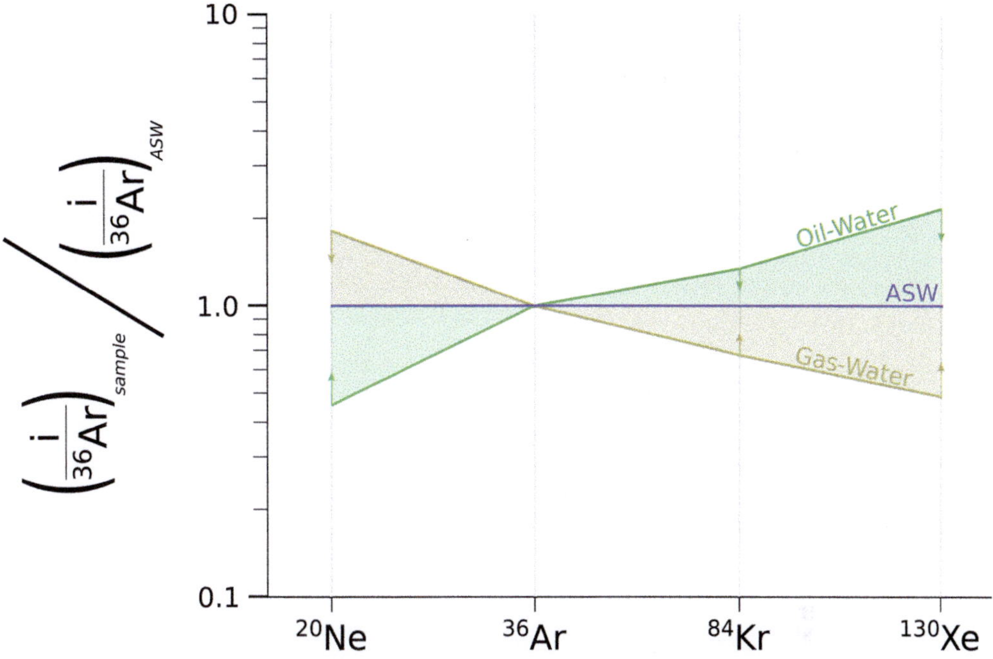

Fig. 3. Model for the distribution of atmosphere-derived noble gases (ANG) in subsurface partitioning between water and oil/gas phases, after Bosch & Mazor (1988). All values are normalized to the air-saturated water (ASW) reference, which is assumed to be the initial composition of groundwater in the subsurface. Upon first interaction with a small gas volume ($V_g/V_w \approx 0$), the different solubilities of the noble gases cause different amounts of exsolution into the gas. The composition of the gas volume at $V_g/V_w = 0$ will be that defined by the gas–water interaction line. As the gas volume and, consequently, V_g/V_w increases, the proportion of noble gases in the gas volume increases, and the composition of the gas will evolve back towards the initial ASW values, as shown by the arrows. In a closed system, the gas remains in contact with the water, and as V_g/V_w increases the noble gases will eventually be *c.* 100% in the gas phase, resulting in the gas phase having the initial ASW composition. In an open system, where gas is able to escape, the remaining water can become highly fractionated, and eventually the noble gases exsolving from this water will evolve to compositions beyond the ASW line. In this way, the ANG can be used to assess to what extent a system is open or closed (i.e. how much gas is potentially escaping from the system). Furthermore, quantitative V_g/V_w ratios can be calculated from this method (see the text). In an oil system, the same method applies. The initial gas–water and oil–water interaction lines were calculated for an initial ASW composition of zero salinity at STP (Kipfer *et al.* 2002), and the phase partitioning occurring at a reservoir temperature of 100°C. Choosing different initial ASW and reservoir conditions can substantially affect the magnitude of fractionation upon partitioning.

subsections, the Henry's law solubility of noble gases is dependent on temperature, pressure, salinity (for water) and API gravity (for oil). These relationships are non-linear and must be empirically derived, with some parameters often needing to be extrapolated from empirical data to match reservoir conditions. Additionally, the initial composition of air-saturated groundwater is similarly dependent on temperature and salinity conditions when it is formed at the surface during aquifer recharge. As groundwater associated with hydrocarbon systems can be millions of years old, it can be difficult to predict what these surface conditions would have been, but the generic assumption is made that perturbations in the noble gas composition caused by petroleum system processing is far larger than any uncertainty in

initial composition. As the increase in quality datasets from case studies increases, it is interesting to note that data inversion techniques are starting to be used to reconstruct surface conditions and demonstrate that some systems can preserve these signals over many millions of years (Barry *et al.* 2016).

Dating by radiogenic ingrowth

The dating of groundwaters using the decay and production of noble gas isotopes is a well-established technique, albeit with many assumptions. Whilst several different methods exist, each appropriate for different timescales or systems, the most widely investigated, and most appropriate for typical basinal fluid timescales, is the accumulation of stable

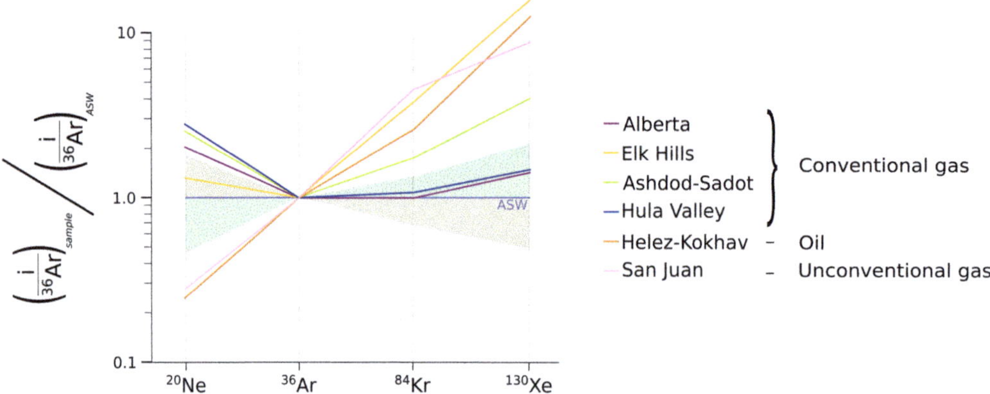

Fig. 4. Selected data displayed on the ANG composition chart described in Figure 5. Shaded areas correspond to the gas–oil and water–oil partitioning lines described in Figure 5, although as these lines are sensitive to temperature, pressure and salinity both in the reservoir and at initial groundwater recharge conditions, the actual partitioning lines will vary for each system. The Hula Valley and Ashdod–Sadot systems (Bosch & Mazor 1988) show similar elevated ^{20}Ne/^{36}Ar ratios suggesting gas–water partitioning with small V_g/V_w, although enrichments in Kr and Xe vary. The Helez-Kokhav oil system (Kennedy *et al.* 1990) shows partitioning in the direction of the oil–water line, suggesting a low V_o/V_w. The San Juan coalbed methane (Zhou *et al.* 2005) shows a gas system fractionated strongly away from the gas–water line past the ASW reference. This is potentially indicative of an open system with significant gas loss, or some other mechanism fractionating the noble gases. The Elk Hills system (Torgersen & Kennedy 1999) shows ^{20}Ne/^{36}Ar values expected for a gas system, although the ^{84}Kr/^{36}Ar and ^{130}Xe/^{36}Ar are highly enriched. The Alberta system (Hiyagon & Kennedy 1992) is described in the caption to Figure 2. A common observation amongst the different systems is an enrichment in atmospheric Kr and Xe compared to the values predicted from the ^{20}Ne/^{36}Ar partitioning. The extent of this enrichment is variable, although it is often suggested to derive from an adsorbed component within the source rock.

radiogenic ^{4}He and ^{40}Ar (Torgersen & Clarke 1985; Marty *et al.* 1993; Tolstikhin *et al.* 1996; Castro *et al.* 1998*a*, *b*; Mahara *et al.* 2009). A similar technique using fissiogenic Xe production is possible, but is only applicable in very old samples (>100 Ma) due to the slow accumulation rate (Holland *et al.* 2013). By using the partitioning laws described in the previous subsection, it is possible to indirectly date the groundwater associated with a petroleum system by measuring the noble gases in the hydrocarbon phase (Zhou & Ballentine 2006). The importance of groundwater involvement in petroleum systems is detailed in the above subsection, and constraining the age is a crucial step in resolving the groundwater–hydrocarbon interaction.

The basis for all of these models is to deconvolve the noble gases produced *in situ* within the system from those present initially in the groundwater, and those brought in from external fluxes. The *in situ* concentration is then compared with the theoretical production rates (calculated from parent isotope concentrations and decay constants) to give the necessary time for this concentration to accumulate:

$$[^4\text{He}]_{\text{total}} = [^4\text{He}]_{\text{ASW}} + [^4\text{He}]_{in\ situ\ \text{production}}$$
$$+ [^4\text{He}]_{\text{external flux}} \quad (7)$$

where 'total' is the total reconstructed concentration in associated groundwater, ASW is the initial concentration in air-saturated groundwater recharge, '*in situ* production' is the amount produced in place over the lifetime of the system and 'external flux' denotes the amount brought in from external sources. The age of the system can then be calculated using the following parameterization, after Torgersen (1980):

$$[^4\text{He}]_{in\ situ\ \text{production}} = \frac{\rho J \Lambda (1 - \varphi)}{\varphi} t \quad (8)$$

where ρ is the density of the rock, J is the *in situ* production of ^{4}He, Λ is a parameter describing the efficiency of transfer of produced ^{4}He from the mineral to the surrounding fluid, where $0 < \Lambda < 1$, and φ is porosity. The residence time of the groundwater in the system is t. Λ is thought to be approximately 1 over geological timescales (Ballentine & Burnard 2002). J is calculated as a function of the concentrations of U and Th in the surrounding rocks, after Craig & Lupton (1976):

$$J = 0.2355 \times 10^{-12} [\text{U}] \left\{ 1 + 0.123 \left(\frac{[\text{U}]}{[\text{Th}]} - 4 \right) \right\} \quad (9)$$

Analogous equations can be formulated for the production of ^{40}Ar, or any other radiogenic isotope.

The principal uncertainty in groundwater dating is the term accounting for external flux of radiogenically produced isotopes. Radiogenic concentrations in excess of those that can be reasonably explained by *in situ* production are frequently observed in groundwater and hydrocarbon systems (Torgersen & Clarke 1985; Ballentine *et al.* 1996; Takahata & Sano 2000). Several studies suggest that this is due to a universal continental degassing flux from the deep crust, whilst others have suggested a more variable input from old isolated fluid bodies or mineral degassing during alteration (Torgersen & Clarke 1985; Solomon *et al.* 1996; Tolstikhin *et al.* 1996; Patriarche *et al.* 2004), or due to thermal perturbation of old continental crust (Ballentine *et al.* 2002; Lowenstern *et al.* 2014). The exact impacts of these various factors are probably dependent on basin-scale hydrogeological behaviours, and, as such, knowledge of the basin structure and history are crucial during these calculations and subsequent interpretations.

Zhou & Ballentine (2006) considered data from three previous studies to investigate ^{4}He ages of hydrocarbon-associated groundwater. In the San Juan Basin biogenic coalbed methane play, ^{4}He ages were found to be dependent on the distance from recharge at the basin margin. Making an argument that the biogenic degradation of the coal is related to groundwater age enabled a biogenic gas production rate to be calculated, and is in reasonable agreement with biogenic gas production rates estimated using a similar approach in the Antrim shale, Illinois Basin, USA (Schlegel *et al.* 2011). ^{4}He ages from the North Sea Magnus oil field gave values of *c.* 2 Ma, which when compared with the age of the reservoir (*c.* 150 Ma) suggest that the influence of formation water has been relatively minor. The He-rich Hugoton–Panhandle gas field in the southern USA yielded slightly older ^{4}He groundwater ages of around 4 Ma. This is still relatively young compared to the estimated age of the petroleum system, and is interpreted to be more representative of the ages of groundwaters bringing in the high He concentrations, indicating a relatively recent injection of commercial He.

Dating of groundwaters associated with natural gases in the Piceance Basin, Colorado, USA, has also been undertaken by McMahon *et al.* (2013). They were able to identify a range of groundwater ages using ^{4}He dating in conjunction with ^{14}C dating. This was used to show the compartmentalization of the field into areas with different ages, with differences in gas composition between the areas. Similarly, the ^{4}He ages determined by Schlegel *et al.* (2011), in the Illinois Basin, showed older groundwater ages associated with thermogenic

methane, whilst younger ages were associated with microbially generated methane. This suggests that noble gas ages have the potential to constrain the onset and extent of microbial methane generation.

Unconventional systems

The rapid expansion of unconventional source-rock reservoir hydrocarbon production over the last decade has created an opportunity for the development of new noble gas techniques to advance our understanding of the mechanisms involved in unconventional oil and gas generation, storage, and production (Curiale & Curtis 2016). As unconventional hydrocarbons are generated and produced *in situ*, with no secondary migration, the pressure–temperature histories of the systems are better constrained, and the noble gas signatures are likely to be less influenced by basin-scale fluid-flow regimes. Furthermore, the retention of hydrocarbons within the source rock allows for the initial noble gas composition of hydrocarbons to be measured directly. This could have important implications for the study of conventional systems, where the initial hydrocarbon noble gas composition is often assumed to be negligible.

Tantalizing insights into the potential behaviour of noble gases in source rocks were found in an investigation into biogenic coalbed methane in the San Juan system (Zhou *et al.* 2005). The produced gases were observed to be highly enriched in atmosphere-derived Xe and, to a lesser extent, Kr. This effect was suggested to be due to preferential sorption of heavy noble gases onto the organic-carbon-rich sediments, a phenomenon that has been observed experimentally in laboratory simulations (Fanale & Cannon 1971; Podosek *et al.* 1981) and observed previously in other hydrocarbon systems (Torgersen & Kennedy 1999). Variable fractionation was also observed in both ^{20}Ne/^{22}Ne and ^{38}Ar/^{36}Ar isotope ratios, consistent with kinetic mass-dependent fractionation. This is proposed to be due to concentration gradients created during gas production imparting a diffusive effect on the produced gas.

An investigation into Marcellus shale gases by Hunt *et al.* (2012), showed that the gases could be separated into distinct groups with different thermal maturities, based on their radiogenic noble gas contents. The ^{4}He/^{40}Ar* and ^{21}Ne*/^{40}Ar* ratios both showed a distinct grouping that correlated with thermal maturity, suggesting that the temperatures experienced by the source rocks affected the release of radiogenically produced noble gases into the surrounding fluids (e.g. Ballentine *et al.* 1996). Whilst there are more practical ways of identifying gas maturity, this shows the possibility for using noble gases to track the extent of gas release with temperature, unaffected by chemical or biological effects.

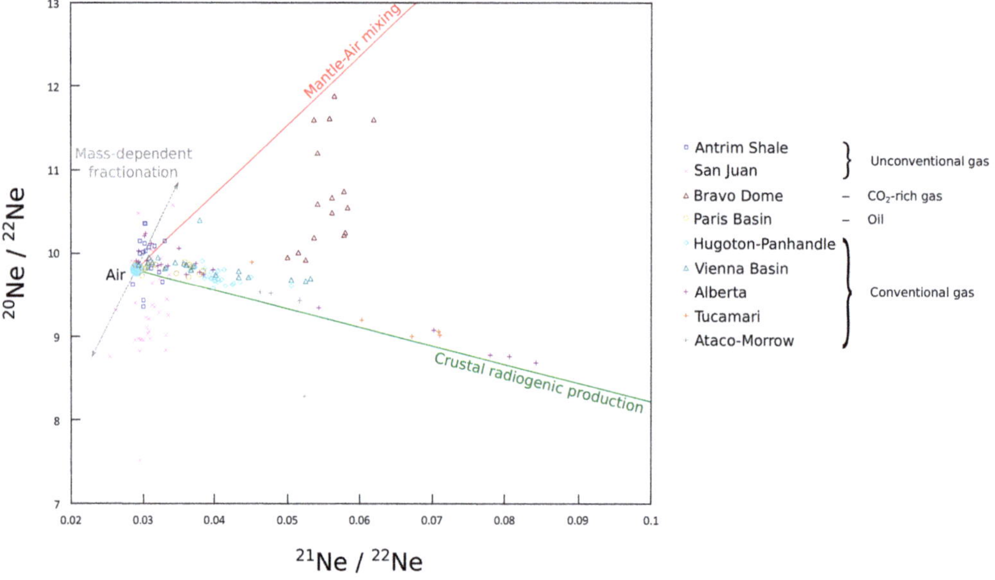

Fig. 5. Neon three-isotope plot for selected noble gas studies of petroleum systems. Neon isotopes are conventionally thought of as a 'three end-member' mixing system, with crustal radiogenic production and mantle mixing lines shown. In crustal systems, both ^{21}Ne and ^{22}Ne are produced radiogenically, resulting in alteration of the isotope ratios along the crustal evolution line defined in Kennedy *et al.* (1990) and Ballentine & Burnard (2002). The Bravo Dome system shows radiogenic ingrowth in addition to mixing with a mantle Ne end member (Ballentine *et al.* 2005). Additionally, mass-dependent fractionation of the system can cause the isotopes to evolve along the mass-fractionation line in either direction, consistent with data observed in both the Antrim and San Juan systems. Both of these systems are unconventional, the San Juan being a coalbed methane system, and the Antrim a more typical shale gas play. The Bravo Dome system is rich in CO_2 that is interpreted to be sourced from mantle fluids; the neon isotopes clearly show mixing with a mantle end member. Data sources: Antrim shale (Wen *et al.* 2015); San Juan coalbed methane (Zhou *et al.* 2005); Bravo Dome (Ballentine *et al.* 2005); Paris Basin (Pinti & Marty 1995); Vienna Basin (Ballentine & O'Nions 1992); Alberta, Tucamari and Ataco-Morrow (Kennedy *et al.* 1990).

The Antrim shale gas play has also been investigated by Wen *et al.* (2015). Trace amounts of mantle fluids were detected using ^{3}He/^{4}He isotope ratios, and ^{4}He groundwater ages were relatively young (<250 ka), coinciding with past glaciations. Both of these findings suggest that despite the low porosity of the rocks, groundwater and other fluids are still able to permeate the system. However, the Antrim shale is naturally fractured to a much greater extent than typical unconventional shale gas systems, which could artificially increase permeability and, hence, local fluid flow (Apotria *et al.* 1994; Ryder 1996). Further observations from this study include variable Ne isotopes exhibiting scatter in the mass-dependent fractionation directions, similar to the San Juan Basin; whether this is a signal common to all unconventional systems is not yet known.

Measured data from unconventional studies are shown alongside those from conventional systems in Figures 2, 5 and 6. Helium isotopes show lower ^{4}He concentrations compared to conventional systems, and no significant mantle contributions. Neon isotopes in both datasets show no significant radiogenic contribution but high levels of scatter along mass-dependent fractionation trajectories. ^{40}Ar/^{36}Ar isotope ratios show little variation in unconventional systems but large ranges in ^{36}Ar concentration compared to conventional systems. This pattern potentially reflects heterogeneity in production when compared to conventional accumulations, caused by lower permeability and connectivity within the source-rock reservoir. However, more datasets are needed to draw any robust conclusions. It is also important to note that both the San Juan and Antrim shale systems would be considered far from typical for unconventional source-rock reservoir production. The San Juan system is a biogenic coalbed methane rather than a more common thermogenic shale gas, and the Antrim system also has a significant biogenic component and has been highly fractured by recent glaciation events. For a detailed discussion of nomenclature and classification of unconventional systems, see Curiale & Curtis (2016).

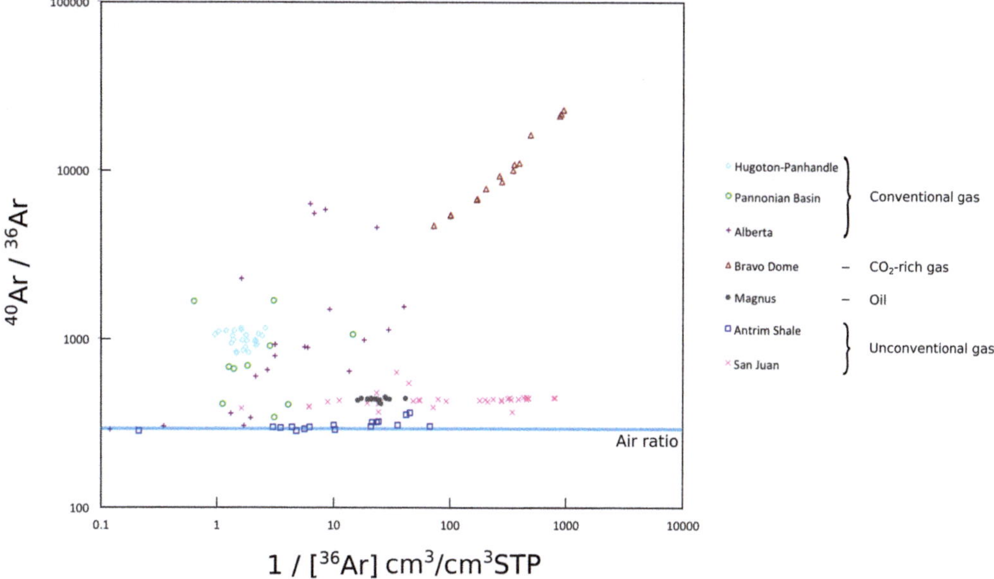

Fig. 6. Selected examples of Ar isotope ratio and concentration measurements from previous studies. ^{40}Ar/^{36}Ar is plotted against $1/[^{36}$Ar] so that mixing trajectories are clearly apparent. The Ar isotope system is a mixing system between the air-like ratio of 298.6 (Lee *et al.* 2006) and addition of radiogenic ^{40}Ar, produced in the crust from ^{40}K decay. As is evident from the data shown, ratios below the air value are not usually observed. Elevated ^{40}Ar/^{36}Ar ratios can be caused both by radiogenic production of ^{40}Ar within the crust, as well as mixing with mantle fluids which have high levels of radiogenic ^{40}Ar. The Ar isotope system is also important for the detection of any air contamination in samples; the relatively high concentration of Ar in air makes even small amounts of contamination drastically affect measured Ar concentrations and isotope ratios. Data sources: Magnus (Ballentine *et al.* 1996); Pannonian Basin (Ballentine & O'Nions 1993); Hugoton–Panhandle gas field (Ballentine & Sherwood Lollar 2002); Alberta Field (Hiyagon & Kennedy 1992); Bravo Dome (Ballentine *et al.* 2005); Antrim shale (Wen *et al.* 2015); San Juan coalbed methane (Zhou *et al.* 2005).

Studies of noble gases in produced unconventional hydrocarbons are still relatively sparse (see also Moore *et al.* this volume, in press), there have been several noble gas studies of the groundwaters associated with unconventional systems, primarily for assessing environmental impact on aquifers. Of these, Jackson *et al.* (2013) and Darrah *et al.* (2014, 2015) have shown the utility of noble gases in discriminating between anthropogenic hydrocarbon contamination of drinking water aquifers and natural hydrocarbon migration in subsurface brines.

Summary

Despite years of investigation and a number of notable publications in the field, noble gases are still not used as a routine geochemical analytical tool in hydrocarbon systems, in the manner of C or H isotopes. Partly, this is due to the expense and difficulty of making the measurements themselves, but it is also due to the complex nature of interpreting noble gas results and the necessity of

having a proper context to do so. Whilst calculating absolute V_g/V_w ratios, and the ^{4}He and ^{40}Ar ages represents significant progress towards quantifying fluid exchange and accumulation within a system, it is still not always intuitive what these numbers actually represent. For V_g/V_w and V_o/V_w ratios, it is not necessarily the ratio of gas to water in the accumulation itself, but rather the entire history of hydrocarbon–water interaction within the system (e.g. during migration from the source rock to the accumulation). In this way, they should be considered not as an alternative to geological observations, but as an additional constraint to complement existing techniques.

A key step in the development of the noble gas toolkit is the continued acquisition and compilation of case studies. Although petroleum systems can be broadly categorized they are all unique, which potentially provides great future utility for identifying the source and history of a hydrocarbon reservoir. Comparison between different systems will allow universal patterns between similar systems to be discerned from local effects particular to that location. This, in turn, will allow more robust conclusions to be drawn in a more straightforward fashion from new studies.

In the study of unconventional systems, it remains unclear which questions noble gases will be most suited to answer. Will the adaptation and application of existing conventional system tools be appropriate, or will new approaches need to be formulated entirely? The isolation of any Xe-enriched sedimentary component and the controls on this process should be a preliminary goal. This may lead to insights into the mechanisms controlling gas storage within organic-rich sediment; the importance of adsorption v. free gas stored in porosity could help explain the extent of gas generation and recoverability. As previously mentioned, constraining the noble gas characteristics of source rocks could also help to refine groundwater models in conventional systems. Indeed, fluid interaction models could also have some utility in unconventional systems themselves. Previous studies have shown that despite their low permeability, groundwater-derived noble gases and mantle fluids are present within produced gases. Unconventional systems are known to expel some generated hydrocarbons, as many simultaneously act as source rocks for conventional accumulations (Robison 1997). Noble gases could be used to shed light on the mechanisms controlling gas release v. retention within the source rock, effectively helping to predict the volumes of hydrocarbons present in a source rock while simultaneously providing constraints on potential volumes in associated conventional petroleum accumulations.

The promise of being able to geochemically decode the physical structure and history of a petroleum system is a tantalizing prospect. Although theoretically straightforward, this goal is complicated by the vagaries of any natural system. Completion of further data-rich studies on a wide range of different systems should be aided by improvements in experimental technology. Past investigations were typically focused on a few specific isotopes, it is now possible to routinely measure all naturally occurring noble gas isotopes in a single sample. The acquisition of accurate data for less-studied isotopes, such as those of Kr and Xe, could lead to development of new investigative techniques, as well as the corroboration of existing ideas. Integration of these data will help investigations of individual systems to be generalized into overarching patterns applicable to all petroleum systems, at which point noble gas analysis can become less descriptive and more predictive.

This work was funded by a NERC PhD studentship as part of the Centre for Doctoral Training in Oil & Gas, as well as ExxonMobil Upstream Research Company. We are grateful to two anonymous reviewers for their constructive and helpful reviews, and Mike Formolo for editorial handling. We would also like to thank Jennifer Mabry and Oliver Warr for their valuable input and insight on the topics discussed here, as well as Melissa Murphy for her assistance in the writing of this manuscript.

References

AESCHBACH-HERTIG, W., PEETERS, F., BEYERLE, U. & KIPFER, R. 2000. Palaeotemperature reconstruction from noble gases in ground water taking into account equilibration with entrapped air. *Nature*, **405**, 1040–1044, https://doi.org/10.1038/35016542

AESCHBACH-HERTIG, W., EL-GAMAL, H., WIESER, M. & PALCSU, L. 2008. Modeling excess air and degassing in groundwater by equilibrium partitioning with a gas phase: modeling gas partitioning. *Water Resources Research*, **44**, W08449, https://doi.org/10.1029/2007WR006454

ALLÈGRE, C.J., HOFMANN, A. & O'NIONS, K. 1996. The argon constraints on mantle structure. *Geophysical Research Letters*, **23**, 3555–3557, https://doi.org/10.1029/96GL03373

ANDERS, E. & GREVESSE, N. 1989. Abundances of the elements: meteoritic and solar. *Geochimica et Cosmochimica Acta*, **53**, 197–214, https://doi.org/10.1016/0016-7037(89)90286-X

APOTRIA, T., KAISER, C.J. & CAIN, B.A. 1994. Fracturing and stress history of the Devonian Antrim Shale, Michigan Basin. *In: 1st North American Rock Mechanics Symposium*. American Rock Mechanics Association, Alexandria, VA.

BACH, W., NAUMANN, D. & ERZINGER, J. 1999. A helium, argon, and nitrogen record of the upper continental crust (KTB drill holes, Oberpfalz, Germany): implications for crustal degassing. *Chemical Geology*, **160**, 81–101, https://doi.org/10.1016/S0009-2541(99)00058-3

BALLENTINE, C.J. & BURNARD, P.G. 2002. Production, release and transport of noble gases in the continental crust. *Reviews in Mineralogy and Geochemistry*, **47**, 481–538, https://doi.org/10.2138/rmg.2002.47.12

BALLENTINE, C.J. & HALL, C.M. 1999. Determining paleotemperature and other variables by using an error-weighted, nonlinear inversion of noble gas concentrations in water. *Geochimica et Cosmochimica Acta*, **63**, 2315–2336, https://doi.org/10.1016/S0016-7037(99)00131-3

BALLENTINE, C.J. & HOLLAND, G. 2008. What CO_2 well gases tell us about the origin of noble gases in the mantle and their relationship to the atmosphere. *Philosophical Transactions of the Royal Society of London A: Mathematical, Physical and Engineering Sciences*, **366**, 4183–4203, https://doi.org/10.1098/rsta.2008.0150

BALLENTINE, C.J. & O'NIONS, R.K. 1992. The nature of mantle neon contributions to Vienna Basin hydrocarbon reservoirs. *Earth and Planetary Science Letters*, **113**, 553–567, https://doi.org/10.1016/0012-821X(92)90131-E

BALLENTINE, C.J. & O'NIONS, R.K. 1993. The use of natural He, Ne and Ar isotopes as constraints on hydrocarbon transport. *In*: PARKER, J.R. (ed.) *Petroleum Geology of Northwest Europe: Proceedings of the 4th Conference*.

Geological Society, London, 1339–1345, https://doi.org/10.1144/0041339

BALLENTINE, C.J. & SHERWOOD LOLLAR, B. 2002. Regional groundwater focusing of nitrogen and noble gases into the Hugoton–Panhandle giant gas field, USA. *Geochimica et Cosmochimica Acta*, **66**, 2483–2497, https://doi.org/10.1016/S0016-7037(02)00850-5

BALLENTINE, C.J., O'NIONS, R.K., OXBURGH, E.R., HORVATH, F. & DEAK, J. 1991. Rare gas constraints on hydrocarbon accumulation, crustal degassing and groundwater flow in the Pannonian Basin. *Earth and Planetary Science Letters*, **105**, 229–246, https://doi.org/10.1016/0012-821X(91)90133-3

BALLENTINE, C.J., O'NIONS, R.K. & COLEMAN, M.L. 1996. A Magnus opus: Helium, neon, and argon isotopes in a North Sea oilfield. *Geochimica et Cosmochimica Acta*, **60**, 831–849, https://doi.org/10.1016/0016-7037(95)00439-4

BALLENTINE, C.J., BURGESS, R. & MARTY, B. 2002. Tracing fluid origin, transport and interaction in the crust. *Reviews in Mineralogy and Geochemistry*, **47**, 539–614, https://doi.org/10.2138/rmg.2002.47.13

BALLENTINE, C.J., MARTY, B., SHERWOOD LOLLAR, B. & CASSIDY, M. 2005. Neon isotopes constrain convection and volatile origin in the Earth's mantle. *Nature*, **433**, 33–38, https://doi.org/10.1038/nature03182

BARRY, P.H., LAWSON, M., MEURER, W.P., WARR, O., MABRY, J.C., BYRNE, D.J. & BALLENTINE, C.J. 2016. Noble gases solubility models of hydrocarbon charge mechanism in the Sleipner Vest Gas Field. *Geochimica et Cosmochimica Acta*, **194**, 291–309, https://doi.org/10.1016/j.gca.2016.08.021

BATTANI, A., SARDA, P. & PRINZHOFER, A. 2000. Basin scale natural gas source, migration and trapping traced by noble gases and major elements: the Pakistan Indus basin. *Earth and Planetary Science Letters*, **181**, 229–249, https://doi.org/10.1016/S0012-821X(00)00188-6

BAXTER, E.F., DEPAOLO, D.J. & RENNE, P.R. 2002. Spatially correlated anomalous ^{40}Ar/^{39}Ar 'age' variations in biotites about a lithologic contact near Simplon Pass, Switzerland: a mechanistic explanation for excess Ar. *Geochimica et Cosmochimica Acta*, **66**, 1067–1083, https://doi.org/10.1016/S0016-7037(01)00828-6

BEN-NAIM, A. & EGEL-THAL, M. 1965. Thermodynamics of aqueous solutions of noble gases. III. Effect of electrolytes. *The Journal of Physical Chemistry*, **69**, 3250–3253, https://doi.org/10.1021/j100894a005

BLANC, P. & CONNAN, J. 1994. Preservation, degradation, and destruction of trapped oil. *In*: MAGOON, L.B. & DOW, W.G. (eds) *The Petroleum System – From Source to Trap*. American Association of Petroleum Geologists, Memoirs, **60**, 237–247.

BOJESEN-KOEFOED, J.A., CHRISTIANSEN, F.G., NYTOFT, H.P. & PEDERSEN, A.K. 1999. Oil seepage onshore West Greenland: evidence of multiple source rocks and oil mixing. *In*: FLEET, A.J. & BOLDY, S.A.R. (eds) *Petroleum Geology of North-West Europe: Proceedings of the 5th Conference*. Geological Society, London, 305–314, https://doi.org/10.1144/0050305

BOSCH, A. & MAZOR, E. 1988. Natural gas association with water and oil as depicted by atmospheric noble gases: case studies from the southeastern Mediterranean Coastal Plain. *Earth and Planetary Science Letters*, **87**, 338–346, https://doi.org/10.1016/0012-821X(88)90021-0

BROOKER, R.A., WARTHO, J.-A., CARROLL, M.R., KELLEY, S.P. & DRAPER, D.S. 1998. Preliminary UVLAMP determinations of argon partition coefficients for olivine and clinopyroxene grown from silicate melts. *Chemical Geology*, **147**, 185–200, https://doi.org/10.1016/S0009-2541(97)00181-2

BROWN, H. 1949. Rare gases and the formation of the Earth's atmosphere. *In*: KUIPER, G.P. (ed.) *The Atmospheres of the Earth and Planets*. Chicago Press, Chicago, IL, 258.

BURNARD, P., GRAHAM, D. & TURNER, G. 1997. Vesicle-specific noble gas analyses of 'popping rock': implications for primordial noble gases in Earth. *Science*, **276**, 568–571, https://doi.org/10.1126/science.276.5312.568

BURNARD, P., ZIMMERMANN, L. & SANO, Y. 2013. The noble gases as geochemical tracers: history and background. *In*: BURNARD, P. (ed.) *The Noble Gases as Geochemical Tracers*. Advances in Isotope Geochemistry. Springer, Berlin, 1–15.

CADY, H.P. & MCFARLAND, D.F. 1906. Helium in natural gas. *Science*, **24**, 344–344.

CASTRO, M.C., GOBLET, P., LEDOUX, E., VIOLETTE, S. & DE MARSILY, G. 1998a. Noble gases as natural tracers of water circulation in the Paris Basin: 2. Calibration of a groundwater flow model using noble gas isotope data. *Water Resources Research*, **34**, 2467–2483, https://doi.org/10.1029/98WR01957

CASTRO, M.C., JAMBON, A., DE MARSILY, G. & SCHLOSSER, P. 1998b. Noble gases as natural tracers of water circulation in the Paris Basin: 1. Measurements and discussion of their origin and mechanisms of vertical transport in the basin. *Water Resources Research*, **34**, 2443–2466, https://doi.org/10.1029/98WR01956

CRAIG, H. & LUPTON, J.E. 1976. Primordial neon, helium, and hydrogen in oceanic basalts. *Earth and Planetary Science Letters*, **31**, 369–385, https://doi.org/10.1016/0012-821X(76)90118-7

CROVETTO, R., FERNÁNDEZ-PRINI, R. & JAPAS, M.L. 1982. Solubilities of inert gases and methane in H_2O and in D_2O in the temperature range of 300 to 600 K. *The Journal of Chemical Physics*, **76**, 1077–1086, https://doi.org/10.1063/1.443074

CROWHURST, P.V., GREEN, P.F. & KAMP, P.J.J. 2002. Appraisal of (U–Th)/He apatite thermochronology as a thermal history tool for hydrocarbon exploration: an example from the Taranaki Basin, New Zealand. *AAPG Bulletin*, **86**, 1801–1819.

CURIALE, J.A. & CURTIS, J.B. 2016. Organic geochemical applications to the exploration for source-rock reservoirs – A review. *Journal of Unconventional Oil and Gas Resources*, **13**, 1–31, https://doi.org/10.1016/j.juogr.2015.10.001

CURTIS, J.B. 2002. Fractured shale-gas systems. *AAPG Bulletin*, **86**, 1921–1938.

DARRAH, T.H., VENGOSH, A., JACKSON, R.B., WARNER, N.R. & POREDA, R.J. 2014. Noble gases identify the mechanisms of fugitive gas contamination in drinking-water wells overlying the Marcellus and Barnett Shales. *Proceedings of the National Academy of Sciences of the United States of America*, **111**, 14076–14081, https://doi.org/10.1073/pnas.1322107111

DARRAH, T.H., JACKSON, R.B. *ET AL.* 2015. The evolution of Devonian hydrocarbon gases in shallow aquifers of the northern Appalachian Basin: insights from integrating noble gas and hydrocarbon geochemistry. *Geochimica et Cosmochimica Acta*, **170**, 321–355, https://doi.org/10.1016/j.gca.2015.09.006

DAUPHAS, N. 2003. The dual origin of the terrestrial atmosphere. *Icarus*, **165**, 326–339, https://doi.org/10.1016/S0019-1035(03)00198-2

DAY, J.M.D., BARRY, P.H., HILTON, D.R., BURGESS, R., PEARSON, D.G. & TAYLOR, L.A. 2015. The helium flux from the continents and ubiquity of low-^{3}He/^{4}He recycled crust and lithosphere. *Geochimica et Cosmochimica Acta*, **153**, 116–133, https://doi.org/10.1016/j.gca.2015.01.008

DEMAISON, G. & HUIZINGA, B.J. 1991. Genetic classification of petroleum systems (1). *AAPG Bulletin*, **75**, 1626–1643.

DEMBICKI, H.J. & ANDERSON, M.J. 1989. Secondary migration of oil: Experiments supporting efficient movement of separate, buoyant oil phase along limited conduits: geologic note. *AAPG Bulletin*, **73**, 1018–1021.

DOW, W.G. 1974. Application of oil-correlation and source-rock data to exploration in Williston Basin. *AAPG Bulletin*, **58**, 1253–1262.

DOW, W.G. 1977. Kerogen studies and geological interpretations. *Journal of Geochemical Exploration*, **7**, 79–99, https://doi.org/10.1016/0375-6742(77)90078-4

DOWNEY, M.W. 1984. Evaluating seals for hydrocarbon accumulations. *AAPG Bulletin*, **68**, 1752–1763.

DYMOND, J.H. & SMITH, E.B. 1980. *Virial Coefficients of Pure Gases and Mixtures. A Critical Compilation.* Oxford Science Research Papers, **2**. Clarendon Press, Oxford.

FANALE, F.P. & CANNON, W.A. 1971. Physical adsorption of rare gas on terrigenous sediments. *Earth and Planetary Science Letters*, **11**, 362–368, https://doi.org/10.1016/0012-821X(71)90195-6

FARLEY, K.A. 2002. (U–Th)/He dating: techniques, calibrations, and applications. *Reviews in Mineralogy and Geochemistry*, **47**, 819–844, https://doi.org/10.2138/rmg.2002.47.18

FERNÁNDEZ-PRINI, R., ALVAREZ, J.L. & HARVEY, A.H. 2003. Henry's constants and vapor–liquid distribution constants for gaseous solutes in H_2O and D_2O at high temperatures. *Journal of Physical and Chemical Reference Data*, **32**, 903–916, https://doi.org/10.1063/1.1564818

GILFILLAN, S.M.V., BALLENTINE, C.J. *ET AL.* 2008. The noble gas geochemistry of natural CO_2 gas reservoirs from the Colorado Plateau and Rocky Mountain provinces, USA. *Geochimica et Cosmochimica Acta*, **72**, 1174–1198, https://doi.org/10.1016/j.gca.2007.10.009

GILFILLAN, S.M.V., LOLLAR, B.S. *ET AL.* 2009. Solubility trapping in formation water as dominant CO_2 sink in natural gas fields. *Nature*, **458**, 614–618, https://doi.org/10.1038/nature07852

GLUYAS, J. & SWARBRICK, R. 2013. *Petroleum Geoscience.* John Wiley & Sons, Chichester, UK.

GÖBEL, R., OTT, U. & BEGEMANN, F. 1978. On trapped noble gases in ureilites. *Journal of Geophysical Research: Solid Earth*, **83**, 855–867, https://doi.org/10.1029/JB083iB02p00855

GRAHAM, D.W. 2002. Noble gas isotope geochemistry of mid-ocean ridge and ocean island basalts: characterization of mantle source reservoirs. *Reviews in Mineralogy and Geochemistry*, **47**, 247–317, https://doi.org/10.2138/rmg.2002.47.8

GRIMBERG, A., BAUR, H. *ET AL.* 2006. Solar wind neon from genesis: implications for the lunar noble gas record.

Science, **314**, 1133–1135, https://doi.org/10.1126/science.1133568

HALLIDAY, A.N. 2013. The origins of volatiles in the terrestrial planets. *Geochimica et Cosmochimica Acta*, **105**, 146–171, https://doi.org/10.1016/j.gca.2012.11.015

HEATON, T.H.E. & VOGEL, J.C. 1981. 'Excess air' in groundwater. *Journal of Hydrology*, **50**, 201–216, https://doi.org/10.1016/0022-1694(81)90070-6

HERZBERG, O. & MAZOR, E. 1979. Hydrological applications of noble gases and temperature measurements in underground water systems: examples from Israel. *Journal of Hydrology*, **41**, 217–231, https://doi.org/10.1016/0022-1694(79)90063-5

HINDLE, A.D. 1997. Petroleum migration pathways and charge concentration: a three-dimensional model. *AAPG Bulletin*, **81**, 1451–1481.

HIYAGON, H. & KENNEDY, B.M. 1992. Noble gases in CH_4-rich gas fields, Alberta, Canada. *Geochimica et Cosmochimica Acta*, **56**, 1569–1589, https://doi.org/10.1016/0016-7037(92)90226-9

HOEFS, J. 1997. *Stable Isotope Geochemistry.* Springer, Berlin.

HOLLAND, G. & GILFILLAN, S.M.V. 2013. Application of noble gases to the viability of CO_2 storage. *In*: BURNARD, P. (ed.) *The Noble Gases as Geochemical Tracers.* Advances in Isotope Geochemistry. Springer, Berlin, 177–223.

HOLLAND, G., CASSIDY, M. & BALLENTINE, C.J. 2009. Meteorite Kr in Earth's mantle suggests a late accretionary source for the atmosphere. *Science*, **326**, 1522–1525, https://doi.org/10.1126/science.1179518

HOLLAND, G., LOLLAR, B.S., LI, L., LACRAMPE-COULOUME, G., SLATER, G.F. & BALLENTINE, C.J. 2013. Deep fracture fluids isolated in the crust since the Precambrian era. *Nature*, **497**, 357–360, https://doi.org/10.1038/nature12127

HOLZNER, C.P., MCGINNIS, D.F., SCHUBERT, C.J., KIPFER, R. & IMBODEN, D.M. 2008. Noble gas anomalies related to high-intensity methane gas seeps in the Black Sea. *Earth and Planetary Science Letters*, **265**, 396–409, https://doi.org/10.1016/j.epsl.2007.10.029

HONDA, M., KURITA, K., HAMANO, Y. & OZIMA, M. 1982. Experimental studies of He and Ar degassing during rock fracturing. *Earth and Planetary Science Letters*, **59**, 429–436, https://doi.org/10.1016/0012-821X(82)90144-3

HOOKER, P.J., O'NIONS, R.K. & OXBURGH, E.R. 1985. Helium isotopes in North Sea gas fields and the Rhine rift. *Nature*, **318**, 273–275, https://doi.org/10.1038/318273a0

HORNAFIUS, J.S., QUIGLEY, D. & LUYENDYK, B.P. 1999. The world's most spectacular marine hydrocarbon seeps (Coal Oil Point, Santa Barbara Channel, California): quantification of emissions. *Journal of Geophysical Research: Oceans*, **104**, 20703–20711, https://doi.org/10.1029/1999JC900148

HORSTAD, I., LARTER, S.R. & MILLS, N. 1992. A quantitative model of biological petroleum degradation within the Brent Group reservoir in the Gullfaks Field, Norwegian North Sea. *Organic Geochemistry*, **19**, 107–117, https://doi.org/10.1016/0146-6380(92)90030-2

HUGHES, W.B., HOLBA, A.G. & DZOU, L.I.P. 1995. The ratios of dibenzothiophene to phenanthrene and pristane to phytane as indicators of depositional environment and lithology of petroleum source rocks. *Geochimica et Cosmochimica Acta*, **59**, 3581–3598, https://doi.org/10.1016/0016-7037(95)00225-O

HUNT, A.G., DARRAH, T.H. & POREDA, R.J. 2012. Determining the source and genetic fingerprint of natural gases using noble gas geochemistry: a northern Appalachian Basin case study. *AAPG Bulletin*, **96**, 1785–1811, https://doi.org/10.1306/03161211093

JACKSON, R.B., VENGOSH, A. *ET AL.* 2013. Increased stray gas abundance in a subset of drinking water wells near Marcellus shale gas extraction. *Proceedings of the National Academy of Sciences of the United States of America*, **110**, 11 250–11 255, https://doi.org/10.1073/pnas.1221635110

KENNEDY, B.M., HIYAGON, H. & REYNOLDS, J.H. 1990. Crustal neon: a striking uniformity. *Earth and Planetary Science Letters*, **98**, 277–286, https://doi.org/10.1016/0012-821X(90)90030-2

KHARAKA, Y.K. & SPECHT, D.J. 1988. The solubility of noble gases in crude oil at 25–100°C. *Applied Geochemistry*, **3**, 137–144, https://doi.org/10.1016/0883-2927(88)90001-7

KIPFER, R., AESCHBACH-HERTIG, W., PEETERS, F. & STUTE, M. 2002. Noble gases in lakes and ground waters. *Reviews in Mineralogy and Geochemistry*, **47**, 615–700, https://doi.org/10.2138/rmg.2002.47.14

LAFARGUE, E. & BARKER, C. 1988. Effect of water washing on crude oil compositions. *AAPG Bulletin*, **72**, 263–276.

LEAHY, J.G. & COLWELL, R.R. 1990. Microbial degradation of hydrocarbons in the environment. *Microbiological Reviews*, **54**, 305–315.

LEE, J.-Y., MARTI, K., SEVERINGHAUS, J.P., KAWAMURA, K., YOO, H.-S., LEE, J.B. & KIM, J.S. 2006. A redetermination of the isotopic abundances of atmospheric Ar. *Geochimica et Cosmochimica Acta*, **70**, 4507–4512, https://doi.org/10.1016/j.gca.2006.06.1563

LEE, K.K.M. & STEINLE-NEUMANN, G. 2006. High-pressure alloying of iron and xenon: 'Missing' Xe in the Earth's core? *Journal of Geophysical Research: Solid Earth*, **111**, B02202, https://doi.org/10.1029/2005JB003781

LEWAN, M.D. 1993. Laboratory simulation of petroleum formation. *In*: ENGEL, M.H. & MACKO, S.A. (eds) *Organic Geochemistry*. Topics in Geobiology, **11**. Springer, New York, 419–442.

LOWENSTERN, J.B., EVANS, W.C., BERGFELD, D. & HUNT, A.G. 2014. Prodigious degassing of a billion years of accumulated radiogenic helium at Yellowstone. *Nature*, **506**, 355–358, https://doi.org/10.1038/nature12992

LUPTON, J. & EVANS, L. 2013. Changes in the atmospheric helium isotope ratio over the past 40 years: Atmospheric Helium Isotope ratio. *Geophysical Research Letters*, **40**, 6271–6275, https://doi.org/10.1002/2013GL057681

MAGOON, L.B. & DOW, W.G. 1994. The petroleum system. *In*: MAGOON, L.B. & DOW, W.G. (eds) *The Petroleum System: From Source to Trap*. American Association of Petroleum Geologists, Memoirs, **60**, 3–24.

MAHARA, Y., HABERMEHL, M.A. *ET AL.* 2009. Groundwater dating by estimation of groundwater flow velocity and dissolved ^{4}He accumulation rate calibrated by ^{36}Cl in the Great Artesian Basin, Australia. *Earth and Planetary Science Letters*, **287**, 43–56, https://doi.org/10.1016/j.epsl.2009.07.034

MARROCCHI, Y., RAZAFITIANAMAHARAVO, A., MICHOT, L.J. & MARTY, B. 2005. Low-pressure adsorption of Ar, Kr, and Xe on carbonaceous materials (kerogen and carbon blacks), ferrihydrite, and montmorillonite: implications for the trapping of noble gases onto meteoritic matter. *Geochimica et Cosmochimica Acta*, **69**, 2419–2430, https://doi.org/10.1016/j.gca.2004.09.016

MARTEL, D.J., O'NIONS, R.K., HILTON, D.R. & OXBURGH, E.R. 1990. The role of element distribution in production and release of radiogenic helium: the Carnmenellis Granite, southwest England. *Chemical Geology*, **88**, 207–221, https://doi.org/10.1016/0009-2541(90)90090-T

MARTY, B. 2012. The origins and concentrations of water, carbon, nitrogen and noble gases on Earth. *Earth and Planetary Science Letters*, **313–314**, 56–66, https://doi.org/10.1016/j.epsl.2011.10.040

MARTY, B., TORGERSEN, T., MEYNIER, V., O'NIONS, R.K. & DE MARSILY, G. 1993. Helium isotope fluxes and groundwater ages in the Dogger Aquifer, Paris Basin. *Water Resources Research*, **29**, 1025–1035, https://doi.org/10.1029/93WR00007

MAZOR, E. 1972. Paleotemperatures and other hydrological parameters deduced from noble gases dissolved in groundwaters; Jordan Rift Valley, Israel. *Geochimica et Cosmochimica Acta*, **36**, 1321–1336, https://doi.org/10.1016/0016-7037(72)90065-8

McMAHON, P.B., THOMAS, J.C. & HUNT, A.G. 2013. Groundwater ages and mixing in the Piceance Basin natural gas province, Colorado. *Environmental Science & Technology*, **47**, 13 250–13 257, https://doi.org/10.1021/es402473c

MERRIHUE, C. & TURNER, G. 1966. Potassium–argon dating by activation with fast neutrons. *Journal of Geophysical Research*, **71**, 2852–2857.

MOORE, M.T., VINSON, D.S., WHYTE, C.J., EYMOLD, W.K., WALSH, T.B. & DARRAH, T.H. In press. Differentiating between biogenic, thermogenic sources of natural gas in coalbed methane reservoirs from the Illinois Basin using noble gas, hydrocarbon geochemistry. *In*: LAWSON, M., FORMOLO, M.J. & EILER, J.M. (eds) *From Source to Seep: Geochemical Applications in Hydrocarbon Systems*. Geological Society, London, Special Publications, **468**, https://doi.org/10.1144/SP468.8

MOREIRA, M., KUNZ, J. & ALLÈGRE, C. 1998. Rare gas systematics in popping rock: isotopic and elemental compositions in the upper mantle. *Science*, **279**, 1178–1181, https://doi.org/10.1126/science.279.5354.1178

OWEN, T., BAR-NUN, A. & KLEINFELD, I. 1992. Possible cometary origin of heavy noble gases in the atmospheres of Venus, Earth and Mars. *Nature*, **358**, 43–46, https://doi.org/10.1038/358043a0

OZIMA, M. & PODOSEK, F.A. 2002. *Noble Gas Geochemistry*. Cambridge University Press, Cambridge.

PATRIARCHE, D., CASTRO, M.C. & GOBLET, P. 2004. Large-scale hydraulic conductivities inferred from three-dimensional groundwater flow and ^{4}He transport modeling in the Carrizo aquifer, Texas. *Journal of Geophysical Research: Solid Earth*, **109**, B11202, https://doi.org/10.1029/2004JB003173

PEPIN, R.O. 1991. On the origin and early evolution of terrestrial planet atmospheres and meteoritic volatiles. *Icarus*, **92**, 2–79, https://doi.org/10.1016/0019-1035(91)90036-S

PEPIN, R.O. 2006. Atmospheres on the terrestrial planets: clues to origin and evolution. *Earth and Planetary Science Letters*, **252**, 1–14, https://doi.org/10.1016/j.epsl.2006.09.014

PEPIN, R.O. & PORCELLI, D. 2006. Xenon isotope systematics, giant impacts, and mantle degassing on the early Earth. *Earth and Planetary Science Letters*, **250**, 470–485, https://doi.org/10.1016/j.epsl.2006.08.014

PETERS, K.E. 1986. Guidelines for evaluating petroleum source rock using programmed pyrolysis. *AAPG Bulletin*, **70**, 318–329.

PETERS, K.E. & CASSA, M.R. 1994. Applied source rock geochemistry: Chapter 5: Part II. Essential elements. *In*: MAGOON, L.B. & DOW, W.G. (eds) *The Petroleum System – From Source to Trap*. American Association of Petroleum Geologists, Memoirs, **60**, 93–120.

PETERS, K.E., BURNHAM, A.K. & WALTERS, C.C. 2015. Petroleum generation kinetics: single v. multiple heating-ramp open-system pyrolysis. *AAPG Bulletin*, **99**, 591–616, https://doi.org/10.1306/11141414080

PHILP, R.P. 1993. Oil–oil and oil–source rock correlations: Techniques. *In*: ENGEL, M.H. & MACKO, S.A. (eds) *Organic Geochemistry*. Topics in Geobiology, **11**. Springer, New York, 445–460.

PIERCE, A.P., GOTT, G.B. & MYTTON, J.W. 1964. *Uranium and Helium in the Panhandle Gas Field, Texas, and Adjacent Areas*. United States Geological Survey, Professional Papers, **454-G**.

PINTI, D.L. & MARTY, B. 1995. Noble gases in crude oils from the Paris Basin, France: implications for the origin of fluids and constraints on oil–water–gas interactions. *Geochimica et Cosmochimica Acta*, **59**, 3389–3404, https://doi.org/10.1016/0016-7037(95)00213-J

PODOSEK, F.A., BERNATOWICZ, T.J. & KRAMER, F.E. 1981. Adsorption of xenon and krypton on shales. *Geochimica et Cosmochimica Acta*, **45**, 2401–2415, https://doi.org/10.1016/0016-7037(81)90094-6

PODOSEK, F.A., WOOLUM, D.S., CASSEN, P. & NICHOLS, R.H. 2000. Solar gases in the Earth by solar wind irradiation. 10th annual Goldschmidt Conference, Oxford. *Journal of Conference Abstracts*, **5**, 804, https://goldschmidtabstracts.info/abstracts/abstractView?id=2000000804

PORCELLI, D. & BALLENTINE, C.J. 2002. Models for distribution of terrestrial noble gases and evolution of the atmosphere. *Reviews in Mineralogy and Geochemistry*, **47**, 411–480, https://doi.org/10.2138/rmg.2002.47.11

PORCELLI, D., BALLENTINE, C.J. & WIELER, R. 2002. An overview of noble gas geochemistry and cosmochemistry. *Reviews in Mineralogy and Geochemistry*, **47**, 1–19, https://doi.org/10.2138/rmg.2002.47.1

PRINZHOFER, A. 2013. Noble gases in oil and gas accumulations. *In*: BURNARD, P. (ed.) *The Noble Gases as Geochemical Tracers*. Advances in Isotope Geochemistry. Springer, Berlin, 225–247.

PUJOL, M., MARTY, B. & BURGESS, R. 2011. Chondritic-like xenon trapped in Archean rocks: a possible signature of the ancient atmosphere. *Earth and Planetary Science Letters*, **308**, 298–306, https://doi.org/10.1016/j.epsl.2011.05.053

RENNE, P.R., MUNDIL, R., BALCO, G., MIN, K. & LUDWIG, K.R. 2010. Joint determination of 40K decay constants and $^{40}Ar*/^{40}K$ for the Fish Canyon sanidine standard, and improved accuracy for $^{40}Ar/^{39}Ar$ geochronology.

Geochimica et Cosmochimica Acta, **74**, 5349–5367, https://doi.org/10.1016/j.gca.2010.06.017

REYNOLDS, J.H. 1963. Xenology. *Journal of Geophysical Research*, **68**, 2939–2956, https://doi.org/10.1029/JZ068i010p02939

ROBISON, C.R. 1997. Hydrocarbon source rock variability within the Austin Chalk and Eagle Ford Shale (Upper Cretaceous), East Texas, U.S.A. *International Journal of Coal Geology*, **34**, 287–305, https://doi.org/10.1016/S0166-5162(97)00027-X

RUTHERFORD, E. 1906. *Radioactive Transformations*. Yale University Press, New Haven, CT.

RYDER, R.T. 1996. *Fracture Patterns and Their Origin in the Upper Devonian Antrim Shale Gas Reservoir of the Michigan Basin; A Review*. United States Geological Survey, Open-File Report, **96-23**.

SANLOUP, C., SCHMIDT, B.C., PEREZ, E.M.C., JAMBON, A., GREGORYANZ, E. & MEZOUAR, M. 2005. Retention of xenon in quartz and Earth's missing xenon. *Science*, **310**, 1174–1177, https://doi.org/10.1126/science.1119070

SARDA, P., STAUDACHER, T. & ALLEGRE, C. 1988. Neon isotopes in submarine basalts. *Earth and Planetary Science Letters*, **91**, 73–88, https://doi.org/10.1016/0012-821X(88)90152-5

SCHLEGEL, M.E., ZHOU, Z., McINTOSH, J.C., BALLENTINE, C.J. & PERSON, M.A. 2011. Constraining the timing of microbial methane generation in an organic-rich shale using noble gases, Illinois Basin, USA. *Chemical Geology*, **287**, 27–40, https://doi.org/10.1016/j.chemgeo.2011.04.019

SCHOWALTER, T.T. 1979. Mechanics of secondary hydrocarbon migration and entrapment. *AAPG Bulletin*, **63**, 723–760.

SMITH, S. 1985. Noble gas solubility in water at high temperature. *Eos, Transactions of the American Geophysical Union*, **66**, 397.

SMITH, S.P. & KENNEDY, B.M. 1983. The solubility of noble gases in water and in NaCl brine. *Geochimica et Cosmochimica Acta*, **47**, 503–515, https://doi.org/10.1016/0016-7037(83)90273-9

SOLOMON, D.K., HUNT, A. & POREDA, R.J. 1996. Source of radiogenic helium 4 in shallow aquifers: implications for dating young groundwater. *Water Resources Research*, **32**, 1805–1813, https://doi.org/10.1029/96WR00600

STAHL, W.J. 1978. Source rock–crude oil correlation by isotopic type-curves. *Geochimica et Cosmochimica Acta*, **42**, 1573–1577, https://doi.org/10.1016/0016-7037(78)90027-3

STAINFORTH, J.G. & REINDERS, J.E.A. 1990. Primary migration of hydrocarbons by diffusion through organic matter networks, and its effect on oil and gas generation. *Organic Geochemistry*, **16**, 61–74, https://doi.org/10.1016/0146-6380(90)90026-V

STAUDACHER, T., SARDA, P., RICHARDSON, S.H., ALLÈGRE, C.J., SAGNA, I. & DMITRIEV, L.V. 1989. Noble gases in basalt glasses from a Mid-Atlantic Ridge topographic high at 14°N: geodynamic consequences. *Earth and Planetary Science Letters*, **96**, 119–133, https://doi.org/10.1016/0012-821X(89)90127-1

STOLPER, D.A., LAWSON, M. *ET AL.* 2014. Formation temperatures of thermogenic and biogenic methane. *Science*, **344**, 1500–1503, https://doi.org/10.1126/science.1254509

STUART, F.M., LASS-EVANS, S., GODFREY FITTON, J. & ELLAM, R.M. 2003. High ^{3}He/^{4}He ratios in picritic basalts from Baffin Island and the role of a mixed reservoir in mantle plumes. *Nature*, **424**, 57–59, https://doi.org/10.1038/nature01711

STUTE, M., FORSTER, M. *ET AL.* 1995. Cooling of Tropical Brazil (5°C) during the Last Glacial Maximum. *Science*, **269**, 379–383, https://doi.org/10.1126/science.269.5222.379

SWEENEY, J.J. & BURNHAM, A.K. 1990. Evaluation of a simple model of Vitrinite Reflectance based on chemical kinetics (1). *AAPG Bulletin*, **74**, 1559–1570.

TAKAHATA, N. & SANO, Y. 2000. Helium flux from a sedimentary basin. *Applied Radiation and Isotopes*, **52**, 985–992, https://doi.org/10.1016/S0969-8043(99)00159-1

TOLSTIKHIN, I., LEHMANN, B.E., LOOSLI, H.H. & GAUTSCHI, A. 1996. Helium and argon isotopes in rocks, minerals, and related ground waters: a case study in northern Switzerland. *Geochimica et Cosmochimica Acta*, **60**, 1497–1514, https://doi.org/10.1016/0016-7037(96)00036-1

TORGERSEN, T. 1980. Controls on pore-fluid concentration of ^{4}He and ^{222}Rn and the calculation of ^{4}He/^{222}Rn ages. *Journal of Geochemical Exploration*, **13**, 57–75, https://doi.org/10.1016/0375-6742(80)90021-7

TORGERSEN, T. 1989. Terrestrial helium degassing fluxes and the atmospheric helium budget: implications with respect to the degassing processes of continental crust. *Chemical Geology: Isotope Geoscience Section*, **79**, 1–14, https://doi.org/10.1016/0168-9622(89)90002-X

TORGERSEN, T. & CLARKE, W.B. 1985. Helium accumulation in groundwater, I: an evaluation of sources and the continental flux of crustal ^{4}He in the Great Artesian Basin, Australia. *Geochimica et Cosmochimica Acta*, **49**, 1211–1218, https://doi.org/10.1016/0016-7037(85)90011-0

TORGERSEN, T. & KENNEDY, B.M. 1999. Air-Xe enrichments in Elk Hills oil field gases: role of water in migration and storage. *Earth and Planetary Science Letters*, **167**, 239–253, https://doi.org/10.1016/S0012-821X(99)00021-7

TORGERSEN, T., HABERMEHL, M.A. & CLARKE, W.B. 1992. Crustal helium fluxes and heat flow in the Great Artesian Basin, Australia. *Chemical Geology*, **102**, 139–152, https://doi.org/10.1016/0009-2541(92)90152-U

TOTH, J. 1980. Cross-formational gravity-flow of groundwater: a mechanism of the transport and accumulation of petroleum (The Generalized Hydraulic Theory of Petroleum Migration). *In*: ROBERTS, W.H., III & CORDELL, R.J. (eds) *Problems of Petroleum Migration*. American Association of Petroleum Geologists, Studies in Geology, **10**, 121–167.

TUCKER, J.M. & MUKHOPADHYAY, S. 2014. Evidence for multiple magma ocean outgassing and atmospheric loss episodes from mantle noble gases. *Earth and Planetary Science Letters*, **393**, 254–265, https://doi.org/10.1016/j.epsl.2014.02.050

VAN KREVELEN, D.W. 1961. *Coal*. Elsevier, Amsterdam.

WAPLES, D.W. 1980. Time and temperature in petroleum formation: application of Lopatin's method to petroleum exploration. *AAPG Bulletin*, **64**, 916–926.

WAPLES, D.W. 1994. Maturity modeling: thermal indicators, hydrocarbon generation, and oil cracking. *In*: MAGOON, L.B. & DOW, W.G. (eds) *The Petroleum System – From Source to Trap*. American Association of Petroleum Geologists, Memoirs, **60**, 285–306.

WARR, O., BALLENTINE, C.J., MU, J. & MASTERS, A. 2015. Optimizing Noble Gas–water interactions via Monte Carlo simulations. *The Journal of Physical Chemistry B*, **119**, 14 486–14 495, https://doi.org/10.1021/acs.jpcb.5b06389

WASSERBURG, G.J., MAZOR, E. & ZARTMAN, R.E. 1963. Isotopic and chemical composition of some terrestrial natural gases. *In*: GEISS, J. & GOLDBERG, E.D. (eds) *Earth Science and Meteoritics*. North-Holland, Amsterdam, 219–240.

WEN, T., CASTRO, M.C., ELLIS, B.R., HALL, C.M. & LOHMANN, K.C. 2015. Assessing compositional variability and migration of natural gas in the Antrim Shale in the Michigan Basin using noble gas geochemistry. *Chemical Geology*, **417**, 356–370, https://doi.org/10.1016/j.chemgeo.2015.10.029

WETHERILL, G.W. 1954. Variations in the isotopic abundances of neon and argon extracted from radioactive minerals. *Physical Review*, **96**, 679–683, https://doi.org/10.1103/PhysRev.96.679

WHITICAR, M.J. 1994. Correlation of natural gases with their sources. *In*: MAGOON, L.B. & DOW, W.G. (eds) *The Petroleum System – From Source to Trap*. American Association of Petroleum Geologists, Memoirs, **60**, 261–283.

WIELER, R. 2002. Noble gases in the solar system. *Reviews in Mineralogy and Geochemistry*, **47**, 21–70, https://doi.org/10.2138/rmg.2002.47.2

WIELER, R. & EIKENBERG, J. 1999. An upper limit on the spontaneous fission decay constant of ^{232}Th derived from xenon in monazites with extremely high Th/U ratios. *Geophysical Research Letters*, **26**, 107–110.

WIPRUT, D. & ZOBACK, M.D. 2000. Fault reactivation and fluid flow along a previously dormant normal fault in the northern North Sea. *Geology*, **28**, 595–598, https://doi.org/10.1130/0091-7613(2000)282.0.CO;2

ZAIKOWSKI, A. & SPANGLER, R.R. 1990. Noble gas and methane partitioning from ground water: an aid to natural gas exploration and reservoir evaluation. *Geology*, **18**, 72–74, https://doi.org/10.1130/0091-7613(1990)0182.3.CO;2

ZARTMAN, R.E., WASSERBURG, G.J. & REYNOLDS, J.H. 1961. Helium, argon, and carbon in some natural gases. *Journal of Geophysical Research*, **66**, 277–306, https://doi.org/10.1029/JZ066i001p00277

ZHOU, Z. & BALLENTINE, C.J. 2006. 4He dating of groundwater associated with hydrocarbon reservoirs. *Chemical Geology*, **226**, 309–327, https://doi.org/10.1016/j.chemgeo.2005.09.030

ZHOU, Z., BALLENTINE, C.J., KIPFER, R., SCHOELL, M. & THIBODEAUX, S. 2005. Noble gas tracing of groundwater/coalbed methane interaction in the San Juan Basin, USA. *Geochimica et Cosmochimica Acta*, **69**, 5413–5428, https://doi.org/10.1016/j.gca.2005.06.027

ZIEGLER, J.F. 1977. *Helium: Stopping Powers and Ranges in All Elemental Matter*. Pergamon, Oxford.

Differentiating between biogenic and thermogenic sources of natural gas in coalbed methane reservoirs from the Illinois Basin using noble gas and hydrocarbon geochemistry

MYLES T. MOORE[1], DAVID S. VINSON[2], COLIN J. WHYTE[1], WILLIAM K. EYMOLD[1], TALOR B. WALSH[3] & THOMAS H. DARRAH[1]*

[1]*Divisions of Solid Earth Dynamics and Water, Climate and the Environment, School of Earth Sciences, The Ohio State University, Columbus, OH 43210, USA*

[2]*Department of Geography and Earth Sciences, University of North Carolina-Charlotte, Charlotte, NC 28223, USA*

[3]*Department of Earth Sciences, Millersville University, Millersville, PA 17511, USA*

**Correspondence: darrah.24@osu.edu*

Abstract: While coalbed methane (CBM) is a significant source of natural gas production globally, uncertainties regarding the proportions of biogenic and thermogenic natural gas in CBM reservoirs still remain. We integrate major gases, hydrocarbon composition, hydrocarbon stable isotopes and noble gases in fluids from 20 producing CBM wells to more accurately constrain the genetic source of natural gases in the eastern Illinois Basin, USA. Previous studies have indicated primarily biogenic production of methane (>99.6%) with negligible contributions from thermogenic natural gases. However, by integrating noble gases, we identify quantifiable (up to 19.2%) contributions of exogenous thermogenic gas in produced gases from the Seelyville and Springfield coal seams. Thermogenic gases are distinguished by a positive relationship between methane, ethane and helium-4, lower C_1/C_2+, heavier $\delta^{13}C$-CH_4, more radiogenic noble gases (^{4}He, ^{21}Ne*, ^{40}Ar*), and lower abundances of atmospherically derived gases (^{20}Ne, ^{36}Ar). Biogenic gases displayed lighter $\delta^{13}C$-CH_4, higher C_1/C_2+, higher levels of atmospheric gases and lower abundances of radiogenic noble gases. Our data suggest that natural gases from a deeper, exogenous thermogenic source likely migrated to the Pennsylvanian-aged coals at an unknown time and later mixed with biogenic methane diluting the geochemical signature of the thermogenic methane within the Springfield and Seelyville coal seams.

Increased demands for cleaner-burning fuels have placed a renewed interest in natural gas extraction from unconventional reservoirs as a substitute for coal-based electricity production throughout the last decade (USEIA 2013). Coalbed methane (CBM) production has been developed in the US as an unconventional source of natural gas since the 1950s, with increased utilization occurring from the late 1980s to the present (Strąpoć *et al.* 2010; Golding *et al.* 2013). The US possesses the third largest reserve of coalbed methane globally following Russia and China and is currently the largest producer of coalbed methane in the world (Ahmed *et al.* 2009; Moore 2012; Pashin *et al.* 2014). In comparison with shale gas development, CBM extraction is significantly less expensive because of the typically shallow nature of CBM coal fields. For example, the average cost to drill a well for CBM extraction is *c.* \$US450 000–550 000 per well as opposed to *c.* \$US5 000 000 for a horizontally drilled and hydraulically fractured shale-gas well (Ritter *et al.* 2015).

In recent decades, CBM development expanded from high-maturity coal fields dominated by thermogenic methane to lower-rank, relatively higher permeability coals with significant biogenic contributions (Zhou *et al.* 2005; Zhou & Ballentine 2006; Bates *et al.* 2011; Schlegel *et al.* 2011*a, b*; Strąpoć *et al.* 2011; McIntosh *et al.* 2012; Golding *et al.* 2013; Vinson *et al.* 2017). Today, natural gas production in the US from CBM is *c.* 110 million cubic metres per day or nearly 8–10% of the total natural gas production domestically with anticipated increases in CBM production globally in the coming years (Golding *et al.* 2013; USEIA 2013).

In CBM reservoirs, natural gas mixtures can be sourced from various genetic pathways including primary biogenic gas from biodegradation of organic matter in the source formation, secondary biodegradation of thermogenic hydrocarbons and thermocatalytic degradation of organic matter. The inherent complexity in thermogenic and biogenic mixtures of natural gas leaves several fundamental questions

From: LAWSON, M., FORMOLO, M. J. & EILER, J. M. (eds) 2018. *From Source to Seep: Geochemical Applications in Hydrocarbon Systems*. Geological Society, London, Special Publications, **468**, 151–188.
First published online January 18, 2018, https://doi.org/10.1144/SP468.8
© 2018 The Author(s). Published by The Geological Society of London. All rights reserved.
For permissions: http://www.geolsoc.org.uk/permissions. Publishing disclaimer: www.geolsoc.org.uk/pub_ethics

about natural gas sources from CBM reservoirs unsettled (Strąpoć *et al.* 2011; Golding *et al.* 2013; Ritter *et al.* 2015; Vinson *et al.* 2017). Some fundamental challenges include: (1) the determination of the proportion of methane (CH_4 or C_1) derived by biogenic v. thermogenic processes; (2) the presence of exogenous fluids in CBM reservoirs; (3) the relationship between formation water residence time, chemistry, microbiology and biogenic gas generation; and (4) better determination of biogeochemical processes that produce methane precursors and control methanogenesis.

Our ability to deconvolve thermogenic and biogenic mixtures of natural gas also has important economic and environmental applications. It is critical to decipher the genetic source of methane in CBM reservoirs in order to optimize drilling and completion strategies, construct accurate gas-in-place estimates, and to enhance natural gas production from CBM reservoirs. Accurate estimates of thermogenic and biogenic mixtures of natural gas are also critical to constrain the role and scale of the natural and anthropogenic hydrocarbon fluxes from and within the deep crust, groundwater, critical zone, surface water and to the atmosphere (Etiope *et al.* 2009*a*; Jackson *et al.* 2013; Darrah *et al.* 2015*b*; Stolper *et al.* 2015; Douglas *et al.* 2016; Kang *et al.* 2016).

Beyond the long-standing problem of separating thermogenic–biogenic mixtures, other questions remain regarding the rates and mechanisms by which coal-derived organics are converted to methane precursors (Strąpoć *et al.* 2011; Furmann *et al.* 2013; Gao *et al.* 2013; Ritter *et al.* 2015; Vinson *et al.* 2017). Methanogenic archaea can degrade organic matter and produce methane by acetoclastic (acetate fermentation), hydrogenotrophic (CO_2 reduction) and methylotrophic pathways in anaerobic conditions at less than 80°C. In many cases, the relative importance of each of these production pathways is not clear (Whiticar *et al.* 1986; Schlegel *et al.* 2011*b*; Head *et al.* 2014; Vinson *et al.* 2017).

Conventionally, the origin of CBM hydrocarbon gases has been mainly inferred through the use of: (1) the molecular composition of gases (e.g. the ratio of methane (C_1) to higher-order hydrocarbons (C_2+) or C_1/C_2+, carbon dioxide (CO_2), and nitrogen (N_2)); (2) the concentrations of dissolved solutes in coexisting waters (e.g. chloride, dissolved inorganic carbon (DIC), sulphate); and (3) the stable isotopic composition of (a) gases (e.g. δ^2H-CH_4, $\delta^{13}C$-CH_4, $\delta^{13}C$-CO_2), (b) water (δ^2H-H_2O; $\delta^{18}O$-H_2O) and (c) dissolved inorganic carbon ($\delta^{13}C$-DIC). However, the application of these traditional molecular and isotopic techniques to delineate thermogenic–biogenic gas mixtures can be complicated by uncertainties in overlapping isotopic composition of endmembers, uncertain mixing, transport/migration effects and/or secondary biodegradation

of hydrocarbons, such as later stages of methanogenesis and aerobic or anaerobic oxidation (Coleman *et al.* 1981; James & Burns 1984; Martini *et al.* 1998; Whiticar 1999; Kinnaman *et al.* 2007; Etiope *et al.* 2009*a*, 2011; Pape *et al.* 2010; Golding *et al.* 2013; Darrah *et al.* 2015*b*; Stolper *et al.* 2015; Anderson *et al.* 2017; Vinson *et al.* 2017).

Recently developed techniques that examine multiply substituted isotopologues of methane show great promise for resolving (1) the relative contributions of biogenic and thermogenic methane, (2) the temperature conditions of gas formation and (3) the rate of biogenic gas generation during which hydrogen isotopes may equilibrate with the coexisting water (Stolper *et al.* 2015; Wang *et al.* 2015; Douglas *et al.* 2016). These improvements better constrain hydrocarbon source apportionment and can lead to more accurate estimates of fluid residence times and methanogenic production rates. However, geochemical signatures that include stable isotopes and isotopologues, namely of carbon, hydrogen and oxygen, can also be affected by oxidation of hydrocarbons (Wang *et al.* 2016; Whitehill *et al.* 2017).

Because they are not altered by chemical reactions or microbial oxidation, noble gases present an advantageous addition to studies of natural gas geochemistry. Still, there are a limited number of studies that have attempted to integrate noble gases with traditional hydrocarbon molecular and isotopic geochemistry towards investigations of natural gas in shallow hydrocarbon reservoirs or in unconventional (non-migrated) CBM systems (e.g. Zhou *et al.* 2005; Zhou & Ballentine 2006; Schlegel *et al.* 2011*b*; Hunt *et al.* 2012; Darrah *et al.* 2015*b*; Wen *et al.* 2015, 2017; Barry *et al.* 2016, 2017). Furthermore, the majority of studies that have integrated these techniques focused on estimating the residence time of groundwater or hydrocarbon gases (Zhou *et al.* 2005; Zhou & Ballentine 2006; Schlegel *et al.* 2011*b*; Barry *et al.* 2017) or resolving fluid transport dynamics (Zhou *et al.* 2005; Darrah *et al.* 2014, 2015*b*; Barry *et al.* 2016), as opposed to apportioning the biogenic and thermogenic contributions to a mixed natural gas.

Recent studies examining the mean residence time of meteoric water recharge to constrain the timing of biogenic methane production in CBM fields have measured the accumulation of terrigenic ^4He in natural gas and groundwater (Zhou & Ballentine 2006; Zhou *et al.* 2005; Schlegel *et al.* 2011*b*; Barry *et al.* 2017). While these works describe the state-of-the-art methods for mean residence time determination, the utility of these methodologies is also complicated by variability in the uranium and thorium concentrations of the source rock, mixing with exogenous sources of natural gas and/or contributions of ^4He from endogenous or exogenous thermogenic sources (Zhou & Ballentine 2006; Zhou *et al.* 2005; Schlegel *et al.* 2011*b*; Darrah *et al.* 2015*b*). More recently developed techniques that estimate residence time based on radiogenic $^{40}Ar*$ and $^{136}Xe*$ or nucleogenic $^{21}Ne*$ (where * denotes the proportion derived by radioactive decay) isotope accumulation rather than relying solely on ^{4}He show great promise for constraining the residence time of older crustal fluids (e.g. Holland *et al.* 2013; Barry *et al.* 2017). However, the application of these newly developed techniques is still limited by mixing effects induced from the introduction of exogenous gases. Therefore, it is critical to determine if CBM reservoirs or other natural gas fields contain exogenous sources of radiogenic noble gases.

Although the release of ^{4}He from crustal mineral grains to fluids in porous media is often thought to be nearly instantaneous on geologically relevant time frames, there are limited data on the ^{4}He transfer rates from coal seam solids (Pepper & Corvi 1995). Additionally, higher temperatures are required for the release of heavier nucleogenic/radiogenic noble gas isotopes, such as $^{21}Ne*$, $^{40}Ar*$ and $^{136}Xe*$, making them better diagnostic tracers for thermogenic gas (Darrah *et al.* 2014, 2015*b*; Hunt *et al.* 2012). The mixing of the natural gas of interest with even relatively small contributions of exogenous thermogenic gases rich in ^{4}He, or the mixing of older biogenic methane (e.g. produced in the Pennsylvanian) with recently produced biogenic methane (e.g. produced in the Pleistocene), can lead to overestimates of fluid residence time. For these reasons, accurate apportionment of biogenic and thermogenic components in mixed gases may be required to reliably estimate formation water residence times when using either the classic ^{4}He or the more recently developed $^{21}Ne*$, $^{40}Ar*$ or $^{136}Xe*$ dating methods.

Here, we examine the major gases (e.g. N_2, CO_2), hydrocarbon composition (C_1–C_5), stable isotopes of hydrocarbon gases and dissolved inorganic carbon ($\delta^{13}C$-CH_4, $\delta^{2}H$-CH_4, $\delta^{13}C$-DIC), and noble gas elemental and isotopic abundances (He, Ne, Ar, Kr, Xe) of natural gases extracted from 20 CBM wells producing from the Seelyville and Springfield coal seams (Pennsylvanian) in the eastern Illinois Basin in Sullivan County, Indiana, USA (Fig. 1). Previous studies of natural gases produced from these shallow coalbeds (screened at depths ranging from 73 to 213 m, Fig. 2; Tables 1–3) in this part of the Illinois Basin inferred an almost exclusively biogenic origin, although stratigraphically equivalent (higher thermal maturity) coalbeds contain thermogenic natural gas in the southeastern portion of the Illinois Basin (McIntosh *et al.* 2002; Mastalerz *et al.* 2004, 2013, 2017; Strąpoć *et al.* 2007; Schlegel *et al.* 2011*a*). The underlying New Albany Shale may also contain mixtures of biogenic and thermogenic gas within the Indiana portion of the Illinois Basin (Schlegel *et al.* 2011*b*).

Integrated methodology utilizing hydrocarbon and noble gas data

Conventional methods to determine source of natural gas

Methanogens almost exclusively produce methane and CO_2 (>99.9%), yielding a dry gas (C_1/C_2+ > 2000) with isotopically light carbon and hydrogen signatures relative to thermogenic gas. The individual pathways of microbial methanogenesis are thought to impart different C and H isotope fractionations, which result in characteristic (but overlapping) ranges of $\delta^{13}C$-CH_4 and $\delta^{2}H$-CH_4 for both hydrogenotrophic and acetoclastic methanogenesis (Whiticar 1999). Hydrogenotrophic methanogenesis is thought to yield CH_4 with a larger fractionation from the source organic matter (i.e. more negative $\delta^{13}C$-CH_4) than CH_4 produced from acetoclastic methanogenesis (Whiticar 1999). While hydrogenotrophic and acetoclastic methanogenesis have been studied extensively in cultures and field-based studies, less is known about methylotrophic methanogenesis, including its environmental significance and isotopic signatures of carbon and hydrogen in CH_4 (e.g. Vinson *et al.* 2017). Secondary methanogenesis by any of these pathways would progressively enrich the natural gas mixture in methane relative to higher-order aliphatic hydrocarbons (higher C_1/C_2+) and lead to more negative $\delta^{13}C$-CH_4 relative to the thermogenic endmember (Schoell 1983; Whiticar *et al.* 1985).

By using a combination of hydrocarbon molecular composition and stable carbon and hydrogen isotopes of methane, biogenic gas (<−55 per mille, denoted as ‰ with C_1/C_2+ > 1×10^3) can be distinguished from thermogenic gas (>−55‰ with C_1/C_2+ < *c.* 200) on a plot of $\delta^{13}C$-CH_4 v. C_1/C_2+ (Bernard *et al.* 1976). Mixing of biogenic methane with a source of thermogenic natural gas could increase C_1/C_2+ and produce more negative $\delta^{13}C$-CH_4 relative to the theoretical thermogenic endmember (e.g. Jenden *et al.* 1993; Tilley & Muehlenbachs 2013), leading to uncertainty in the source of some natural gases, especially in the range of −55‰ to −40‰ (Vinson *et al.* 2017).

In comparison with biogenic methane, thermogenic hydrocarbons are produced by thermocatalytic decomposition of kerogen with increasing temperatures, which are generally associated with greater burial depths (termed catagenesis) (Whiticar *et al.* 1985; Tissot & Welte 2012). The hydrocarbon compositions of thermogenic gases change as the organic sources (e.g. kerogen or liquid hydrocarbons)

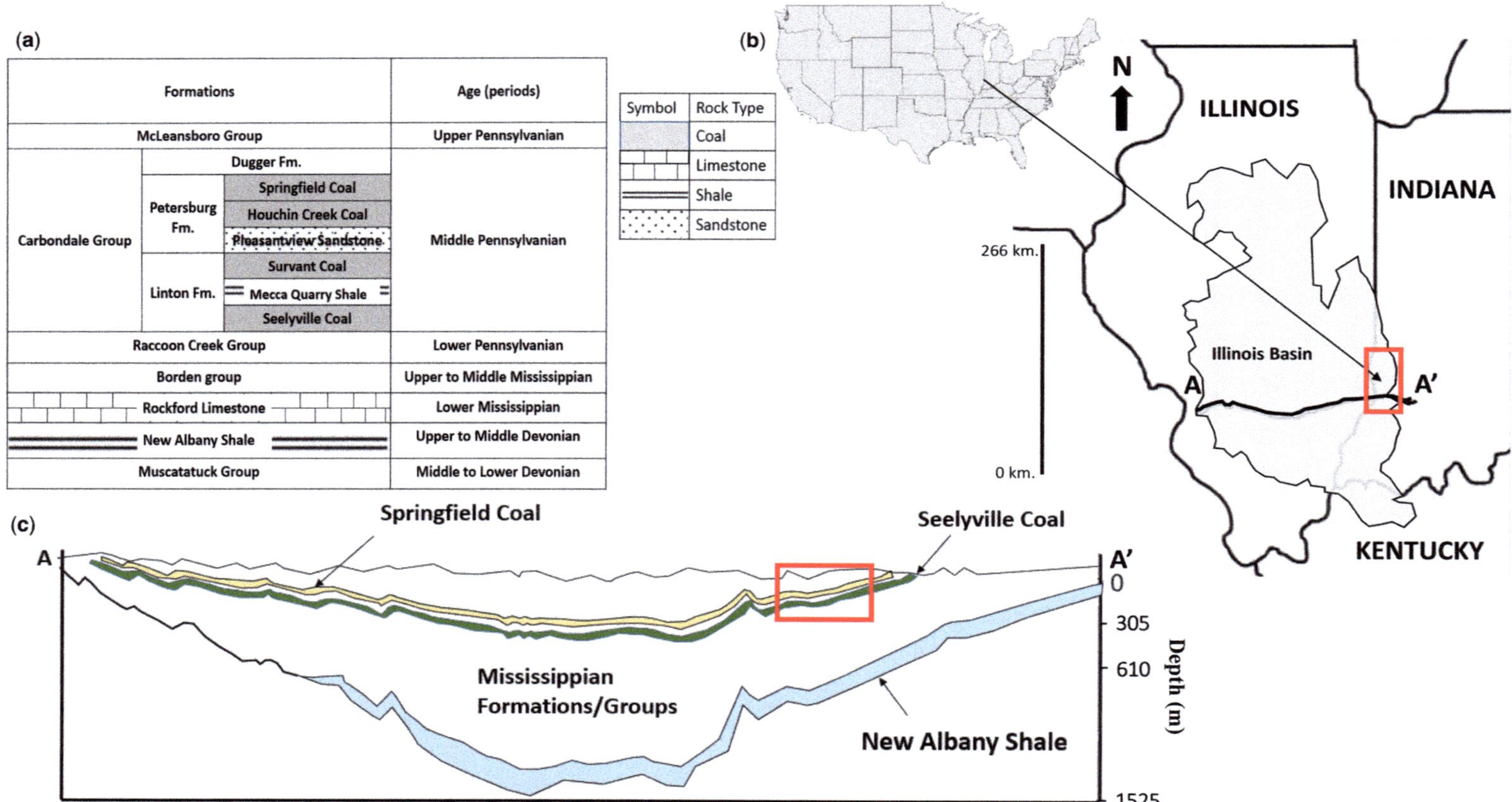

Fig. 1. Generalized stratigraphic column (**a**), areal extent of the Illinois Basin (**b**), and a simplified cross-section (**c**) (reproduced from Strąpoć *et al.* 2007) of the synclinal Illinois Basin. The stratigraphic column displays the Middle Pennsylvanian–aged Springfield and Seelyville coal seams (of the Carbondale Group) and the Upper to Middle Devonian–aged New Albany Shale unit. Samples were collected from producing CBM wells in the area denoted by the red box on the eastern edge of the Illinois Basin (1c).

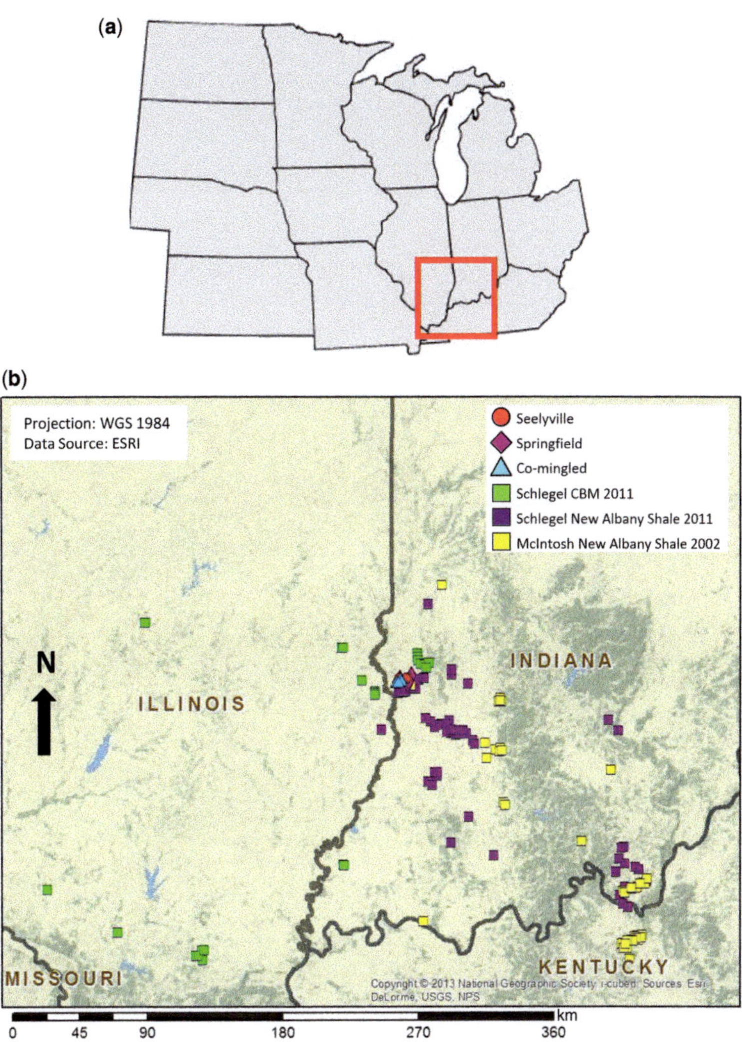

Fig. 2. Inset map of study area denoted by the red box (**a**) and a digital terrain map (**b**) showing the locations of producing CBM wells sampled as part of this study in Sullivan County, Indiana denoted as red, pink and cyan symbols ($n = 20$). CBM well locations from previous studies (Schlegel *et al.* 2011*a*) are shown as green squares and New Albany Shale well locations from previous studies (McIntosh *et al.* 2002; Schlegel *et al.* 2011*a*) are depicted as purple and yellow squares, respectively. Previous publications that did not provide latitude and longitude coordinates or that re-sampled previous wells are not shown.

degrade. In general, there is an approximately linear relationship between the hydrogen (δ^2H) and carbon (δ^{13}C) isotopes of CH_4, C_2H_6, C_3H_8, etc., in which the δ^2H and δ^{13}C values increase in each compound with increasing thermal maturity (Chung *et al.* 1988; Tang *et al.* 2000). If a single generation of thermogenic natural gas were produced, one would anticipate a linear relationship between δ^{13}C of CH_4, C_2H_6, C_3H_8 and the δ^{13}C of the source rock. The fractionation between the source of carbon and the associated hydrocarbon gas following thermogenic

maturation is smaller than that from microbial processes (Faber *et al.* 1992; Whiticar *et al.* 1994). Microbial gases have the most negative δ^{13}C–CH_4, followed by fairly immature thermogenic gases, and highly mature thermogenic gases have the highest δ^{13}C–CH_4. Therefore, the δ^{13}C–CH_4 of biogenic gas is most likely to overlap with early-maturity thermogenic gases. The increases in thermal maturity for natural gas formation by thermogenic processes is also associated with an increase in the C_1/C_2+ from an early-maturity thermogenic gas often

Table 1. *Major gas composition*

Sample ID	Coal gas production unit	Well maximum depth screened (m)	CH$_4$ cm^3 STP cm^{-3}	($\pm 1\sigma$)	C$_2$H$_6$ cm^3 STP cm^{-3}	($\pm 1\sigma$)	C$_3$H$_8$ cm^3 STP cm^{-3}	N$_2$ cm^3 STP cm^{-3}	($\pm 1\sigma$)	O$_2$ cm^3 STP cm^{-3}	CO$_2$ cm^3 STP cm^{-3}	($\pm 1\sigma$)
IN-1	Springfield	76	0.922	0.011	1.27×10^{-4}	1.87×10^{-6}	b.d.l.	0.066	7.91×10^{-4}	<0.002	0.012	1.23×10^{-4}
IN-2	Springfield	73	0.961	0.013	2.05×10^{-4}	3.05×10^{-6}	b.d.l.	0.031	4.13×10^{-4}	<0.002	0.008	7.47×10^{-5}
IN-3	Springfield	91	0.965	0.011	4.50×10^{-4}	6.72×10^{-6}	b.d.l.	0.023	3.19×10^{-4}	<0.002	0.013	1.26×10^{-4}
IN-4	Springfield	107	0.953	0.013	5.07×10^{-4}	7.63×10^{-6}	b.d.l.	0.043	5.46×10^{-4}	<0.002	0.004	4.01×10^{-5}
IN-5	Springfield	76	0.950	0.010	1.73×10^{-4}	2.34×10^{-6}	b.d.l.	0.044	4.99×10^{-4}	<0.002	0.006	6.72×10^{-5}
IN-6	Seelyville	173	0.939	0.013	4.85×10^{-4}	4.44×10^{-6}	b.d.l.	0.048	5.35×10^{-4}	<0.002	0.008	7.87×10^{-5}
IN-7	Seelyville	177	0.959	0.012	5.04×10^{-4}	6.83×10^{-6}	b.d.l.	0.034	3.77×10^{-4}	<0.002	0.007	5.64×10^{-5}
IN-8	Seelyville	213	0.943	0.010	4.38×10^{-4}	5.89×10^{-6}	b.d.l.	0.043	4.42×10^{-4}	<0.002	0.008	6.83×10^{-5}
IN-9	Seelyville	171	0.963	0.012	3.73×10^{-4}	4.20×10^{-6}	b.d.l.	0.021	2.22×10^{-4}	<0.002	0.017	1.69×10^{-4}
IN-10	Seelyville	176	0.959	0.010	2.47×10^{-4}	3.03×10^{-6}	b.d.l.	0.032	3.70×10^{-4}	<0.002	0.010	9.82×10^{-5}
IN-11	Seelyville	153	0.961	0.012	3.82×10^{-4}	5.41×10^{-6}	b.d.l.	0.030	3.15×10^{-4}	<0.002	0.009	7.50×10^{-5}
IN-12	Seelyville	167	0.954	0.011	2.38×10^{-4}	3.26×10^{-6}	b.d.l.	0.040	4.30×10^{-4}	<0.002	0.005	5.68×10^{-5}
IN-13	Seelyville	175	0.899	0.011	9.25×10^{-5}	1.35×10^{-6}	b.d.l.	0.085	9.44×10^{-4}	<0.002	0.006	6.11×10^{-5}
IN-14	Seelyville	175	0.942	0.011	1.09×10^{-3}	1.58×10^{-5}	b.d.l.	0.041	4.16×10^{-4}	<0.002	0.007	7.30×10^{-5}
IN-15	Co-mingled	158	0.948	0.012	7.15×10^{-5}	9.68×10^{-7}	b.d.l.	0.049	5.80×10^{-4}	<0.002	0.004	4.09×10^{-5}
IN-16	Co-mingled	174	0.973	0.012	4.73×10^{-4}	6.41×10^{-6}	b.d.l.	0.021	2.45×10^{-4}	<0.002	0.006	6.53×10^{-5}
IN-17	Co-mingled	No information	0.902	0.010	4.25×10^{-4}	6.18×10^{-6}	b.d.l.	0.075	7.99×10^{-4}	<0.002	0.010	1.10×10^{-4}
IN-18	Co-mingled	185	0.875	0.009	7.84×10^{-5}	9.84×10^{-7}	b.d.l.	0.097	1.09×10^{-3}	<0.002	0.009	9.11×10^{-5}
IN-19	Co-mingled	No information	0.938	0.011	3.12×10^{-4}	4.86×10^{-6}	b.d.l.	0.052	6.13×10^{-4}	<0.002	0.011	1.06×10^{-4}
IN-20	Co-mingled	No information	0.971	0.011	5.40×10^{-4}	8.91×10^{-6}	b.d.l.	0.020	2.38×10^{-4}	<0.002	0.009	8.17×10^{-5}
Method Detection Limit			1.00×10^{-5}		5.00×10^{-6}		5.00×10^{-6}	1.00×10^{-3}		2.00×10^{-3}	1.00×10^{-3}	

b.d.l., below detection limits.

Table 2. *Noble gas composition*

Sample ID	Coal gas production unit	Well maximum depth screened (m)	^{4}He $\times 10^{-6}$ cm^3 STP cm^{-3}	$(\pm 1\sigma)$	^{22}Ne $\times 10^{-6}$ cm^3 STP cm^{-3}	$(\pm 1\sigma)$	Ne $\times 10^{-6}$ cm^3 STP cm^{-3}	^{36}Ar $\times 10^{-6}$ cm^3 STP cm^{-3}	$(\pm 1\sigma)$	Ar $\times 10^{-6}$ cm^3 STP cm^{-3}	^{84}Kr $\times 10^{-9}$ cm^3 STP cm^{-3}	$(\pm 1\sigma)$	Kr $\times 10^{-9}$ cm^3 STP cm^{-3}	^{130}Xe $\times 10^{-9}$ cm^3 STP cm^{-3}	$(\pm 1\sigma)$	Xe $\times 10^{-9}$ cm^3 STP cm^{-3}
IN-1	Springfield	76	29.6	0.20	0.107	1.20×10^{-3}	1.153	2.717	8.69×10^{-3}	833.4	98.49	1.62	173.10	0.540	1.40×10^{-3}	12.68
IN-2	Springfield	73	74.2	0.42	0.017	2.10×10^{-4}	0.182	0.506	1.62×10^{-3}	166.1	8.49	0.14	14.93	0.050	1.29×10^{-4}	1.12
IN-3	Springfield	91	18.9	0.12	0.012	1.44×10^{-4}	0.127	0.397	1.27×10^{-3}	123.2	8.70	0.14	15.28	0.056	1.46×10^{-4}	1.33
IN-4	Springfield	107	423.4	2.10	0.029	2.95×10^{-4}	0.310	0.439	1.40×10^{-3}	144.6	5.54	0.09	9.74	0.033	8.62×10^{-5}	0.69
IN-5	Springfield	76	244.7	1.31	0.078	9.78×10^{-4}	0.848	1.082	3.46×10^{-3}	334.4	15.82	0.26	27.81	0.070	1.82×10^{-4}	1.57
IN-6	Seelyville	173	421.3	2.01	0.061	7.60×10^{-4}	0.664	0.602	1.93×10^{-3}	201.9	6.61	0.11	11.62	0.032	8.29×10^{-5}	0.69
IN-7	Seelyville	177	460.8	2.38	0.028	3.48×10^{-4}	0.302	0.511	1.64×10^{-3}	166.9	7.70	0.13	13.52	0.039	1.00×10^{-4}	0.82
IN-8	Seelyville	213	23.7	0.12	0.043	4.80×10^{-4}	0.460	2.000	6.40×10^{-3}	581.6	65.91	1.08	115.83	0.384	9.99×10^{-4}	8.17
IN-9	Seelyville	171	174.8	0.90	0.015	1.76×10^{-4}	0.167	0.287	9.18×10^{-4}	96.0	3.26	0.05	5.74	0.024	6.23×10^{-5}	0.47
IN-10	Seelyville	176	1464.6	6.40	0.159	1.83×10^{-3}	1.715	0.528	1.69×10^{-3}	175.9	5.03	0.08	8.85	0.018	4.78×10^{-5}	0.36
IN-11	Seelyville	153	149.6	0.86	0.019	1.95×10^{-4}	0.205	0.376	1.20×10^{-3}	124.8	4.96	0.08	8.71	0.017	4.41×10^{-5}	0.39
IN-12	Seelyville	167	70.3	0.18	0.016	1.95×10^{-4}	0.172	0.343	1.10×10^{-3}	112.8	4.16	0.07	7.31	0.017	4.33×10^{-5}	0.36
IN-13	Seelyville	175	31.3	0.15	0.048	6.80×10^{-4}	0.515	1.597	5.11×10^{-3}	475.6	61.81	1.01	108.62	0.292	7.59×10^{-4}	6.54
IN-14	Seelyville	175	286.7	1.95	0.030	3.05×10^{-4}	0.321	0.705	2.26×10^{-3}	225.9	10.68	0.18	18.77	0.085	2.21×10^{-4}	1.86
IN-15	Co-mingled	158	47.0	0.31	0.036	4.02×10^{-4}	0.386	1.203	3.85×10^{-3}	371.4	37.88	0.62	66.57	0.192	5.00×10^{-4}	4.29
IN-16	Co-mingled	174	192.0	1.22	0.015	1.95×10^{-4}	0.159	0.405	1.29×10^{-3}	133.6	5.42	0.09	9.52	0.035	9.15×10^{-5}	0.81
IN-17	Co-mingled	No information	11.6	0.09	0.107	1.20×10^{-3}	1.153	3.688	1.18×10^{-2}	1092.4	145.35	2.38	255.45	0.749	1.95×10^{-3}	16.30
IN-18	Co-mingled	185	37.2	0.17	0.102	1.04×10^{-3}	1.099	3.420	1.09×10^{-2}	1015.9	139.82	2.29	245.73	0.680	1.77×10^{-3}	15.04
IN-19	Co-mingled	No information	62.5	0.32	0.016	1.80×10^{-4}	0.172	0.343	1.10×10^{-3}	112.7	4.07	0.07	7.15	0.027	7.13×10^{-5}	0.58
IN-20	Co-mingled	No information	96.1	0.59	0.018	2.03×10^{-4}	0.194	0.380	1.21×10^{-3}	126.2	5.15	0.08	9.05	0.032	8.26×10^{-5}	0.71

Table 3. *Statistical summary of major and noble gas compositions*

	CH_4	C_2H_6	C_3H_8	N_2	O_2	CO_2	^4He	^{22}Ne	Ne	^{36}Ar	Ar	^{84}Kr	Kr	^{130}Xe	Xe
	cm^3 STP cm^{-3}	cm^3 STP cm^{-3}	cm^3 STP cm^{-3}	cm^3 STP cm^{-3}	cm^3 STP cm^{-3}	cm^3 STP cm^{-3}	cm^3 STP cm^{-3}	cm^3 STP cm^{-3}	cm^3 STP cm^{-3}	cm^3 STP cm^{-3}	cm^3 STP cm^{-3}	cm^3 STP cm^{-3}	cm^3 STP cm^{-3}	cm^3 STP cm^{-3}	cm^3 STP cm^{-3}
All samples							$\times10^{-6}$	$\times10^{-6}$	$\times10^{-6}$	$\times10^{-6}$	$\times10^{-6}$	$\times10^{-9}$	$\times10^{-9}$	$\times10^{-9}$	$\times10^{-9}$
Average	0.944	3.60×10^{-4}	b.d.l.	0.045	<0.002	0.009	216.0	0.048	0.515	1.08	330.8	32.2	56.7	0.17	3.74
Minimum	0.875	7.15×10^{-5}	b.d.l.	0.020	<0.002	0.004	11.6	0.012	0.127	0.29	96.0	3.3	5.7	0.02	0.36
Maximum	0.973	1.09×10^{-3}	b.d.l.	0.097	<0.002	0.017	1464.6	0.159	1.715	3.69	1092.4	145.4	255.5	0.75	16.30
Standard deviation	0.026	2.32×10^{-4}	b.d.l.	0.021	<0.002	0.003	327.9	0.041	0.448	1.06	311.6	45.9	80.6	0.23	5.20
Springfield															
Average	0.950	2.92×10^{-4}	b.d.l.	0.041	<0.002	0.009	158.2	0.048	0.524	1.03	320.3	27.4	48.2	0.15	3.48
Minimum	0.922	1.27×10^{-4}	b.d.l.	0.023	<0.002	0.004	18.9	0.012	0.127	0.40	123.2	5.5	9.7	0.03	0.69
Maximum	0.965	5.07×10^{-4}	b.d.l.	0.066	<0.002	0.013	423.4	0.107	1.153	2.72	833.4	98.5	173.1	0.54	12.68
Standard deviation	0.017	1.73×10^{-4}	b.d.l.	0.017	<0.002	0.004	173.8	0.042	0.453	0.98	298.8	39.9	70.2	0.22	5.16
Seelyville															
Average	0.947	4.27×10^{-4}	b.d.l.	0.041	<0.002	0.009	342.6	0.047	0.502	0.77	240.1	18.9	33.2	0.10	2.18
Minimum	0.899	9.25×10^{-5}	b.d.l.	0.021	<0.002	0.005	23.7	0.015	0.167	0.29	96.0	3.3	5.7	0.02	0.36
Maximum	0.963	1.09×10^{-3}	b.d.l.	0.085	<0.002	0.017	1464.6	0.159	1.715	2.00	581.6	65.9	115.8	0.38	8.17
Standard deviation	0.020	2.80×10^{-4}	b.d.l.	0.018	<0.002	0.003	450.0	0.045	0.485	0.60	170.8	25.6	45.0	0.14	3.00
Co-mingled															
Average	0.934	3.17×10^{-4}	b.d.l.	0.052	<0.002	0.008	74.4	0.049	0.527	1.57	475.4	56.3	98.9	0.29	6.29
Minimum	0.875	7.15×10^{-5}	b.d.l.	0.020	<0.002	0.004	11.6	0.015	0.159	0.34	112.7	4.1	7.2	0.03	0.58
Maximum	0.973	5.40×10^{-4}	b.d.l.	0.097	<0.002	0.011	192.0	0.107	1.153	3.69	1092.4	145.4	255.5	0.75	16.30
Standard deviation	0.039	2.02×10^{-4}	b.d.l.	0.030	<0.002	0.003	64.1	0.044	0.472	1.57	459.1	68.1	119.7	0.34	7.41
ANOVA comparison of groups (*p*-value)	0.569	0.671	n.a.	0.516	n.a.	0.936	0.284	0.993	0.994	0.377	0.379	0.307	0.307	0.342	0.341

starting out with >20% ethane (Bernard *et al.* 1976). As a result, the C_1/C_2+ is often critical to determining the proportion of biogenic and thermogenic gases in a mixture of natural gas. The challenge of using $\delta^{13}C$–CH_4 and C_1/C_2+ is exacerbated by the large range of possibilities and uncertainty for the biogenic and thermogenic endmembers.

Following the formation of biogenic and thermogenic hydrocarbon gases, post-genetic processes can also alter their molecular and isotopic compositions. These processes include: (1) secondary biodegradation that consumes wet ($C_2+ > 2\%$) gas components, while yielding CH_4 via methanogenesis; (2) mixing with other biogenic or thermogenic natural gases; (3) anaerobic oxidation, which is mediated by methanotrophs and paired with a terminal electron acceptor (e.g. sulphate); (4) aerobic oxidation in the presence of oxygen (O_2) inorganically or microbially through the activity of methane monooxygenase mediated by the Methylococcaceae family of bacteria; (5) diffusive fractionation; and (6) fractionation by solubility partitioning between the liquid and gas phases during fluid migration (James & Burns 1984; Rowe & Muehlenbachs 1999*b*; Whiticar 1999; Martini *et al.* 2003; Kinnaman *et al.* 2007; Etiope *et al.* 2009*a*; Pape *et al.* 2010; Valentine *et al.* 2010; Darrah *et al.* 2015*b*). These post-genetic alteration processes further obscure the interpretation of natural gas origins based on C and H stable isotopes and C_1/C_2+ alone.

Additional constraints yielded by integrating hydrocarbons and noble gases

Noble gases occur naturally in low (but readily quantifiable) abundances relative to hydrocarbons within the crust (Ballentine *et al.* 2002; Ballentine & Burnard 2002). As a result of their inert nature, noble gas compositions in the crust are not altered by microbial respiration, chemical reactions (e.g. sulphate reduction) or oxidation (Ballentine *et al.* 2002; Ballentine & Burnard 2002; Sherwood Lollar & Ballentine 2009). The elemental concentration and isotopic composition of noble gases are also well constrained in all terrestrial reservoirs, including the atmosphere, hydrosphere, crust and mantle, which further enhances their functionality as geochemical tracers of crustal fluids and their interactions (Ballentine *et al.* 2002; Ballentine & Burnard 2002).

Noble gas compositions of most crustal fluids represent a binary mixture of atmospherically sourced isotopes derived from meteoric water recharge (e.g. ^{20}Ne, ^{36}Ar, ^{84}Kr) and radiogenic/nucleogenic isotopes derived from radioactive decay (e.g. ^{4}He, $^{21}Ne^*$, $^{40}Ar^*$, $^{136}Xe^*$) (Ballentine *et al.* 2002; Ballentine & Burnard 2002). Atmospheric noble gases (termed air-saturated water or ASW) dissolve into water following Henry's Law

of solubility, in which the solubility of a given gas increases with increasing atomic mass (He < Ne < Ar < Kr < Xe) (Ballentine & Burnard 2002). These gases are subsequently incorporated into subsurface fluids during recharge or can be trapped in fluids that occupy sediment pores prior to lithification (i.e. groundwater and formation water, respectively). Because the meteoric source for ASW is considered constant globally, the concentrations of ASW components are a reasonably well-constrained function of temperature, salinity and atmospheric pressure (elevation of recharge) (Ballentine & Burnard 2002).

Radiogenic and nucleogenic noble gases are produced in the Earth's crust during the decay of U (e.g. ^{4}He and $^{21}Ne^*$, $^{136}Xe^*$), Th (e.g. ^{4}He and $^{21}Ne^*$) and ^{40}K ($^{40}Ar^*$) (Ballentine & Burnard 2002). In the continental crust, typical ranges for radioactive components are U (*c.* 0.5–10 mg kg^{-1}), Th (*c.* 1–15 mg kg^{-1}) and ^{40}K (*c.* 1 mg kg^{-1} of ^{40}K, which undergoes branched decay whereby 11% of ^{40}K decays to $^{40}Ar^*$) (Taylor & McLennan 1995). As fluid–rock interactions occur within the Earth's crust, the noble gas compositions of crustal fluids change according to the radiogenic nature and temperature conditions of the rock in which the fluids form and through which they migrate (Ballentine *et al.* 2002). Because the process of fluid migration (e.g. diffusion, single-phase advection, multiple-phase advection) fractionates hydrocarbons and noble gases in a predictable fashion (Etiope *et al.* 2009*a*; Gilfillan *et al.* 2009; Anderson *et al.* 2017), noble gas isotopic ratios (e.g. $^{3}He/^{4}He$, $^{21}Ne/^{22}Ne$, $^{40}Ar/^{36}Ar$, $^{20}Ne/^{36}Ar$) can be used to distinguish between ASW, crustal, and mantle sources, resolve the relative contributions of thermogenic and biogenic methane, and identify post-genetic processes that may alter hydrocarbon composition (Zhou *et al.* 2005; Zhou & Ballentine 2006; Schlegel *et al.* 2011*b*; Hunt *et al.* 2012; Darrah *et al.* 2014, 2015*b*; Wen *et al.* 2015; Barry *et al.* 2016, 2017; Harkness *et al.* 2017).

The presence of primordial ^{3}He and elevated ratios of helium to atmospherically derived neon (i.e. He/Ne) can be used to apportion small contributions (*c.* 1%) of mantle-derived fluids in shallow groundwater, natural gas or other crustal fluids (Oxburgh *et al.* 1986; Poreda *et al.* 1986). Air-saturated water and excess atmospheric air exhibit relatively low He/Ne ranging from 0.219 to 0.247 and *c.* 0.288, respectively (Ballentine *et al.* 2002; Ballentine & Burnard 2002). By comparison, crustal-derived and mantle fluids typically exhibit He/Ne >1000, but can be distinguished by the $^{3}He/^{4}He$ (shown as R/R_A) (Craig *et al.* 1978; Hilton 1996). In general, elevated R/R_A (where R is the $^{3}He/^{4}He$ ratio measured in a sample and R_A is the globally constrained $^{3}He/^{4}He$ ratio of air; 1.384×10^{-6}) values have been associated with active tectonic margins, thinned lithosphere or areas of high heat flow (Oxburgh *et al.* 1986; Oxburgh & Onions 1987; Ballentine & Burnard 2002).

Study area

Geological setting: Pennsylvanian coals of the Eastern Illinois Basin

The Illinois Basin is a 1.55×10^4 km^2 oval-shaped basin extending from central Illinois to western Indiana and northwestern Kentucky (Fig. 1; Buschbach & Kolata 1990; Drobniak *et al.* 2004; Mastalerz *et al.* 2013; Karacan *et al.* 2014). It is bounded by the Mississippi River Arch to the NW, the Kankakee Arch to the NE, the Ozark dome to the SW and the New Madrid Rift Complex to the south (Buschbach & Kolata 1990). These structural features separate the Illinois Basin from adjacent basins such as the Appalachian Basin to the east, the Forest City Basin to the west and the Michigan Basin to the NE (Buschbach & Kolata 1990).

The Illinois Basin is dominantly a foredeep, intracratonic basin that, like the Michigan Basin, formed during the evolution of Appalachian tectonics *c.* 530–280 Ma (McIntosh *et al.* 2002; Strąpoć *et al.* 2007). Sedimentation in the Illinois Basin occurred during both transgressive and regressive phases and led to infills up to 3700 m thick near the depocentre, which is located *c.* 100 km SW of the study area in Clay County, Illinois (Strąpoć *et al.* 2007; Fig. 1).

During the Middle to Upper Devonian, sedimentation occurred within a semi-restricted basin that produced a stratified, anoxic, marine environment leading to the accumulation of organic-rich sediments (Cluff 1980) that eventually formed the New Albany Shale (Strąpoć *et al.* 2010). Currently, depths to the New Albany Shale range from surface outcrop at the edges of the Illinois Basin to near 1500 m at the depocentre. Thermal maturity of the New Albany Shale ranges from a vitrinite reflectance of *c.* 0.5% R_o on the margins of the basin to *c.* 1.5% R_o near the depocentre (Cluff 1980; Strąpoć *et al.* 2010).

The overlying Pennsylvanian coal-bearing strata in the Illinois Basin were deposited from *c.* 318 to 299 Ma and are divided into three major groups: the Raccoon Creek, the Carbondale and the McLeansboro Groups (Strąpoć *et al.* 2007). Two coal intervals for CBM production in the Illinois Basin today are the Springfield and Seelyville coal seams, which are members of the Petersburg and Linton Formations, respectively, and are contained within the Carbondale Group (Strąpoć *et al.* 2007; Figs 1 & 2).

In our study area located at the eastern edge of the basin, the Springfield and Seelyville coal seams have thicknesses of 1.37–1.83 and 0.3–3.0 m, respectively (Ruppert *et al.* 2002; Drobniak *et al.* 2004). Both were deposited in a near-shore to marginal marine environment characterized by tidal coastal plains (Mastalerz *et al.* 1999, 2004; Strąpoć *et al.* 2007). Both the Springfield and Seelyville coal seams crop out on the western and eastern portions of the Illinois Basin and reach depths of 270 and 305 m, respectively, near the depocentre of the basin (Fig. 1; Ruppert *et al.* 2002; Drobniak *et al.* 2004). In general, the highest thermal maturities occur in the deepest portion of the basin, with vitrinite reflectance values decreasing towards the basin margins; exceptions include areas that experienced thermal metamorphism related to dyke intrusions (Green *et al.* 2003; Stewart *et al.* 2005; Strąpoć *et al.* 2007, 2010; Schimmelmann *et al.* 2009). These vitrinite reflectance values are consistent with high-volatility bituminous A coals (Green *et al.* 2003; Strąpoć *et al.* 2007), which transition to lower-ranked coals (e.g. high-volatility bituminous B and C) towards the basin margins as observed elsewhere (Green *et al.* 2003; Strąpoć *et al.* 2007). The highest thermal maturities in the Springfield and Seelyville coal seams are located near the basin centre with vitrinite reflectance values of 0.7–0.8% R_o. In our study area, the observed R_o values range from 0.5% to 0.65% R_o (Mastalerz *et al.* 2004; Strąpoć *et al.* 2007, 2008*a*, 2010).

Natural gas production

Production of natural gas from CBM reservoirs in the Illinois Basin began in 2000 and has increased significantly within the last decade. Currently, CBM production in the Illinois Basin is *c.* 11 million cubic metres per day, with total estimated reserves of 1.5–6 billion cubic metres (Drobniak *et al.* 2004; Mastalerz *et al.* 2004, 2013; Karacan *et al.* 2014).

Several lines of evidence indicate that gases produced from the Springfield and Seelyville coal seams are predominantly microbial in origin. Previous work suggests that >99% of the natural gas in the current study area was produced by microbial methanogenesis; these conclusions were based on high cell counts of methanogenic archaea (Strąpoć *et al.* 2008*b*; Schlegel *et al.* 2011*a*), loss of aliphatic and aromatic compounds from the coals by biodegradation (Furmann *et al.* 2013; Gao *et al.* 2013; Schlegel *et al.* 2013), elevated C_1/C_2+, more negative δ^{13}C–CH$_4$ (<−55‰), and more positive values of δ^{13}C-DIC (Strąpoć *et al.* 2008*a*; Schlegel *et al.* 2011*b*; Tables 4–7). Based on these findings, it appears that a substantial biogenic contribution is irrefutable. Furthermore, the evidence of coal biodegradation implies that some or all of the organic source material supporting microbial

Table 4. *Noble gas and hydrogen and carbon stable isotopic composition in natural gas*

Sample ID	Coal gas production unit	Well maximum depth screened (m)	^{3}He/^{4}He R/R$_A$	($\pm1\sigma$)	^{20}Ne/^{22}Ne	($\pm1\sigma$)	^{21}Ne/^{22}Ne	($\pm1\sigma$)	^{38}Ar/^{36}Ar	($\pm1\sigma$)	^{40}Ar/^{36}Ar	($\pm1\sigma$)	δ^2H-CH$_4$ (‰)	($\pm1\sigma$)	δ^{13}C-CH$_4$ (‰)	($\pm1\sigma$)	δ^{13}C-DIC (‰)	($\pm1\sigma$)
IN-1	Springfield	76	0.110	9.83×10^{-4}	9.785	6.11×10^{-2}	0.0279	1.74×10^{-4}	0.1887	7.44×10^{-4}	305.61	1.23	−209.74	4.22×10^{-1}	−64.1	9.71×10^{-2}	n.r.	n.a.
IN-2	Springfield	73	0.052	4.64×10^{-4}	9.776	6.20×10^{-2}	0.0314	1.93×10^{-4}	0.1830	6.85×10^{-4}	327.34	1.35	n.r.	n.a.	−58.1	8.69×10^{-2}	17	2.24×10^{-2}
IN-3	Springfield	91	0.051	4.49×10^{-4}	9.771	6.02×10^{-2}	0.0304	1.93×10^{-4}	0.1897	7.64×10^{-4}	308.73	1.16	−202.3	4.29×10^{-1}	−53.9	8.11×10^{-2}	n.r.	n.a.
IN-4	Springfield	107	0.051	4.51×10^{-4}	9.772	6.08×10^{-2}	0.0301	1.85×10^{-4}	0.1849	7.11×10^{-4}	328.61	1.30	−206.4	4.21×10^{-1}	−54.6	8.06×10^{-2}	26	3.40×10^{-2}
IN-5	Springfield	76	0.057	5.12×10^{-4}	9.772	5.91×10^{-2}	0.0312	2.00×10^{-5}	0.1864	6.79×10^{-4}	307.84	1.09	−214.9	3.91×10^{-1}	−59.58	8.80×10^{-2}	n.r.	n.a.
IN-6	Seelyville	173	0.044	3.92×10^{-4}	9.778	6.13×10^{-2}	0.0297	1.71×10^{-4}	0.1934	7.79×10^{-4}	334.35	1.35	−210.6	3.62×10^{-1}	−55.31	5.06×10^{-2}	27	3.44×10^{-2}
IN-7	Seelyville	177	0.043	3.89×10^{-4}	9.780	6.04×10^{-2}	0.0294	1.75×10^{-4}	0.1920	7.73×10^{-4}	325.14	1.18	−203.9	3.92×10^{-1}	−55.2	5.31×10^{-2}	24	3.30×10^{-2}
IN-8	Seelyville	213	0.901	7.92×10^{-3}	9.771	6.06×10^{-2}	0.0298	1.72×10^{-4}	0.1863	7.16×10^{-4}	289.61	1.00	−208.66	4.20×10^{-1}	−63.14	8.31×10^{-2}	n.r.	n.a.
IN-9	Seelyville	171	0.045	4.02×10^{-4}	9.776	5.93×10^{-2}	0.0310	1.67×10^{-4}	0.2060	8.12×10^{-4}	333.38	1.21	−202.6	3.89×10^{-1}	−55.87	7.13×10^{-2}	26	3.29×10^{-2}
IN-10	Seelyville	176	0.044	3.83×10^{-4}	9.765	6.00×10^{-2}	0.0316	1.86×10^{-4}	0.1874	7.74×10^{-4}	331.80	1.34	n.r.	n.a.	−57.3	7.94×10^{-2}	27	3.79×10^{-2}
IN-11	Seelyville	153	0.058	5.01×10^{-4}	9.761	5.89×10^{-2}	0.0303	1.60×10^{-4}	0.1902	7.31×10^{-4}	331.18	1.31	−201.4	3.81×10^{-1}	−54.68	6.92×10^{-2}	21	2.87×10^{-2}
IN-12	Seelyville	167	0.061	5.30×10^{-4}	9.769	5.43×10^{-2}	0.0292	1.57×10^{-4}	0.1828	6.26×10^{-4}	327.40	1.23	−203.5	3.65×10^{-1}	−58.64	6.31×10^{-2}	24	3.25×10^{-2}
IN-13	Seelyville	175	0.103	9.28×10^{-4}	9.786	5.05×10^{-2}	0.0289	1.46×10^{-4}	0.1837	7.58×10^{-4}	296.66	1.17	−203.1	3.90×10^{-1}	−63.24	9.29×10^{-2}	n.r.	n.a.
IN-14	Seelyville	175	0.055	5.03×10^{-4}	9.777	5.24×10^{-2}	0.0313	1.98×10^{-4}	0.1859	9.80×10^{-4}	319.33	1.23	−203.6	3.85×10^{-1}	−49.87	6.85×10^{-2}	n.r.	n.a.
IN-15	Co-mingled	158	0.094	8.53×10^{-4}	9.787	5.64×10^{-2}	0.0291	1.68×10^{-4}	0.1895	6.87×10^{-4}	307.50	1.21	−209.8	3.55×10^{-1}	−62.32	8.95×10^{-2}	23	3.30×10^{-2}
IN-16	Co-mingled	174	0.045	4.01×10^{-4}	9.778	4.85×10^{-2}	0.0305	1.79×10^{-4}	0.1870	7.53×10^{-4}	328.98	1.00	n.r.	n.a.	−54.77	8.05×10^{-2}	27	3.96×10^{-2}
IN-17	Co-mingled	No information	0.318	2.77×10^{-3}	9.785	5.05×10^{-2}	0.0282	1.52×10^{-4}	0.1885	7.03×10^{-4}	294.99	1.22	−213.55	3.59×10^{-1}	−65.67	9.95×10^{-2}	n.r.	n.a.
IN-18	Co-mingled	185	0.115	9.85×10^{-4}	9.786	5.91×10^{-2}	0.0289	1.34×10^{-4}	0.1855	6.71×10^{-4}	295.87	1.19	−208.7	4.01×10^{-1}	−69.8	1.02×10^{-1}	n.r.	n.a.
IN-19	Co-mingled	No information	0.050	4.36×10^{-4}	9.762	6.28×10^{-2}	0.0317	1.38×10^{-4}	0.1861	6.41×10^{-4}	327.58	1.23	−208.63	4.20×10^{-1}	−56.8	8.15×10^{-2}	n.r.	n.a.
IN-20	Co-mingled	No information	0.049	4.29×10^{-4}	9.766	6.00×10^{-2}	0.0320	1.80×10^{-4}	0.1889	6.12×10^{-4}	331.29	1.21	−208.9	4.18×10^{-1}	−54.9	8.05×10^{-2}	n.r.	n.a.
ASW values			0.985		9.780		0.0289		0.188		295.50							

Table 5. *Statistical summary of noble gas and hydrogen and carbon stable isotopic compositions in methane*

All samples	^{3}He/^{4}He	^{20}Ne/^{22}Ne	^{21}Ne/^{22}Ne	^{38}Ar/^{36}Ar	^{40}Ar/^{36}Ar	δ^2H-CH$_4$	δ^{13}C-CH$_4$	δ^{13}C-DIC
	R/R_A					(‰)	(‰)	(‰)
Average	0.117	9.775	0.0301	0.1883	317.66	−207.08	−58.39	24.20
Minimum	0.043	9.761	0.0279	0.1828	289.61	−214.90	−69.80	17.00
Maximum	0.901	9.787	0.0320	0.2060	334.35	−201.40	−49.87	27.00
Standard deviation	0.194	0.008	0.0012	0.0050	15.10	4.09	4.89	3.22
Springfield								
Average	0.064	9.775	0.0302	0.1865	315.63	−208.34	−58.06	21.50
Minimum	0.051	9.771	0.0279	0.1830	305.61	−214.90	−64.10	17.00
Maximum	0.110	9.785	0.0314	0.1897	328.61	−202.30	−53.90	26.00
Standard deviation	0.026	0.006	0.0014	0.0027	11.34	5.33	4.13	6.36
Seelyville								
Average	0.150	9.774	0.0301	0.1897	320.98	−204.67	−57.03	24.83
Minimum	0.043	9.761	0.0289	0.1828	289.61	−210.60	−63.24	21.00
Maximum	0.901	9.786	0.0316	0.2060	334.35	−201.40	−49.87	27.00
Standard deviation	0.282	0.008	0.0010	0.0070	16.55	3.20	4.23	2.32
Co-mingled								
Average	0.112	9.777	0.0301	0.1876	314.37	−209.92	−60.71	25.00
Minimum	0.045	9.762	0.0282	0.1855	294.99	−213.55	−69.80	23.00
Maximum	0.318	9.787	0.0320	0.1895	331.29	−208.63	−54.77	27.00
Standard deviation	0.105	0.011	0.0016	0.0016	16.97	2.08	6.23	2.83
ANOVA comparison of groups (*p*-value)	0.748	0.728	0.989	0.500	0.690	0.167	0.375	0.468

methanogenesis is derived from primary biodegradation of the coal rather than migrated hydrocarbons.

Nonetheless, recently published clumped isotope palaeothermometry and noble gas data from the Antrim Shale of the Michigan Basin (Stolper *et al.* 2015; Wen *et al.* 2015) have identified the presence of significantly larger thermogenic contributions than previously estimated. These original interpretations for a nearly pure biogenic endmember were based on traditional hydrocarbon molecular and isotopic compositions and geomicrobiology (e.g. Martini *et al.* 1996, 1998; Schlegel *et al.* 2013). Thus, while we anticipated a significant biogenic component for natural gases from the current study area, we further examine the molecular and isotopic composition of hydrocarbons and the elemental and isotopic composition of noble gases to test the hypothesis that >99% of natural gas within the Springfield and Seelyville coal seams is biogenic in origin.

Methods

Sample collection and data compilation

Gas samples were collected with negligible air contamination using thick-walled 0.95 cm (3/8 inch) outside diameter and 40.6 cm (16 inch) long refrigeration-grade copper tubes from 20 actively producing CBM wells in Sullivan County, Indiana. Copper tubes were connected in-line of the CBM well and purged with production fluids for *c.* 10 min (≫50 tube volumes). The copper tubes were then sealed using brass refrigeration clamps with a 0.762 mm (0.030 inch) gap (Darrah *et al.* 2015*b*; Kang *et al.* 2016).

Gas samples were collected along the eastern margin of the Illinois Basin from CBM wells at screened intervals ranging from 73 to 213 m in depth (Fig. 1c; Tables 1–2). Five wells were producing natural gas from the Springfield coal seam and nine from the Seelyville coal seam; six wells had comingled natural gas production from both coal seams (denoted as 'comingled' production in the figures; Tables 1–2).

Data from this study are shown in red, pink and cyan symbols, with symbol shape and colour denoting the coal seam from which samples are produced. Pink diamonds and red circles represent samples collected from CBM wells producing from Springfield and Seelyville coal seams, respectively, while comingled wells are denoted by cyan triangles in Figures 2–12.

We also present data from our study area within the context of published reports of natural gas data from actively producing CBM wells (*n* = 41) (Strąpoć *et al.* 2007; Schlegel *et al.* 2011*b*; Mastalerz *et al.* 2017) and shale gas wells from the New Albany Shale (*n* = 62) (McIntosh *et al.* 2002, 2004), located within the Illinois Basin. Data from published reports on the Springfield and Seelyville coal

Table 6. *Molecular and isotopic gas ratios*

Sample ID	Coal gas production unit	Well maximum depth screened (m)	$CO_2/^3He$	$CH_4/^3He$	$CH_4/^{36}Ar$ ×10^6	$^4He/CH_4$ ×10^{-6}	N_2/Ar	CH_4/C_2H_6+	CH_4/CO_2	CO_2/CH_4	$^4He/^{20}Ne$	$^{20}Ne/^{36}Ar$	$^4He/^{36}Ar$	$^{84}Kr/^{36}Ar$	$^{130}Xe/^{84}Kr$	$^4He/^{21}Ne*$ ×10^6	$^4He/^{40}Ar*$
IN-1	Springfield	76	1.24×10^{10}	2.03×10^{11}	0.34	32.1	79.6	7256.6	74.3	0.0135	28.4	0.384	10.9	0.036	5.48×10^{-3}	n.a.	1.1
IN-2	Springfield	73	8.38×10^{9}	1.81×10^{11}	1.90	77.2	185.2	4687.6	114.6	0.0087	451.4	0.325	146.7	0.017	5.86×10^{-3}	1.77	4.6
IN-3	Springfield	91	1.30×10^{10}	7.21×10^{11}	2.43	19.5	184.5	2145.6	74.0	0.0135	164.3	0.289	47.4	0.022	6.47×10^{-3}	1.10	3.6
IN-4	Springfield	107	3.95×10^{9}	3.19×10^{10}	2.17	444.3	299.2	1879.6	241.5	0.0041	1510.4	0.639	965.5	0.013	5.97×10^{-3}	12.17	29.2
IN-5	Springfield	76	6.48×10^{9}	4.91×10^{10}	0.88	257.6	131.3	5489.4	146.5	0.0068	319.2	0.709	226.1	0.015	4.42×10^{-3}	1.38	18.3
IN-6	Seelyville	173	8.10×10^{9}	3.69×10^{10}	1.56	448.7	235.4	1935.1	115.9	0.0086	700.8	0.999	700.1	0.011	4.82×10^{-3}	9.12	18.0
IN-7	Seelyville	177	6.95×10^{9}	3.48×10^{10}	1.88	480.6	201.0	1901.1	137.9	0.0073	1685.9	0.535	901.2	0.015	5.01×10^{-3}	35.41	30.4
IN-8	Seelyville	213	7.50×10^{9}	3.18×10^{10}	0.47	25.1	74.2	2154.6	125.7	0.0080	56.9	0.208	11.8	0.033	5.83×10^{-3}	0.60	n.a.
IN-9	Seelyville	171	1.67×10^{10}	8.74×10^{10}	3.35	181.6	217.4	2578.9	57.7	0.0173	1158.2	0.526	609.2	0.011	7.34×10^{-3}	5.34	16.1
IN-10	Seelyville	176	1.00×10^{10}	1.08×10^{10}	1.82	1527.6	179.6	3879.6	95.8	0.0104	944.4	2.936	2773.1	0.010	3.65×10^{-3}	3.37	76.4
IN-11	Seelyville	153	9.25×10^{9}	8.01×10^{10}	2.56	155.6	237.8	2513.4	103.9	0.0096	807.2	0.494	398.5	0.013	3.42×10^{-3}	5.56	11.2
IN-12	Seelyville	167	5.46×10^{9}	1.60×10^{11}	2.78	73.7	358.5	4002.1	174.8	0.0057	452.8	0.452	204.8	0.012	4.01×10^{-3}	13.38	6.4
IN-13	Seelyville	175	5.90×10^{9}	2.01×10^{11}	0.56	34.8	178.4	9725.6	152.4	0.0066	67.0	0.292	19.6	0.039	4.72×10^{-3}	n.a.	16.9
IN-14	Seelyville	175	7.11×10^{9}	4.28×10^{10}	1.34	304.3	179.6	867.6	132.5	0.0075	986.0	0.413	406.8	0.015	7.96×10^{-3}	4.04	17.1
IN-15	Co-mingled	158	3.91×10^{9}	1.54×10^{11}	0.79	49.6	130.9	13256.6	242.3	0.0041	134.5	0.290	39.1	0.031	5.07×10^{-3}	7.31	3.3
IN-16	Co-mingled	174	6.30×10^{9}	8.18×10^{10}	2.41	197.4	155.2	2057.9	154.3	0.0065	1335.8	0.355	474.6	0.013	6.49×10^{-3}	7.94	14.2
IN-17	Co-mingled	No-information	1.04×10^{10}	1.76×10^{11}	0.24	12.9	68.9	2124.5	86.7	0.0115	11.1	0.283	3.1	0.039	5.15×10^{-3}	n.a.	n.a.
IN-18	Co-mingled	185	8.70×10^{9}	1.48×10^{11}	0.26	42.5	95.0	11164.4	100.6	0.0099	37.4	0.291	10.9	0.041	4.86×10^{-3}	n.a.	29.4
IN-19	Co-mingled	No-information	1.05×10^{10}	2.15×10^{11}	2.74	66.7	464.3	3001.6	89.2	0.0112	401.6	0.454	182.4	0.012	6.74×10^{-3}	1.39	5.7
IN-20	Co-mingled	No-information	8.97×10^{9}	1.49×10^{11}	2.56	99.0	161.0	1796.6	108.2	0.0092	547.0	0.463	253.1	0.014	6.16×10^{-3}	1.73	7.1
ASW (10°C)							37.03					0.255	0.156	0.036	0.045	0.010		

Table 7. *Statistical summary of molecular and isotopic gas ratios*

	$CO_2/^3He$	$CH_4/^3He$	$CH_4/^{36}Ar$	$^4He/CH_4$	N_2/Ar	CH_4/C_2H_6+	CH_4/CO_2	CO_2/CH_4	$^4He/^{20}Ne$	$^{20}Ne/^{36}Ar$	$^4He/^{36}Ar$	$^{84}Kr/^{36}Ar$	$^{130}Xe/^{84}Kr$	$^4He/^{21}Ne*$	$^4He/^{40}Ar*$
All samples			$\times10^6$	$\times10^{-6}$										$\times10^6$	
Average	8.50×10^9	1.40×10^{11}	1.65	226.5	190.8	4220.9	126.5	9.01×10^{-3}	590.0	0.567	419.3	0.021	0.005	6.57	16.25
Minimum	3.91×10^9	1.08×10^{10}	0.24	12.9	68.9	867.6	57.7	4.13×10^{-3}	11.1	0.208	3.1	0.010	0.003	0.00	0.00
Maximum	1.67×10^{10}	7.21×10^{11}	3.35	1527.6	464.3	13256.6	242.3	1.73×10^{-2}	1685.9	2.936	2773.1	0.041	0.008	35.41	76.38
Standard deviation	3.13×10^9	1.53×10^{11}	0.98	342.2	97.2	3474.1	49.7	3.29×10^{-3}	529.2	0.587	630.2	0.011	0.001	8.47	17.40
Springfield															
Average	8.85×10^9	2.37×10^{11}	1.54	166.2	176.0	4291.7	130.2	9.33×10^{-3}	494.7	0.469	279.3	0.020	0.006	4.10	11.35
Minimum	3.95×10^9	3.19×10^{10}	0.34	19.5	79.6	1879.6	74.0	4.14×10^{-3}	28.4	0.289	10.9	0.013	0.004	1.10	1.08
Maximum	1.30×10^{10}	7.21×10^{11}	2.43	444.3	299.2	7256.6	241.5	1.35×10^{-2}	1510.4	0.709	965.5	0.036	0.006	12.17	29.17
Standard deviation	3.88×10^9	2.81×10^{11}	0.89	182.4	81.6	2280.7	69.3	4.13×10^{-3}	589.7	0.192	392.8	0.009	0.001	5.38	12.01
Seelyville															
Average	8.55×10^9	7.62×10^{10}	1.81	359.1	206.9	3284.2	121.8	9.01×10^{-3}	762.1	0.762	669.4	0.018	0.005	9.60	24.05
Minimum	5.46×10^9	1.08×10^{10}	0.47	25.1	74.2	867.6	57.7	5.72×10^{-3}	56.9	0.208	11.8	0.010	0.003	0.60	6.42
Maximum	1.67×10^{10}	2.01×10^{11}	3.35	1527.6	358.5	9725.6	174.8	1.73×10^{-2}	1685.9	2.936	2773.1	0.039	0.008	35.41	76.38
Standard deviation	3.37×10^9	6.46×10^{10}	0.97	469.4	74.8	2605.3	34.0	3.44×10^{-3}	521.5	0.845	843.8	0.011	0.002	11.12	22.22
Co-mingled															
Average	8.13×10^9	1.54×10^{11}	1.50	78.0	179.2	5566.9	130.2	8.75×10^{-3}	411.2	0.356	160.6	0.025	0.006	4.59	11.92
Minimum	3.91×10^9	8.18×10^{10}	0.24	12.9	68.9	1796.6	86.7	4.13×10^{-3}	11.1	0.283	3.1	0.012	0.005	1.39	3.26
Maximum	1.05×10^{10}	2.15×10^{11}	2.74	197.4	464.3	13256.6	242.3	1.15×10^{-2}	1335.8	0.463	474.6	0.041	0.007	7.94	29.42
Standard deviation	2.57×10^9	4.36×10^{10}	1.19	65.0	144.1	5204.3	60.1	2.90×10^{-3}	499.9	0.084	184.2	0.014	0.001	3.51	10.59
ANOVA comparison of groups (p-value)	0.94	0.164	0.81	0.281	0.817	0.483	0.939	0.963	0.429	0.408	0.275	0.478	0.393	0.217	0.304

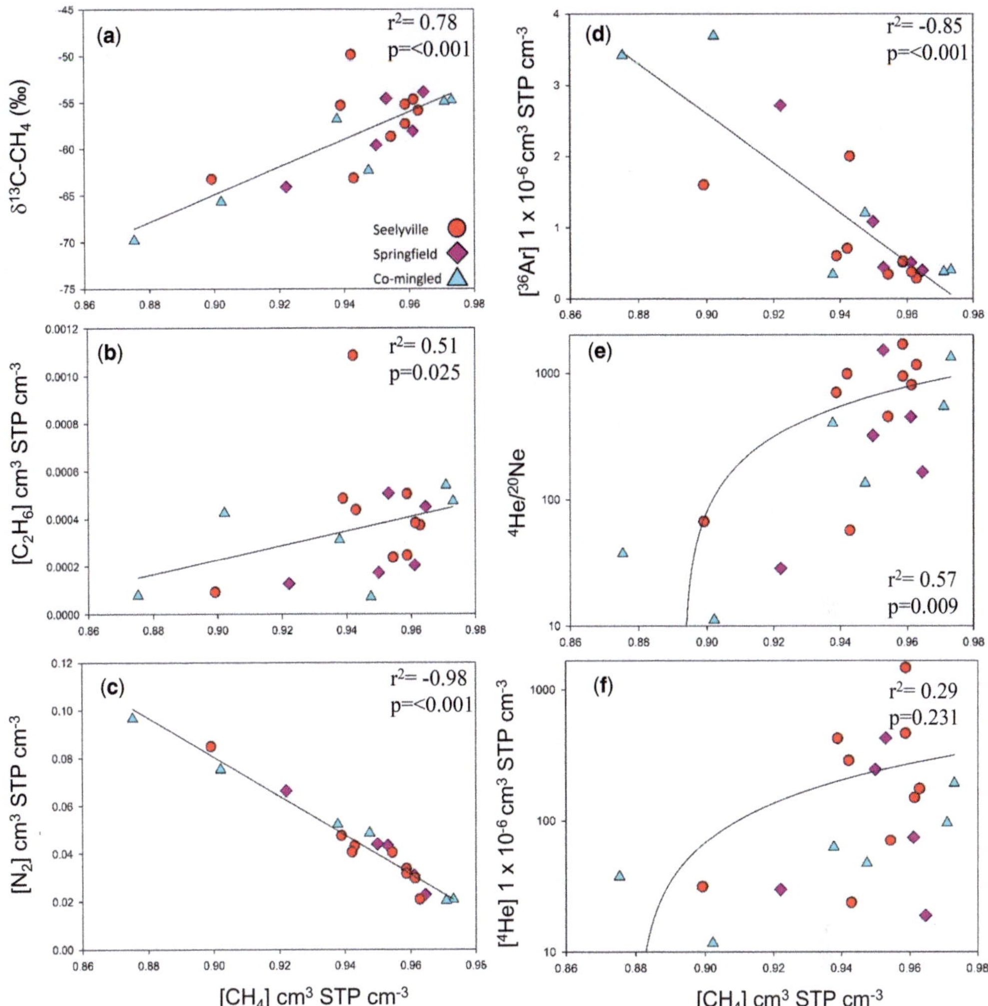

Fig. 3. The stable isotopic values of carbon in methane (δ^{13}C-CH$_4$) (**a**), the concentration of ethane [C$_2$H$_6$] (**b**), the concentration of nitrogen [N$_2$] (**c**), the concentration of argon [^{36}Ar] (**d**), the ratio of ^{4}He/^{20}Ne (**e**; note log scale for visibility), and the concentration of helium [^{4}He] (**f**; note log scale for visibility) v. the concentration of methane [CH$_4$] in producing CBM wells sampled in this study. Linear regression lines, r^2 and p-values of the regression for all data are shown in the plots.

seams are denoted by blue squares (Strąpoć *et al.* 2007), green squares (Schlegel *et al.* 2011*a*) and black squares (Mastalerz *et al.* 2017). Published gas data from the New Albany Shale natural gas wells are denoted by yellow squares (McIntosh *et al.* 2002) and purple squares (Schlegel *et al.* 2011*a*) (Fig. 2).

Sample analyses

Gas samples. Gas samples in the copper tubes were prepared for analysis by cold welding *c.* 2.5 cm

(*c.* 1 inch) splits of the copper tubing using stainless steel clamps. The copper tube was then attached to an ultra-high-vacuum steel line (total pressure = 1–3 × 10^{-9} torr), which was monitored continuously using a 0–20 torr MKS capacitance monometer (accurate to the nearest thousandth) and an isolated ion gauge, using a 0.64 cm (1/4 inch) VCR connection. An aliquot of the gas sample was introduced into the vacuum line for noble gas analyses; this process was repeated sequentially on similar vacuum introduction lines for gas chromatography and isotope ratio mass spectrometry analyses.

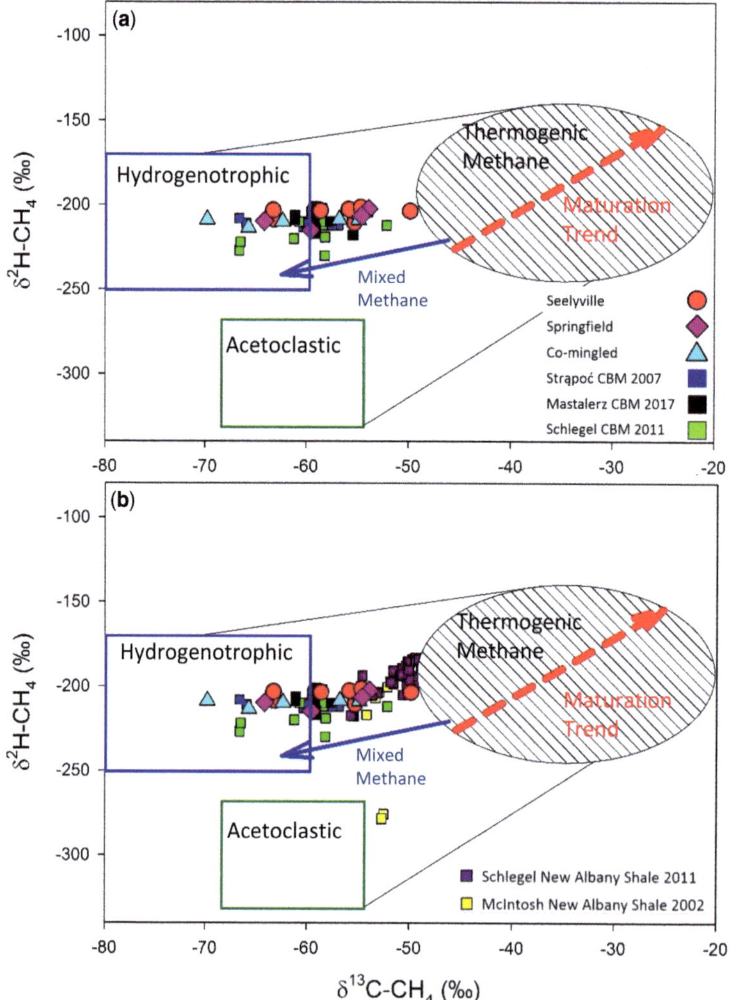

Fig. 4. A comparison of the stable isotopic values of hydrogen (δ^2H-CH$_4$) and carbon (δ^{13}C-CH$_4$) in methane from producing CBM wells from this study (red, pink and cyan symbols) and those from previously published CBM studies in the Illinois Basin (Strąpoć *et al.* 2007, blue squares; Mastalerz *et al.* 2017, black squares; Schlegel *et al.* 2011*a*, green squares) (**a**). Previously published data from gases produced from the New Albany Shale (McIntosh *et al.* 2002; Schlegel *et al.* 2011*a* in purple and yellow squares, respectively) are also displayed for comparison (**b**). The hatched oval represents the typical range of thermogenic natural gases, with the dashed red line indicating the trend of increasing thermal maturity up and to the right (Schoell 1980; Whiticar *et al.* 1986). The outlined blue box represents hydrogenotrophic methane and the outlined green box represents acetoclastic methane (Schoell 1980; Whiticar *et al.* 1986).

The isotopic analyses of noble gases were performed using a Thermo Fisher Helix SFT mass spectrometer at The Ohio State University Noble Gas Laboratory. Noble gas procedures for analysis and purification of samples are summarized in Darrah & Poreda (2012). The average external precision for noble gas isotope concentrations based on 'known–unknown' standards was within ±1.72%, with values reported in parentheses (^{4}He concentrations (0.69%), ^{22}Ne concentrations (1.27%), ^{36}Ar concentrations (0.32%), ^{84}Kr concentrations (1.64%), ^{130}Xe concentrations (1.72%)). These values were determined by measuring referenced and cross-validated laboratory standards including an established atmospheric air standard (Lake Erie, Ohio Air), the Yellowstone MM standard and a series of synthetic natural gas standards obtained from Praxair, including known and cross-validated concentrations of

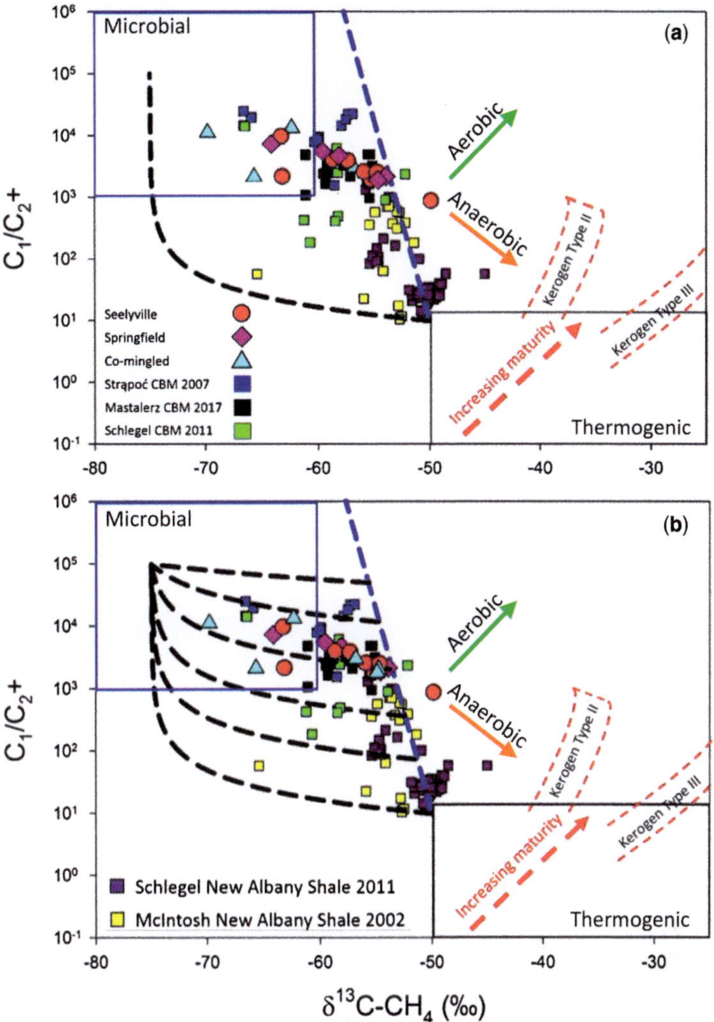

Fig. 5. The relative abundances of methane to higher-order aliphatic hydrocarbons (C_1/C_2+) v. the stable isotopic composition of carbon in methane (δ^{13}C-CH$_4$) from producing CBM wells from this study (red, pink and cyan symbols) and those from previously published CBM studies in the Illinois Basin (Strąpoć *et al.* 2007, blue squares; Mastalerz *et al.* 2017, black squares; Schlegel *et al.* 2011a, green squares) (**a**). Previously published data from gases produced from the New Albany Shale (McIntosh *et al.* 2002; Schlegel *et al.* 2011a in purple and yellow squares, respectively) are also displayed for comparison (**b**). The typical ranges for the composition of thermogenic and biogenic methane are shown in the black and blue boxes, respectively (Bernard *et al.* 1976). Black trend lines demonstrate hypothetical mixing between biogenic and New Albany Shale endmembers. The dashed red arrow in the thermogenic box reflects the trend of increasing thermal maturation and the dashed red boxes denote the thermal maturation trend for gas generated from a Type II or Type III kerogen. The green and orange arrows show post-genetic, microbial alteration by aerobic and anaerobic oxidation, respectively. The blue trend line displays post-genetic alteration from the New Albany Shale following fractionation by solubility partitioning of methane, ethane and propane by GGS-R models (adapted from Gilfillan *et al.* 2009; Darrah *et al.* 2014). The black mixing lines in Figure 5b depict hypothetical mixing between (1) a New Albany Shale-like gas that was modified by phase partitioning starting at various points along the solubility fractionation trend line and (2) a biogenic gas.

C_1–C_5 hydrocarbons, N_2, CO_2, O_2 and each of the noble gases (Tedesco *et al.* 2010; Harkness *et al.* 2017). Noble gas isotopic standard errors were *c.* ±0.0091 times the ratio of air (1.384×10^{-6}) for the ^{3}He/^{4}He ratio, less than ±0.402% and ±0.689% for ^{20}Ne/^{22}Ne and ^{21}Ne/^{22}Ne ratios, respectively,

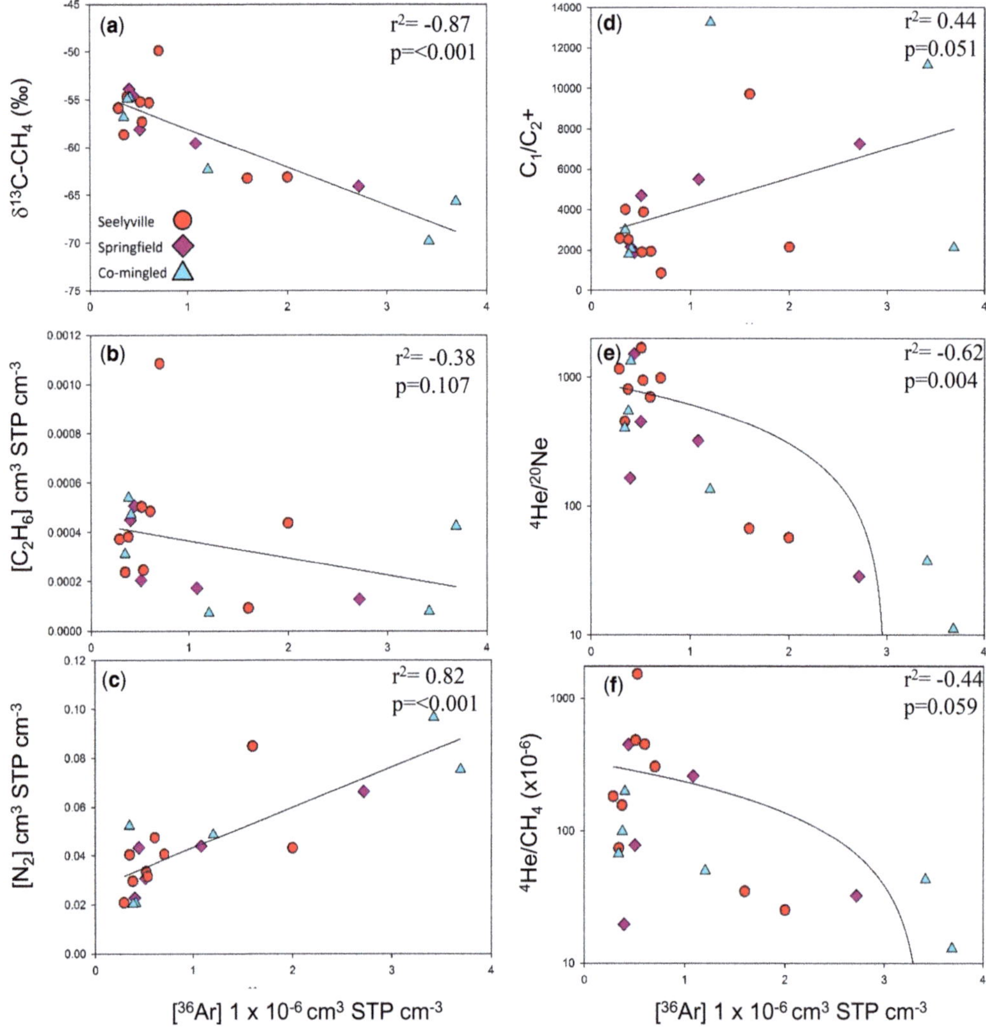

Fig. 6. The stable isotopic values of carbon in methane (δ^{13}C-CH$_4$) (**a**), the concentration of ethane [C$_2$H$_6$] (**b**), the concentration of nitrogen [N$_2$] (**c**), the ratio of methane over higher order hydrocarbons (C$_1$/C$_2$+) (**d**), the ratio of ^{4}He/^{20}Ne (**e**; note the log scale), the ratio of ^{4}He/CH$_4$ (**f**; note log scale) v. the concentration of argon [^{36}Ar] of producing CBM wells sampled in this study. [^{36}Ar] concentrations serve as a proxy for the contributions from ASW in the earth's crust. Linear regression lines, r^2 and p-values of the regression for all data are shown in the plots.

and less than ±0.643% and ±0.427% for ^{38}Ar/^{36}Ar and ^{40}Ar/^{36}Ar ratios, respectively (higher than typical because of interferences from C$_3$H$_8$ on mass = 36 and 38).

Using the atmospheric ratios of ^{21}Ne/^{22}Ne (0.0289) and ^{40}Ar/^{36}Ar (295.5), excess ^{21}Ne (^{21}Ne*, derived mainly from nucleogenic reactions) and excess ^{40}Ar (^{40}Ar*, derived mainly from radiogenic production, although small mantle contributions cannot necessarily be discounted) were calculated by subtracting the atmospheric contributions from the measured values in equations (1) and (2):

$$^{21}\text{Ne}^* = \left(^{21}\text{Ne}/^{22}\text{Ne}_{\text{measured}} - 0.0289\right) \times {}^{22}\text{Ne}_{\text{measured}} \tag{1}$$

$$^{40}\text{Ar}^* = \left(^{40}\text{Ar}/^{36}\text{Ar}_{\text{measured}} - 295.5\right) \times {}^{36}\text{Ar}_{\text{measured}} \tag{2}$$

Major gas components were measured on a SRS Quadrupole MS and the SRI 8610C Multi-Gas 3+ gas chromatograph equipped with a flame ionization detector and thermal conductivity detector at the

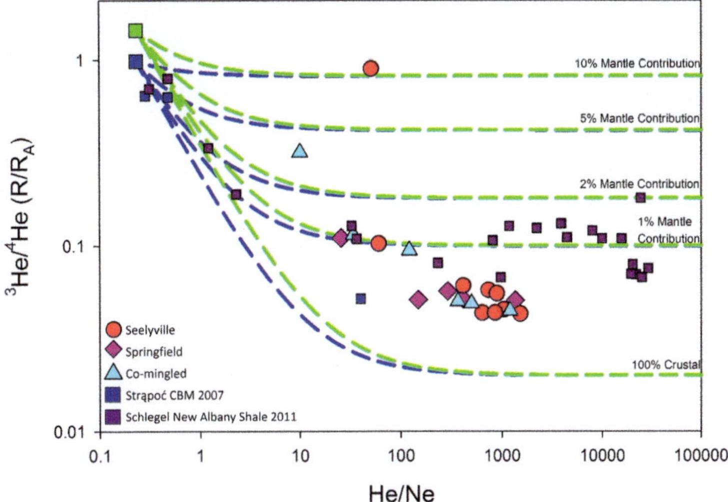

Fig. 7. The helium isotopic ratio ^{3}He/^{4}He (shown as R/R_A, where R_A is the ratio of ^{3}He/^{4}He in atmospheric air; $R_A = 1.384 \times 10^{-6}$) v. the ratio of He/Ne from producing CBM wells in this study. The blue and green boxes denote the globally constrained values for ASW and tritiated ASW, respectively. The dashed blue and green lines represent mixing of ASW and tritiated ASW fluids, respectively, with crustal endmembers with increasing amounts of mantle contributions.

Ohio State University Noble Gas Laboratory (Darrah & Poreda 2012; Hunt *et al.* 2012; Kang *et al.* 2016). Standard analytical errors were all less than ±3.41% for major gas concentrations above the detection limit. The average external precision was determined by measuring an established atmospheric air standard (Lake Erie, Ohio Air) and a series of synthetic natural gas standards obtained from Praxair. The results of the average external precision analyses are as follows: CH$_4$ (1.46%), C$_2$H$_6$ (1.61%), C$_3$H$_8$ (1.97%), N$_2$ (1.25%), CO$_2$ (1.06%), O$_2$ (1.39%) and Ar (0.59%) based on daily replicate measurements during analyses.

Procedures for analyses of stable isotopic values of carbon and hydrogen in methane were described previously (Darrah *et al.* 2013, 2015b; Jackson *et al.* 2013; Harkness *et al.* 2017). Using a Thermo Finnigan Trace Ultra gas chromatographic separation, followed by combustion and dual-inlet isotope ratio mass spectrometry using a Thermo Fisher Delta V Plus, the detection limits for δ^{13}C-CH$_4$ and δ^2H-CH$_4$ were 0.001 and 0.005 cm^3 STP cm^{-3} (STP, standard temperature and pressure 22°C, 1 atm), respectively. The δ^2H-CH$_4$ values are expressed in per mille v. Vienna Standard Mean Ocean Water, with a reproducibility of ±0.5‰. The δ^{13}C-CH$_4$ values are expressed in per mille v. Vienna Peedee belemnite, with a standard deviation of ±0.1‰.

Coal seam solid samples. Six coal seam solid samples, including two from the Springfield ($n = 2$) and four Seelyville ($n = 4$) coal seams were analysed

for U, Th, ^{4}He and ^{21}Ne abundances. Springfield and Seelyville coal seam solids were collected from core samples obtained from Sullivan County, Indiana at modern depths ranging from *c.* 73 to 213 m below the surface, respectively.

Samples analysed for U and Th concentrations were prepared by cutting pieces of coal seam solid samples using a diamond-blade saw. Analyses of U and Th concentrations were completed using a Thermo Finnigan Element 2 ICP-Sector Field MS and a Photon Machine Excite Laser at the Trace Element Research Laboratory at the Ohio State University following methods developed previously (Cuoco *et al.* 2013).

Samples analysed for ^{4}He and ^{21}Ne abundance were prepared by taking *c.* 50 g of each sample, lightly crushing, drying at 25°C in an oven overnight and then sieving through a 63 μm sieve. The >63 μm fraction was retained, dried, weighed, then rinsed fully with deionized water and then dried again at 25°C overnight. Samples were then placed in a vacuum chamber and heated using an external vacuum oven at 400°C for 90 min following methods reported previously (Hunt 2000). Released noble gases were analysed using a Thermo Fisher Helix SFT noble gas mass spectrometer by peak height comparison with a calibrated air standard (Lake Erie, Ohio Air). In order to correct helium concentrations for air leakage, it was assumed that all ^{3}He came from air, and the air component was subtracted from the overall helium concentration (Hunt 2000; Darrah *et al.* 2015b).

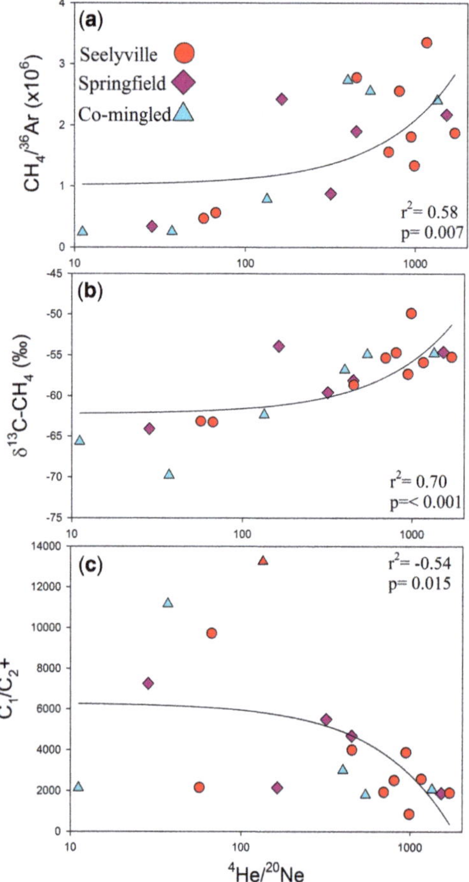

Fig. 8. The ratio $CH_4/^{36}Ar$ (a proxy for gas-water ratio) (**a**), stable isotopic values of carbon in methane (δ^{13}C-CH$_4$) (**b**), and the ratio of methane over higher-order hydrocarbons (C$_1$/C$_2$+) (**c**) v. the ratio $^4He/^{20}Ne$ (a proxy for radiogenic He to ASW; note log scale) in producing CBM wells sampled in this study. Linear regression lines, r^2 and p-values of the regression for all data are shown in the plots.

Numerical modelling

We used numerical models to investigate and constrain the subsurface conditions and viable mechanisms of gas transport to shallow aquifers including: (1) single-phase advection of methane-rich brine (gas dissolved in water); (2) 'buoyant' advection of a gas-phase natural gas or two-phase advection (i.e. free gas + brine); and (3) mixing with a helium-poor, biogenic endmember. Mechanisms 1 (grey arrow), 2 (blue dashed lines) and 3 (purple arrow) are shown in Figures 5, 11 & 12.

All models make two assumptions. The first is that the initial composition of ASW noble gases is consistent with recently recharged meteoric water.

Groundwater recharged under these conditions is assumed to have a salinity of zero and to have equilibrated at *c.* 10°C at an elevation of 200 m (i.e. present day equilibration conditions for shallow groundwater in this area). Under these conditions, the ^{20}Ne and ^{36}Ar in ASW are 201 × 10^{-6} cm^3 STP kg^{-1} and 1394 × 10^{-6} cm^3 STP kg^{-1}, respectively. The second is that the starting natural gas composition of migrated hydrocarbon gases is consistent with New Albany Shale production gases: ^{20}Ne/^{36}Ar = 0.156, ^{4}He/CH$_4$ = 125 (×10^{-6}), C$_1$/C$_2$+ = 16, and δ^{13}C-CH$_4$ = −49‰ generated at an initial reservoir depth of *c.* 1000 m and *c.* 1 M NaCl (Figs 11 & 12). We use a New Albany-like starting composition because of the *ad hoc* observation of the need for an exogenous source of ^{4}He and thermogenic natural gas. Owing to a lack of empirical or field-based evidence for oil-phase hydrocarbons in our study, we did not incorporate oil–gas partitioning into our numerical models. Similarly, we did not incorporate hydrothermal circulation into our numerical or physical transport models.

We model Mechanism 1 (single-phase advection; grey trend in Fig. 11) assuming gas and brine migrate as a single-phase fluid (i.e. no free gas-phase is present) and the resulting fluid does not experience any gas-phase partitioning. In this scenario (i.e. $V_{gas}/V_{liquid} = 0$), the ASW noble gases (e.g. ^{20}Ne/^{36}Ar) would not experience any quantifiable fractionation of ASW components in the absence of multiple phases (e.g. gas + brine) (Ballentine *et al.* 2002). However, the migrating fluid would accumulate radiogenic ^{4}He from the country rock, resulting in an increase in the ^{4}He/CH$_4$.

Mechanism 2 (dashed blue lines) assumes that a free gas-phase or dual-phase (i.e. free gas + brine) fluid migrates within a crustal matrix that contains static, capillary bound or connate pore fluids (e.g. crustal water). In this scenario, gases would partition according to their respective solubilities (Smith & Kennedy 1982; Ballentine *et al.* 2002). The Bunsen solubility (β) constants for the noble gases are determined using Smith & Kennedy (1982) and Weiss (1971*a*, *b*), while the partition coefficients (termed alpha (α), where $\alpha = \beta_X/\beta_Y$) are calculated following Ballentine *et al.* (2002) and Smith & Kennedy (1982). Because the timing and thermal history for natural gas migration are uncertain, we model a range of temperature conditions (a proxy for burial depth), assuming a standard geothermal gradient (i.e. 25°C km^{-1}). The modelled temperatures range from 10°C to 200°C. Since all natural gas wells were sampled at depths <213 m, Bunsen solubility constants (β values, e.g. β_{Ne} or β_{Ar}) were not corrected for geothermal gradients. The solubility fractionation model adapts the groundwater gas stripping and re-dissolution (GGS-R) model

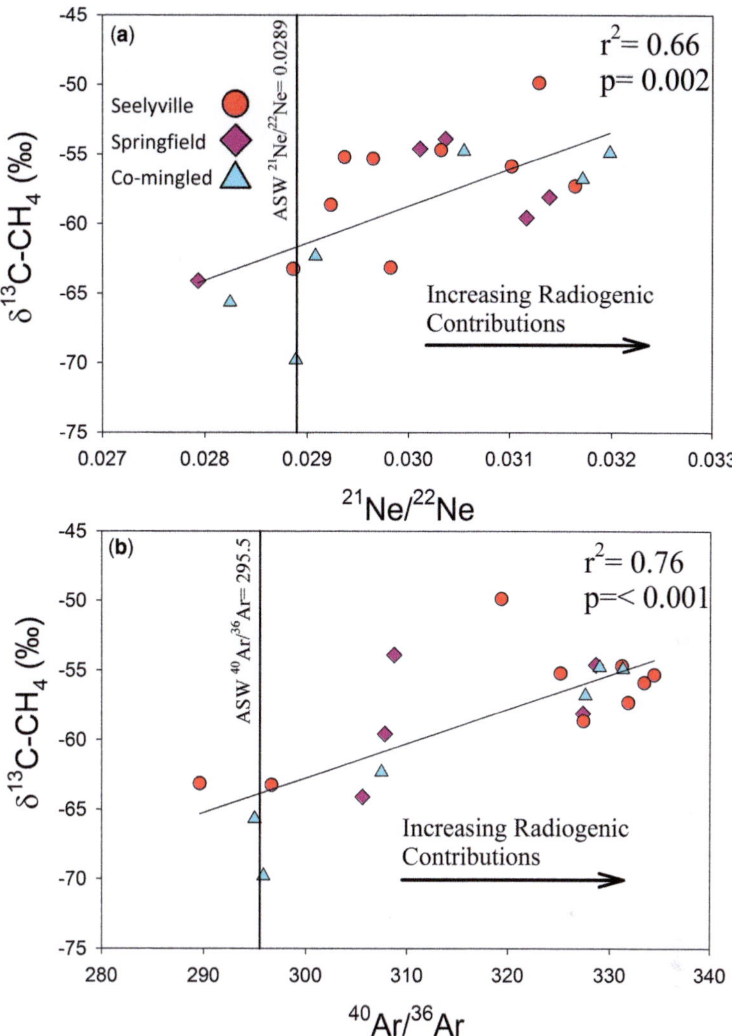

Fig. 9. The carbon isotopic composition in methane (δ^{13}C-CH$_4$) v. the ratio of ^{21}Ne/^{22}Ne (a) and ^{40}Ar/^{36}Ar (b) from producing CBM wells in this study. Note the correlation between increasing radiogenic contributions (a proxy for deeper crustal fluids) with the more thermogenic in origin producing wells (relatively heavier carbon isotopes in methane). The r^2 and p-values of the regression for all data are shown in the plots.

developed for CO_2 (Gilfillan *et al.* 2008, 2009; Zhou *et al.* 2012), and previously adapted for hydrocarbons (Darrah *et al.* 2014), to evaluate the fractionation of atmospheric (e.g. ^{20}Ne/^{36}Ar) and crustally derived components (e.g. ^{4}He/CH$_4$, C$_1$/C$_2$+, and δ^{13}C-CH$_4$).

The GGS-R model postulates that trace gas components are extracted from the water phase in equilibrium with a gas phase through 'gas stripping'. The nucleation of a separate gas phase in groundwater occurs when the sum of the partial pressures of all dissolved gases (but generally dominated by a few major components, such as CO_2 or CH_4) exceeds the hydrostatic pressure, commonly called the saturation point or 'bubble point'. In cases where the partial pressure of dissolved methane is dominant, it controls the partitioning of trace gases (He, Ne, Ar) because these trace components alone cannot attain (or sustain) sufficient partial pressures to nucleate or remain in the bubble phase even at the relatively low hydrostatic pressures present near the vadose zone.

Our adaptation of the GGS-R model for hydrocarbons postulates that catagenesis produces

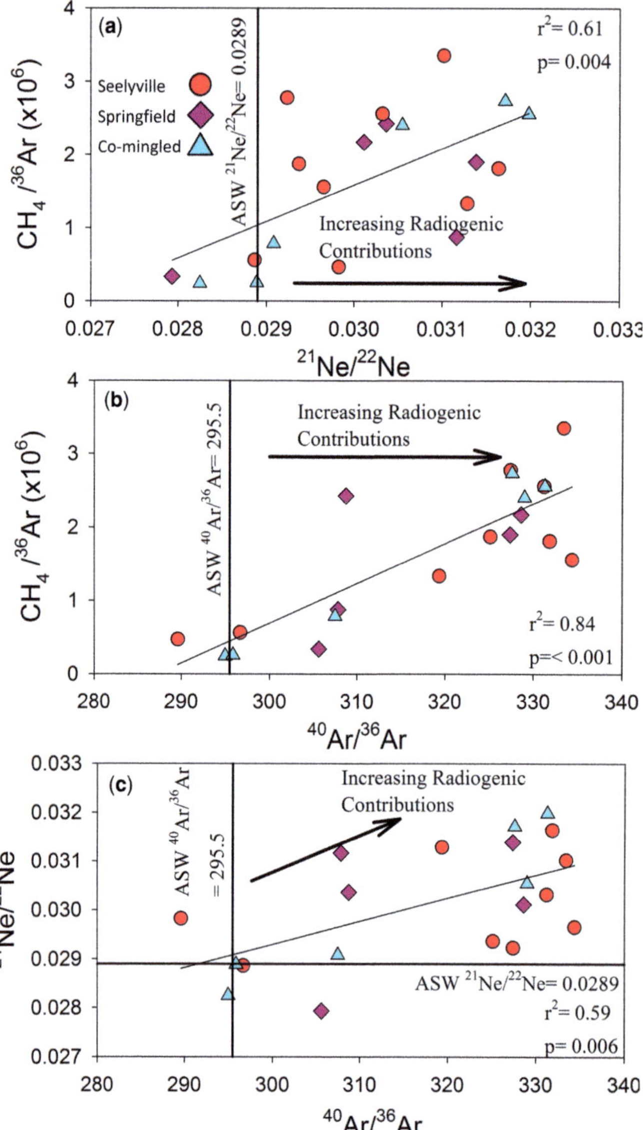

Fig. 10. The ratio of $CH_4/^{36}Ar$ v. the ratio of $^{21}Ne/^{22}Ne$ (**a**) and $^{40}Ar/^{36}Ar$ (**b**) and the ratio of $^{21}Ne/^{22}Ne$ v. $^{40}Ar/^{36}Ar$ (**c**) of producing CBM wells sampled in this study. Note the increasing proportion of crustal noble gases in samples with an increasing ratio of gas to water (i.e. $CH_4/^{36}Ar$). The r^2 and p-values of the regression for all data are shown in the plots.

sufficiently high natural gas concentrations to generate a free gas-phase or 'bubble', leading to the *primary* and later *secondary* migration of hydrocarbon gases. The gas-phase (or two-phase) fluid would then contain crustal and ASW noble gases with the relative partitioning of the trace gas components between the gas and water phases dependent upon: (1) the respective Bunsen solubility coefficients for

each phase (β_X, the ratio at equilibrium of the volume of dissolved gas X (at STP conditions) per unit volume of solution, when the partial pressure of X is 1 atmosphere); and (2) the *in situ* volume ratio of gas to water (i.e. V_{gas}/V_{water}) (Aeschbach-Heritg *et al.* 1999; Ballentine *et al.* 2002; Holocher *et al.* 2003; Aeschbach-Hertig *et al.* 2008). This initial stage would significantly enrich the migrated

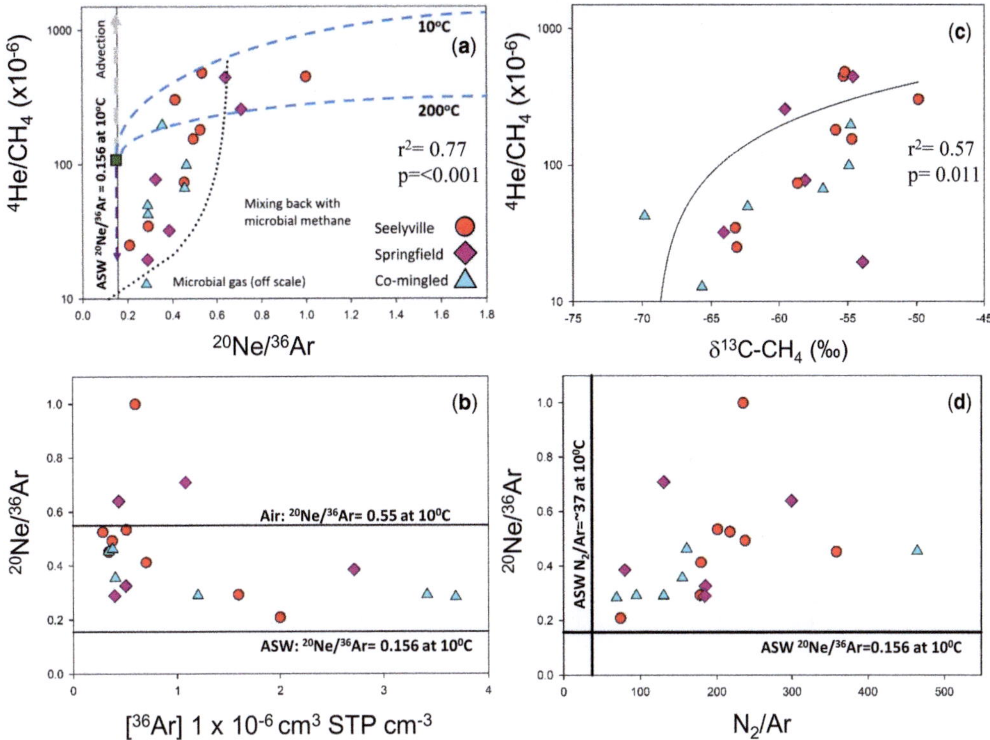

Fig. 11. The ratio of ^{4}He/CH$_4$ v. the ratio of ^{20}Ne/^{36}Ar (**a**), the ratio of ^{20}Ne/^{36}Ar v. the concentration of argon [^{36}Ar] (**b**), the ratio of ^{4}He/CH$_4$ v. the carbon isotopic composition in methane (δ^{13}C-CH$_4$) (**c**), and the ratio of ^{20}Ne/^{36}Ar v. the ratio of N$_2$/Ar (**d**) from producing CBM wells in this study. In Figure 8a, the dark green square represents the previously reported average composition of produced gases from the New Albany Shale (Schlegel *et al.* 2011*a*), the dashed grey arrow represents one-phase advection of a New Albany-like gas source, the purple dashed arrow represents mixing of New Albany-like production gas with a 'young' biogenic source, the dashed blue trend lines represent modelled fractionation by two-phase advection of New Albany Shale-like gas to the Springfield and Seelyville coal seams, and the dotted black line represents a hypothetical mixing line between the migrated endmember that previously underwent phase partitioning and a 'young' microbial source. Two-phase solubility fractionation and advection trends are based off the two-stage groundwater gas stripping and re-dissolution (GGS-R) model developed in Gilfillan *et al.* (2008, 2009) and Zhou *et al.* (2012). Linear regression lines, r^2 and *p*-values of the regression for all data are shown in the plots.

hydrocarbon fluid in [^{4}He] and [^{20}Ne] and increase the ^{4}He/CH$_4$ and ^{20}Ne/^{36}Ar ratios. As the gas migrates into overlying groundwater containing gas levels below methane saturation, there is a partial re-dissolution of methane and the previously exsolved components back into groundwater (Gilfillan *et al.* 2008, 2009). In this scenario, the more soluble trace gas components (e.g. solubility of C$_2$H$_6$ > CH$_4$, which is equal to ^{36}Ar) preferentially re-dissolve into the static, capillary-bound or connate pore fluids according to their relative partition coefficients (α), while less soluble components (e.g. ^{4}He and ^{20}Ne) preferentially remain in the gas phase.

In low gas-to-water conditions (i.e. as V_{gas}/V_{water} approaches 0), trace gases with different solubilities will strongly fractionate as they partition from groundwater into the gas phase (Gilfillan *et al.* 2009). In this scenario, the degree of molecular fractionation increases and can approach a maximum fractionation value of the ratio of the respective Bunsen coefficients. This assumes single-stage fractionation because multi-stage fractionation can greatly exceed α. For example, the Bunsen coefficient ratio for ^{20}Ne v. ^{36}Ar (β_{Ne}/β_{Ar}) is *c.* 3.7 at 10°C, and for ^{4}He v. CH$_4$ (β_{He}/β_{CH4}) is *c.* 4.4 at the same temperature. On the other hand, for components with similar solubilities such as CH$_4$ v. ^{36}Ar (β_{CH4}/β_{Ar}), α is about 1 at 10°C (Ballentine *et al.* 1991; Sherwood Lollar & Ballentine 2009) and for ^{4}He v. ^{20}Ne (β_{He}/β_{Ne}) α is about 1.2 at the same temperature (Weiss 1971*a, b*). These pairs of gases will partition

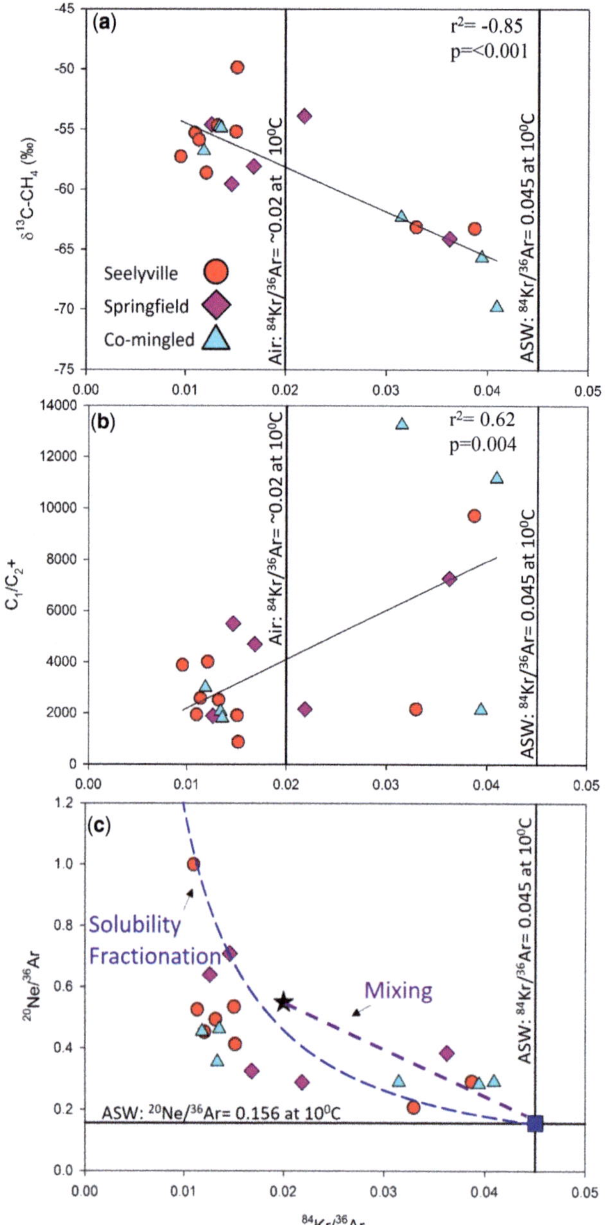

Fig. 12. The carbon isotopic composition in methane (δ^{13}C-CH$_4$) (**a**), the ratio of methane over higher-ordered aliphatic hydrocarbons (C$_1$/C$_2$+) (**b**), and the ratio of ^{20}Ne/^{36}Ar (**c**) v. the ratio of ^{84}Kr/^{36}Ar of producing CBM wells sampled in this study. The blue box and the black star in (c) represent the globally constrained ratios of ^{20}Ne/^{36}Ar and ^{84}Kr/^{36}Ar for ASW and air, respectively. The blue trend line demonstrates the change in a gases composition as it fractionates by two-phase solubility migration. The dashed purple trend line demonstrates mixing between an ASW fluid and air.

into the gas phase according to the $V_{\text{gas}}/V_{\text{water}}$, but will not experience significant molecular or isotopic fractionation relative to each other. As the $V_{\text{gas}}/V_{\text{water}}$ increases, the amount of dissolved gas that partitions into the gas phase will increase, while the degree of molecular (or isotopic) fractionation between gases with different solubilities decreases. Thus, in high V_{gas}/V_{water} conditions nearly all of the trace gases will partition into the gas phase. In this case, even components with relatively large differences in solubility will not experience changes in molecular or isotopic ratios in either the gas or the residual water phases. Our model uses maximum α fractionation values because we assume that gas emplacement occurs when relatively low volumes of gas migrate through the water-saturated crust (i.e. low V_{gas}/V_{water}). Note also that our empirical observations require extensive fractionation (near maximum α) to accommodate the observed $^{20}Ne/^{36}Ar$ in the methane-rich samples.

Results

Data obtained from produced gases in Springfield and Seelyville coal seams, as well as comingled gases, are treated as a single group for three reasons. First, there is noted hydraulic communication between both formations (pers. coms. Mr. Larry Neely, Maverick Energy Drilling Co.). Secondly, past work from this region has treated these formations as hydraulically connected units (Mastalerz *et al.* 2004; Schlegel *et al.* 2011*a*; Mastalerz *et al.* 2017). Finally, despite a comprehensive statistical evaluation of our study area, we identify no statistically meaningful differences among any of our three subgroups (Tables 1–7). For example, neither the differences in the means ($p = 0.569$) or medians ($p = 0.613$) nor the 95% confidence intervals ($p = 0.514$) of the methane abundances or other parameters were statistically significant by simultaneous analysis of variance (ANOVA).

Hydrocarbon and major gases

Methane is the dominant gas component in all samples with concentrations ranging from 0.875 to 0.973 cm^3 STP cm^{-3} (Tables 1–3). Less abundant gases include nitrogen (0.020–0.097 cm^3 STP cm^{-3}), carbon dioxide (0.004–0.017 cm^3 STP cm^{-3}) and ethane (0.15 × 10^{-5} to 1.09 × 10^{-3} cm^3 STP cm^{-3}). Oxygen, propane, normal-butane (C_4n), iso-butane (C_4i), normal-pentane (C_5n) and iso-pentane (C_5i) were below detection limits (<1 × 10^{-5} cm^3 STP cm^{-3}) in all samples (Table 1). The ratio of C_1/C_2+ in our study ranged from 867 to 13 256 (Tables 6 & 7).

While nearly all of the gas samples are composed of predominantly CH_4 and N_2, there are significant correlations between the abundances of CH_4 and ethane ($r^2 = 0.51$; $p = 0.025$), N_2 ($r^2 = -0.98$; $p = <0.001$), ^{36}Ar ($r^2 = -0.85$; $p = <0.001$) and ^4He (shown as $^4He/^{20}Ne$: $r^2 = 0.57$; $p = 0.009$), and the ratio of C_1/C_2+ ($r^2 = -0.54$; $p = 0.014$), and

negative correlations between ^{36}Ar and N_2 (Fig. 3). We do not identify significant correlations between [CO_2] and other gas components (e.g. CH_4, C_2H_6, N_2, ^4He), potentially indicating that CO_2 gas abundances are buffered by dissolution into formation waters as DIC or precipitation of carbonate minerals, or modulated by microbial activity.

Isotopic composition of hydrocarbon gases

The stable isotopic compositions of methane (i.e. $\delta^2H\text{-}CH_4$ and $\delta^{13}C\text{-}CH_4$) ranged from −69.80 to −49.87‰ (Figs 3–5; Tables 4 & 5). Stable isotopic values of hydrogen in methane ranged from −214.9‰ to −201.4‰ (Fig. 4; Tables 4 & 5). The $\delta^{13}C\text{-}CH_4$ and $\delta^2H\text{-}CH_4$ show positive correlations with methane concentration ($r^2 = 0.78$; $p = <0.001$ and $r^2 = 0.32$ $p = 0.213$, respectively), ethane concentration ($r^2 = 0.66$; $p = 0.002$ and $r^2 = 0.16$; $p = 0.563$, respectively) and each other ($r^2 = 0.48$; $p = 0.051$) (Figs 3 & 4). Importantly, these values are statistically indistinguishable ($p = >0.694$) from previously published data from the same field, while analyses of the same wells were also statistically indistinguishable between each of the studies (paired Student's t-test $p = >0.576$; Strąpoć *et al.* 2007, 2008*a*; Mastalerz *et al.* 2017).

Negative correlations were observed between $\delta^{13}C\text{-}CH_4$ and N_2 concentrations ($r^2 = -0.80$; $p = <0.001$), ^{36}Ar concentrations ($r^2 = -0.87$; $p = <0.001$) and C_1/C_2+ ($r^2 = -0.70$; $p = 0.001$) (Tables 1–5; Figs 5–6), suggesting an increasing contribution from more depleted $\delta^{13}C\text{-}CH_4$ with increasing contributions from ASW gases. These results are also consistent with an increasing proportion of biogenic methane with relatively higher abundances of ASW, suggesting that biogenic methane production may be associated with greater proportions of ASW fluids in the Springfield and Seelyville coal seams.

Noble gases

Helium-4 concentrations ranged from 11.6 to 1460 × 10^{-6} cm^3 STP cm^{-3}. The $^3He/^4He$, shown as R/R_A ratios, displayed a range from 0.043 to 0.901 R_A, with corresponding $^4He/^{20}Ne$ and $^4He/^{36}Ar$ values that ranged from 11.1 to 1686 and from 3.1 to 2773, respectively. These data reflect predominantly crustal sources of helium with resolvable excesses of ^3He as compared with crustal production values (*c.* 0.02 R_A; Craig *et al.* 1978; Oxburgh *et al.* 1986). These excesses of ^3He indicate small percentages of mantle-derived helium (up to *c.* 10% MORB, where pure MORB has a $^3He/^4He$ ratio of *c.* 8 R_A; Poreda *et al.* 1992; Tables 4 & 5; Fig. 7).

The ^{22}Ne and ^{36}Ar concentrations ranged from 0.01 to 0.16 × 10^{-6} cm^3 STP cm^{-3}, and from 0.29

to 3.69×10^{-6} cm^3 STP cm^{-3} (Tables 1–3), respectively. The ^{20}Ne/^{22}Ne and ^{21}Ne/^{22}Ne values ranged from 9.761 to 9.787 and from 0.0286 to 0.0317, respectively (Tables 4 & 5). The ^{38}Ar/^{36}Ar values ranged from 0.1828 to 0.2060, while the ^{40}Ar/^{36}Ar values ranged from 294.61 to 334.36, respectively (Tables 4 & 5). These ranges suggest that both neon and argon isotopes are dominated by ASW contributions (^{20}Ne/^{22}Ne = 9.78, ^{21}Ne/^{22}Ne = 0.0289 and ^{40}Ar/^{36}Ar = 295.5), but contain resolvable excesses in nucleogenic (^{21}Ne*) and radiogenic (^{40}Ar*) noble gases.

The dataset displayed a large range of atmospheric noble gas (e.g. ^{36}Ar, ^{22}Ne) and N$_2$ compositions. Atmospheric noble gas abundances positively correlate with N$_2$ concentrations ($r^2 = 0.82$; $p = $ <0.001 and $r^2 = 0.51$; $p = 0.021$ for ^{36}Ar and ^{22}Ne, respectively; Tables 4–7). Both ^{36}Ar concentrations ($r^2 = 0.44$; $p = 0.051$) and N$_2$ concentrations ($r^2 = 0.62$; $p = 0.004$) show significant positive correlations with C$_1$/C$_2$+ (Fig. 6; Tables 4–7). Given that [N$_2$] and [CH$_4$] are strongly negatively correlated (Fig. 3), the abundance of ASW components is also negatively correlated to CH$_4$ concentrations ($r^2 = $ −0.85; $p = $ <0.001 and $r^2 = $ −0.98; $p = $ <0.001 for ^{36}Ar and N$_2$, respectively), C$_2$H$_6$ concentrations ($r^2 = $ −0.38; $p = 0.107$ and $r^2 = $ −0.57; $p = 0.010$ for ^{36}Ar and N$_2$, respectively), δ^{13}C-CH$_4$ ($r^2 = $ −0.87; $p = $ <0.001 and $r^2 = $ −0.80; $p = $ <0.001 for ^{36}Ar and N$_2$, respectively) and ^{4}He/CH$_4$ ($r^2 = $ −0.44; $p = 0.059$ and $r^2 = $ −0.29; $p = 0.223$ for ^{36}Ar and N$_2$, respectively) (Figs 3 & 6).

Radiogenic noble gas isotopes correlate significantly with the proportion of hydrocarbon gases (e.g. CH$_4$, C$_2$H$_6$), N$_2$ and hydrocarbon stable isotopes of carbon (δ^{13}C-CH$_4$) (Figs 8–10). Methane and ethane concentrations correlate positively with the ^{4}He/^{20}Ne ($r^2 = 0.57$; $p = 0.009$ and $r^2 = 0.56$; $p = 0.012$, respectively), ^{4}He/^{36}Ar ($r^2 = 0.33$; $p = 0.150$ and $r^2 = 0.18$; $p = 0.450$, respectively), ^{21}Ne/^{22}Ne ($r^2 = 0.63$; $p = 0.003$ and $r^2 = 0.29$; $p = 0.226$, respectively) and ^{40}Ar/^{36}Ar ($r^2 = 0.70$; $p = 0.001$ and $r^2 = 0.42$; $p = 0.072$, respectively) (Tables 1–7). Similarly, N$_2$ concentrations are negatively correlated with ^{4}He/^{20}Ne ($r^2 = $ −0.57; $p = 0.009$), ^{21}Ne/^{22}Ne ($r^2 = $ −0.66; $p = 0.001$) and ^{40}Ar/^{36}Ar ($r^2 = $ −0.69; $p = 0.001$) (Tables 1–7). The δ^{13}C-CH$_4$ data display a significant positive correlation with ^{4}He/^{20}Ne ($r^2 = 0.70$; $p = 0.001$), ^{21}Ne/^{22}Ne ($r^2 = 0.66$; $p = 0.002$), and ^{40}Ar/^{36}Ar ($r^2 = 0.76$; $p = 0.001$) (Figs 5, 8 & 9; Tables 4–7).

There are other important gas isotope ratios that provide important clues about the transport mechanisms and gas–water interactions that occur in the subsurface. For example, the value of CH$_4$/^{36}Ar serves as a proxy for the relative abundance of natural gas and ASW (i.e. natural gas to water ratio) irrespective of any gas–water partitioning that may have

occurred (Fig. 10), the ^{4}He/CH$_4$ ratio is a sensitive indicator of biogenic v. thermogenic contributions and water–gas interactions, fractionation of the ^{20}Ne/^{36}Ar and ^{84}Kr/^{36}Ar ratios records diagnostic partitioning between the gas and water phases during fluid transport (Figs 11 & 12), and the ^{130}Xe/^{84}Kr ratio records evidence for extensive interaction with organic-rich sediments (Zhou *et al.* 2005; Gilfillan *et al.* 2009; Darrah *et al.* 2015*b*). The CH$_4$/^{36}Ar (gas–water ratio) ranges from 2.4×10^5 to 3.35×10^6 and correlates negatively with [N$_2$] ($r^2 = $ −0.76; $p = $ <0.001) and C$_1$/C$_2$+ ($r^2 = $ −0.54; $p = 0.014$; Tables 1–3 & 6–7; Fig. 8). The ^{4}He/CH$_4$ ratios ranges from 12.9 to 1528×10^{-6} and correlate significantly with δ^{13}C-CH$_4$ ($r^2 = 0.57$; $p = 0.011$) and ^{20}Ne/^{36}Ar ($r^2 = 0.77$; $p = 0.001$). There is a general trend of increasing ^{4}He/CH$_4$ with lower abundances of ASW ($r^2 = $ −0.44; $p = 0.059$ for [^{36}Ar] and $r^2 = $ −0.29; $p = 0.223$ for [N$_2$]; Fig. 6) and heavier δ^{13}C-CH$_4$ ($r^2 = 0.57$; $p = 0.011$; Fig. 11). The ^{20}Ne/^{36}Ar ratios ranged from 0.208 to 0.999 after the removal of a noted outlier, which had an ^{20}Ne/^{36}Ar = 2.936 (Fig. 11; Tables 6 & 7). The gas well noted as an outlier was excluded because this well had been shut in for more than two years. The ^{84}Kr/^{36}Ar ratios ranged from 0.0095 to 0.0409 (Fig. 12), while the ^{130}Xe/^{84}Kr ratios ranged from 0.00342 to 0.00796 (Tables 6 & 7).

Coal seam solids

The uranium (U) and thorium (Th) concentrations measured from coal seam solids ranged from 0.85 to 1.54 and from 2.95 to 4.15 µg g^{-1}, respectively (Table 8); these values produced ratios of Th/U that ranged from 2.53 to 4.16. The concentrations of ^{4}He and ^{21}Ne in the coal seam solids, which represent previously accumulated products of U and Th decay, ranged from 2.17 to $44\,100 \times 10^{-6}$ cm^3 STP kg^{-1} and from 1.21 to 3.04×10^{-12} cm^3 STP kg^{-1}, respectively. Using equation (1), the ^{4}He/^{21}Ne* ratios ranged from 9.61 to 21.2 (Table 8).

Discussion

The integration of hydrocarbon molecular and isotopic compositions, major gases and noble gases provides important constraints on the genetic sources and proportions of biogenic and thermogenic natural gas found in CBM reservoirs from the Springfield and Seelyville formations in the eastern Illinois Basin, USA. Moreover, our results are important for future efforts to determine the timing of meteoric water recharge and fluid migration histories associated with the stimulation of biogenic methane production in the Illinois Basin and other sedimentary basins.

Table 8. *Uranium, thorium, helium, and neon concentrations in coal seam solids*

Well core sampled	Producing coal seam	U (mg kg^{-1})	($\pm 1\sigma$)	Th (mg kg^{-1})	($\pm 1\sigma$)	Th/U	Measured ^{4}He (cm^3 STP kg^{-1}) ($\times 10^{-6}$)	($\pm 1\sigma$)	Measured ^{21}Ne* (cm^3 STP kg^{-1}) ($\times 10^{-12}$)	($\pm 1\sigma$)	^{4}He/^{21}Ne*
IN-2	Springfield	1.06	0.027	3.65	0.090	3.44	27861.42	2.64×10^3	1548	2.22×10^1	18
IN-5	Springfield	1.24	0.029	4.15	0.094	3.34	34335.41	2.60×10^3	1619.9	1.48×10^1	21.2
IN-7	Seelyville	0.85	0.022	2.95	0.064	3.48	21748.97	2.22×10^3	1211.7	1.61×10^1	17.95
IN-8	Seelyville	0.98	0.022	4.06	0.096	4.16	35079.36	3.33×10^3	2081.9	2.94×10^1	16.85
IN-10	Seelyville	1.47	0.036	5.35	0.136	3.65	23649.14	2.86×10^3	2460.9	3.77×10^1	9.61
IN-14	Seelyville	1.54	0.035	3.9	0.093	2.53	44060.81	4.14×10^3	3040.8	3.77×10^1	14.49
Average		1.19		4.01		3.43	31122.52		1993.86		16.35

Genetic source of natural gases in the Springfield and Seelyville formations

The majority of production gas data from CBM wells in the Illinois Basin display higher C_1/C_2+ ratios compared with most primary thermogenic gases and also have intermediate δ^{13}C-CH$_4$ values (<−49‰) that can be interpreted as early-maturity thermogenic gases or mixtures of biogenic–thermogenic gases (Figs 4 & 5; Schoell 1988; Whiticar 1999). For these reasons, previous researchers who used only traditional hydrocarbon techniques (e.g. C_1/C_2+, δ^{13}C, δ^2H-CH$_4$) and did not consider the potential for exogenous thermogenic endmembers interpreted these natural gases as being primarily biogenic with less than 0.4% thermogenic contributions (Strąpoć *et al.* 2007, 2008*a*; Mastalerz *et al.* 2017).

The most confounding issue with the interpretation of a nearly pure biogenic source for natural gases in this study area is the low, but persistent occurrence of ethane (with ethane concentrations up to 1.09×10^{-3} cm^3 STP cm^{-3}; Tables 1–3), even in very shallow CBM wells (depth range 73–213 m; Tables 1–3). Because ethane is not produced by biogenic processes in any sufficient quantity in the subsurface, regardless of the methanogenic pathway (Bernard *et al.* 1976; Schoell 1980; Rowe & Muehlenbachs 1999*a*; Jackson *et al.* 2013), these data suggest that a resolvable component of thermogenic gas is present in all samples. Given that early-maturity thermogenic gases typically contain >20% ethane, the ethane abundances observed here could be consistent with insignificant thermogenic contributions. Notably, ethane concentrations decrease with increasing proportions of atmospherically derived gases (e.g. [N$_2$], [^{20}Ne] and [^{36}Ar]; Fig. 6; Tables 1–3). Elevated contents of atmospherically derived gases are an indicator of increasing influence from meteoric water (Ballentine *et al.* 2002; Ballentine & Burnard 2002). Samples with higher levels of atmospherically derived gases also appear to correspond to an increasing proportion of biogenic natural gas (i.e. higher C_1/C_2+ and more negative δ^{13}C–CH$_4$; Figs 3 & 6). Although the current study does not address the mean residence time of groundwater in the Springfield and Seelyville coal seams, the association of atmospherically derived gases with the biogenic component is broadly consistent with the hypothesis that fresh groundwater recharge and subsequent dilution of high-salinity formation waters may stimulate methanogenesis in coalbeds (Scott *et al.* 1994; Martini *et al.* 1996; McIntosh *et al.* 2002; Schlegel *et al.* 2011*b*; Pashin *et al.* 2014).

In addition to ethane concentrations, most hydrocarbon stable isotope data do not fall *exclusively* within the expected ranges for microbial origin (Figs 4 & 5). Instead, on plots of stable carbon v. hydrogen isotopes (Fig. 4), the data fall along an apparent mixing trend between a hydrogenotrophic microbially sourced component and a quantifiable thermogenic natural gas component. Data from the current study area also display more ethane, shown by C_1/C_2+ as low as 867.6, and more enriched δ^{13}C-CH$_4$ values (δ^{13}C-CH$_4$ up to −49.87‰) than expected for a pure (un-mixed) microbial endmember gas in natural gas reservoirs (δ^{13}C-CH$_4$ <−55‰, C_1/C_2+ >2000; Bernard *et al.* 1976; Fig. 5). It is important to note that the interpretation of these data is complicated by the potential for complex mixtures and/or modification of natural gases by secondary oxidation processes and/or solubility partitioning. For example, oxidation alone can either increase (aerobic oxidation) or decrease (anaerobic oxidation) the C_1/C_2+ and lead to more positive values for residual δ^{13}C–CH$_4$.

Constraining source and migration using noble gases

While molecular and stable isotopes of hydrocarbons alone may not allow gas origins to be fully delineated, radiogenic (^{4}He, ^{40}Ar*) and nucleogenic (^{21}Ne*) noble gases can impart important clues about the thermal maturity of natural gases and/or the presence of exogenous fluids in the Springfield and Seelyville formations. Like ethane, helium-4 and heavier crustal noble gases (^{21}Ne*, ^{40}Ar*) occur at elevated levels in thermogenic natural gases, whereas recently formed microbial gases are nearly devoid of ethane and radiogenic noble gases (Hunt *et al.* 2012; Darrah *et al.* 2015*b*; Wen *et al.* 2015; Harkness *et al.* 2017).

Our data reveal a corresponding increase in the levels of C_2H_6 (i.e. lower C_1/C_2+), more positive δ^{13}C-CH$_4$, lower levels of atmospherically derived gases (N$_2$, ^{20}Ne, ^{36}Ar) and higher gas–water ratios (CH$_4$/^{36}Ar) with increasing levels of the crustal noble gases (^{4}He, ^{21}Ne*, and ^{40}Ar*; Figs 6 & 8–10; Tables 1–3). Each of the above trends is consistent with inputs of thermogenic natural gas, which has recently been observed in the nearby Appalachian and Michigan Basins and the Dallas–Fort Worth Basin (Hunt *et al.* 2012; Darrah *et al.* 2015*a, b*; Wen *et al.* 2015, 2016, 2017; Harkness *et al.* 2017).

The relative abundances of radiogenic noble gases also provide important clues about temperature conditions of natural gas formation (Hunt *et al.* 2012; Darrah *et al.* 2015*b*). The production of ^{4}He and ^{21}Ne* in oxygen-rich silicate/carbonate minerals is linked and nearly constant (^{4}He/^{21}Ne* = 22 × 10^6; Kennedy *et al.* 1990; Yatsevich & Honda 1997; Ballentine & Burnard 2002). As a result, one would anticipate a relatively narrow range of ^{4}He/^{21}Ne* in most crustal minerals and associated fluids that have not experienced extensive fluid migration (Kennedy *et al.* 1990; Hunt *et al.* 2012). Similarly, the production ratio of ^{4}He/^{40}Ar* occurs within the fairly small range (5–8) based on the K/U ratio of source rocks in the bulk continental crust (Ballentine & Burnard 2002; Hunt *et al.* 2012). Once radiogenic gases are formed in the crust, they are released from different lithologies at predictable and quantifiable rates as a function of formation temperature (Ballentine & Burnard 2002; Hunt *et al.* 2012; Darrah *et al.* 2014).

Because of its small atomic radius, helium can diffuse through quartz on geological timescales as short as decades, particularly at the elevated temperatures of hydrocarbon formations, and thus equilibrate with crustal fluids (Cook *et al.* 1996; Hunt *et al.* 2012). As a result, although microbial methane emplaced in the crust would acquire small quantities of ^{4}He from α-decay or from the release of ^{4}He that has previously accumulated in sedimentary grains, microbial production of natural gas is not itself associated with ^{4}He generation (Darrah *et al.* 2014, 2015*b*; Wen *et al.* 2015). Like ^{4}He, nucleogenic ^{21}Ne* and radiogenic ^{40}Ar* production are also not associated with biogenic natural gas formation. However, in comparison with ^{4}He, nucleogenic ^{21}Ne* or radiogenic ^{40}Ar* is only released from crustal silicate mineral grains into migrating fluids at higher temperatures (*c.* 80 and 220°C, respectively; Hunt *et al.* 2012) and thus provides more diagnostic indicators for the presence of thermogenic natural gas that forms at temperatures >80°C. Since the microbial consortium thought to produce microbial CH$_4$ is unable to function above 80°C (Schlegel *et al.* 2011*a*; Head *et al.* 2014), the release of ^{4}He, ^{21}Ne* and ^{40}Ar* provides a fingerprint of formation temperature as well as potentially gas generation temperature.

The crustal noble gas ratios suggest reservoir temperatures higher than the vitrinite reflectance values (R_o = 0.5–0.8%, which corresponds to a temperature of *c.* 60–90°C) reported for the Springfield and Seelyville coalbeds in the current study area (Green *et al.* 2003; Strapoć *et al.* 2007). The reported ^{4}He/^{21}Ne* and ^{4}He/^{40}Ar* ratios from the current study span from 0.6×10^6 to 35.4×10^6 and from 1.1 to 76.4, respectively (Tables 6 & 7). These ranges overlap the crustal production ratios (22×10^6 and 5–8, respectively; Ballentine & Burnard 2002), which indicates that in many samples the majority of the crustal noble gases were transferred from the mineral phases to fluids without significant mass-dependent elemental fractionation (Ballentine *et al.* 1994; Hunt *et al.* 2012). A lack of observable fractionation from the crustal production ratios for the crustal noble gases, specifically between ^{4}He and ^{40}Ar*, suggests that the transfer of radiogenic isotopes occurred at elevated temperatures, apparently in excess of *c.* 150°C temperatures based on the rates of He, Ne and Ar diffusion from siliclastic minerals (Ballentine *et al.* 1994; Hunt *et al.* 2012;

Darrah *et al.* 2014). These values are consistent with vitrinite reflectance values greater than R_o = 2.0%, which is not observed in the Springfield or Seelyville coals. Given that the ^{4}He/^{40}Ar* values are in equilibrium with crustal production ratios in the thermogenic endmember and that increasing abundances of crustal noble gases correspond to lower C$_1$/C$_2$+, less negative δ^{13}C-CH$_4$, lower levels of atmospherically derived gases (N$_2$, ^{20}Ne, ^{36}Ar) and higher gas–water ratios (CH$_4$/^{36}Ar), we infer that these data suggest the presence of an apparently higher temperature thermogenic natural gas than expected based on reported vitrinite reflectance values of the Springfield or Seelyville coals, an exogenous source of natural gas, or both.

Endogenous v. exogenous thermogenic gases

While ethane is indicative of contributions from thermogenic natural gas, its presence alone does not distinguish between the endogenous generation of thermogenic hydrocarbons or an exogenous source of thermogenic gas that is persistent in all of our samples. Noble gases can also be a powerful suite of tracers for determining the timing and migration of natural gases in CBM reservoirs and/or whether natural gases evolved endogenously within coal seams or were sourced from exogenous formations. Before apportioning the potential thermogenic contributions to the mixture of natural gases in the study area, we must better constrain the possible sources of thermogenic gas to the Springfield and Seelyville coal seams. While elevated ^{4}He primarily represents older, deeper fluids, several unique sources of ^{4}He are possible. The ^{4}He in any crustal fluid results from the combination of: (1) atmospheric inputs; (2) excess air-bubble entrainment; (3) *in situ* production of ^{4}He from α-decay of U–Th series nuclides; (4) the release of ^{4}He that previously accumulated in detrital grains; and (5) the flux from exogenous sources (Solomon *et al.* 1996; Ballentine *et al.* 2002; Ballentine & Burnard 2002; Zhou & Ballentine 2006).

Since ^{20}Ne is exclusively atmospherically derived and has a similar solubility to ^{4}He (β_{He}/β_{Ne} = 1.2 at 10°C), we can use the abundance of ^{20}Ne to estimate the atmospheric contributions of ^{4}He. Even in the sample with the lowest [^{4}He], ASW can conservatively account for less than 0.042% of the total [^{4}He]. Using the average measured [U] = 1.2 μg g^{-1} and [Th] = 4.0 μg g^{-1} of the Springfield and Seelyville coal seams (Table 8) to estimate the potential [^{4}He]$_{in\ situ}$, we estimate steady-state production and accumulation at less than 1.030×10^{-9} cm^3 STP l^{-1} of water per year, which, even assuming a fluid residence time equivalent to the depositional age of coals, would account for an insignificant proportion of the measured ^{4}He concentrations in the current study. The transfer of ^{4}He previously trapped in detrital mineral grains from coal seam solids into coal formation waters can significantly exceed steady-state production (Solomon *et al.* 1996). However, based on measured diffusion coefficients for coals and the average initial [^{4}He] in coals of $31\,122 \times 10^{-6}\,\text{cm}^3$ STP kg^{-1} of rock (Table 8), the calculated maximum release rates only range up to *c.* $0.2 \times 10^{-7}\,\text{cm}^3$ STP l^{-1} of water per year, which is proportionally $<10^{-3}\%$ of the maximum measured ^{4}He concentrations in the current study (methods as reported in Darrah *et al.* 2015*b*).

Therefore, the levels of ^{4}He that we observe (0.15% in the gas phase, which accounts for up to 0.33 cm^3 STP l^{-1} in the water phase) greatly exceed the viable combined concentrations from [^{4}He]$_{\text{ASW}}$, the maximum [^{4}He]$_{in\ situ}$ production and the release from ^{4}He that previously accumulated in coal seam solids from the Springfield or Seelyville coals for time spans exceeding tens of million years. In fact, these data imply an apparent residence time for groundwater in these systems of up to 18.9 Myr in order to accumulate such levels of ^{4}He from the release of endogenous helium from detrital grains in coal seams. This groundwater residence time seems implausible given the shallow depth of the sampled wells (*c.* 73–213 m below land surface) and previous reports for extensive recharge of glacial meltwater to the underlying New Albany Shale within the last 1.2 Ma (McIntosh *et al.* 2004, 2012; Schlegel *et al.* 2011*b*). Thus, based on the correlations between ^{4}He, CH$_4$, C$_2$H$_6$, δ^{13}C-CH$_4$ and heavier crustal noble gases, we suggest that the excess ^{4}He concentrations observed from producing CBM wells in the current study area are not endogenous to the Springfield and Seelyville coal seams, but are instead consistent with the migration of an exogenous source of thermogenic gas into the Springfield and Seelyville coal seams. As a result, we hypothesize that elevated helium levels probably indicate that ^{4}He (and potentially ^{21}Ne* and ^{40}Ar*) migrated to and was entrapped within the Springfield and Seelyville coal seams. Following this gas migration, the samples in this study have a mixed composition with biogenic methane that was relatively devoid of ^{4}He.

Ratios that compare deep-sourced v. atmospheric components of crustal fluids (e.g. ^{4}He/CH$_4$, ^{20}Ne/^{36}Ar, ^{84}Kr/^{36}Ar; Figs 11 & 12) are particularly useful for evaluating the processes of gas migration in the subsurface (e.g. via diffusion or solubility fractionation; Darrah *et al.* 2014, 2015*a, b*). The relative concentration of exogenous ^{4}He in crustal fluids (e.g. ^{4}He/CH$_4$) results from a combination of the ^{4}He released within the original thermogenic natural gas source, mixing and the subsequent modification that occurs during post-genetic alteration processes (e.g. fluid migration, oxidation, methanogenic inputs; Ballentine *et al.* 2002; Hunt *et al.* 2012; Darrah *et al.* 2014, 2015*b*). For example, an increase in ^{4}He/CH$_4$, ^{20}Ne/^{36}Ar and C$_1$/C$_2$+, accompanied by a decrease in ^{84}Kr/^{36}Ar, is diagnostic of two-phase (free gas + brine) solubility fractionation, while one-phase (gas dissolved in brine) migration or mixing between thermogenic and biogenic gases would be associated exclusively with enrichment of ^{4}He and hence ^{4}He/CH$_4$ (Darrah *et al.* 2014, 2015*b*). These fractionation processes can significantly impact the composition of migrating natural gases and lead to marked increases in sensitive parameters such as ^{4}He or ^{20}Ne that are diagnostic of fluid migration in the subsurface (Zhou *et al.* 2005; Zhou & Ballentine 2006; Darrah *et al.* 2014, 2015*b*).

Many samples display ^{20}Ne/^{36}Ar significantly above the ASW equilibrium value of *c.* 0.156 (up to 0.999) and ^{84}Kr/^{36}Ar significantly below (as low at 0.01) the ASW equilibrium value (0.045) in the apparently more thermogenic (i.e. ethane-rich, elevated δ^{13}C-CH$_4$) endmember (Figs 11 & 12; Tables 1–3 & 6–7). We note that the increase in ^{20}Ne/^{36}Ar relates almost exclusively to an increase in [^{20}Ne] concomitantly with [CH$_4$], [C$_2$H$_6$] and [^{4}He], but without increases in [^{36}Ar]. Based on the elevated ^{20}Ne/^{36}Ar and lower ^{84}Kr/^{36}Ar ratios relative to the ASW equilibrium value, we conclude that the source of thermogenic natural gas is exogenous and has experienced multiple-stage, two-phase partitioning during fluid transport (secondary hydrocarbon migration) from its source to the Pennsylvanian-aged coal seam traps (Figs 11 & 12).

Although phase partitioning during multiple-stage fluid migration can account for the elevated ^{4}He/CH$_4$ and ^{20}Ne/^{36}Ar in the apparently thermogenic endmember (Fig. 11), it cannot account for the subset of samples with relatively low ^{4}He/CH$_4$. One important difference between our data and other studies that have examined the ^{4}He/CH$_4$ v. ^{20}Ne/^{36}Ar of migrated thermogenic gases (Darrah *et al.* 2014, 2015*a, b*) is the presence of a ^{4}He/CH$_4$ ratio lower than any hypothetical sources of thermogenic natural gas in the basin modelled from the measured U and Th content of coal and shale (Table 8). Samples from the microbial-rich endmember with the most negative δ^{13}C-CH$_4$ and higher C$_1$/C$_2$+ values also display relatively low ^{4}He/CH$_4$ and exhibit ^{20}Ne/^{36}Ar near that of ASW (Figs 4, 6, 8, 11 & 12). Therefore, the very low values of ^{4}He/CH$_4$ documented here are inconsistent with migration of a thermogenic source of natural gas, but are instead consistent with mixing with a pure source of biogenic methane devoid of ^{4}He. To evaluate this process, we fit a hypothetical mixing line between the migrated thermogenic-rich endmember and a microbial source rich in atmospherically derived gases and lacking ^{4}He (dotted black line in Fig. 11*a*).

While the chosen thermogenic-rich endmember in Figure 11a is *ad hoc*, we note that the mixing line approximates the trajectory of most of our data. We conclude that, complementary to the hydrocarbon data and trends displayed in Figures 4 & 5, the data trends displayed in Figure 11a also suggest mixing between a migrated thermogenic natural gas and endogenous biogenic methane.

The new Albany Shale as a potential source of exogenous thermogenic gases

In an *ad hoc* attempt to find a viable source of migrated thermogenic natural gas within the context of the petroleum systems of the Illinois Basin broadly, we compared published data and data produced as part of this study from the Springfield and Seelyville coal seams with previously published results from the underlying New Albany Shale (located *c.* 1200 m below the Springfield and Seelyville coal seams; McIntosh *et al.* 2002; Schlegel *et al.* 2011*a*). Based on this comparison, we note that the thermogenic-rich CBM samples fall along an apparent mixing line between data from the New Albany Shale and a biogenic endmember (Figs 4 & 5). An examination of published New Albany Shale data indicates an ambiguous genetic source for the shale gas. Based on the published $\delta^{13}C$–CH_4, δ^2H–CH_4 and C_1/C_2+ data from the New Albany Shale (McIntosh *et al.* 2002; Schlegel *et al.* 2011*b*), we recognize that the New Albany Shale may itself contain unquantified microbial components (Figs 4 & 5; McIntosh *et al.* 2002; Schlegel *et al.* 2011*b*). Independent of this hypothesis, it is apparent that the New Albany Shale gases have more thermogenic-looking characteristics, specifically higher ethane abundance, than natural gases produced from the shallower Pennsylvanian CBM. While the genetic source of the New Albany Shale gas remains uncertain, it is interesting to note that application of the GGS-R gas fractionation model adapted for hydrocarbons by Darrah *et al.* (2014, 2015*b*) accounts for the observed trends in the New Albany Shale gas data (despite the relatively large range in composition) and allows us to characterize a plausible thermogenic source of natural gas that contributes to the mixed biogenic–thermogenic gases found in the Springfield and Seelyville coal seams (Figs 4 & 5). Based on geochemistry of the currently unidentified exogenous source of natural gas observed in the Springfield and Seelyville formations and the relative stratigraphic positions of both the Devonian/ Mississippian-aged New Albany Shale and the Pennsylvanian-aged coal seams in the petroleum systems of the Illinois Basin, we suggest that the New Albany Shale is a potential exogenous source of thermogenic gas to the Springfield and Seelyville coal seams. While the New Albany Shale gases are

consistent with hypothetical migration to the Pennsylvanian coals, it should also be noted that there could be additional, uncharacterized source formations in the Illinois Basin that might have similar attributes to the New Albany Shale.

Post-migration modification of the thermogenic endmember

Before concluding that the New Albany Shale contributed exogenous thermogenic natural gases to the Springfield and Seelyville coal seams, one must consider how post-genetic processes can alter the original composition of natural gas. Solubility partitioning during dual-phase migration, anaerobic oxidation and aerobic oxidation are known to lead to significant post-genetic modification in various hydrocarbon reservoirs (Whiticar 1999; Kinnaman *et al.* 2007; Etiope *et al.* 2009*a*; Darrah *et al.* 2015*b*). In addition to mixing (discussed above), we suggest two additional processes that could affect $^4He/CH_4$ and C isotope ratios of a thermogenic gas in predictable ways.

Anaerobic oxidation. Anaerobic oxidation couples methane with a terminal electron acceptor such as sulphate, yielding CO_2 and H_2 in proportions linked to the hydrocarbons that were biodegraded. It has been most significantly documented in seafloor environments where sulphate is abundant (e.g. Alperin & Hoehler 2009, 2010). Anaerobic oxidation by methanotrophs would only affect methane and therefore decrease the C_1/C_2+ in the residual natural gas mixture (i.e. methane is oxidized preferentially to higher-order aliphatic hydrocarbons), resulting in more positive $\delta^{13}C$ and δ^2H values of methane (Whiticar 1999; Kinnaman *et al.* 2007; Pape *et al.* 2010), and ultimately yielding an apparently 'wetter' natural gas that is referred to as secondary biogenic gas (Scott *et al.* 1994; Pallasser 2000; Etiope *et al.* 2009*b*).

Anaerobic oxidation of an originally ethane-poor biogenic natural gas would preferentially convert the methane to CO_2, leaving the residual un-oxidized methane enriched in ^{13}C (Whiticar 1999; Etiope *et al.* 2009*b*; Pape *et al.* 2010). While this process could account for the elevated $^4He/CH_4$ (Fig. 11), lower C_1/C_2+ ratios and heavier $\delta^{13}C$-CH_4 than would be anticipated for biogenic gas (down and to the right in Fig. 5), anaerobic methane oxidation cannot explain the correlation between $\delta^{13}C$-CH_4 and $^4He/^{20}Ne$ v. gas–water ratios ($CH_4/^{36}Ar$) and the correlation between elevated $^4He/CH_4$ v. $^{20}Ne/$ ^{36}Ar or $\delta^{13}C$-CH_4, nor the proportion of methane in the same samples. Therefore, we conclude that anaerobic methane oxidation alone cannot account for all trends in the data.

Aerobic oxidation v. mixing. Aerobic oxidation can degrade methane, ethane, propane and other higher aliphatic hydrocarbons. Because heavier aliphatic hydrocarbons are oxidized preferentially to methane (James & Burns 1984; Etiope *et al.* 2009*a*; Harkness *et al.* 2017), aerobic oxidation could account for increased C_1/C_2+, less negative $\delta^{13}C$ and δ^2H values in methane and higher $^4He/CH_4$ in the residual natural gas mixture. Aerobic oxidation is not considered likely in the subsurface without the recent introduction of oxic groundwaters. The persistence of oxic groundwaters in coals is even more unlikely as dissolved oxygen would be consumed rapidly in organic-rich environments such as coal seams (Martini *et al.* 2003; Vinson *et al.* 2017).

We envision one scenario that could lead to extensive aerobic oxidation of hydrocarbons in the Springfield and Seelyville coal seams. The Illinois Basin is known to have experienced periods of extensive glacioisostatic rebound, neotectonic fracturing and glacial meltwater recharge within the last 1.2 Ma, specifically near the basin margins (McIntosh *et al.* 2004, 2012; Schlegel *et al.* 2011*b*). In this scenario, the voluminous recharge of glacial meltwater during interglacial periods could induce aerobic oxidation. While it is nearly impossible to quantify the influence of an episodic infiltration of oxic glacial meltwater into the Springfield and Seelyville coal seams (or the penetration depth of O_2 associated with a recharge pulse) solely on gas composition, the corresponding increase in the levels of ASW gases (N_2, ^{36}Ar, ^{20}Ne) in samples with the highest C_1/C_2+ is broadly consistent with this scenario. The shallow locations of the Springfield and Seelyville coal seams (*c.* 73–213 m) in this area may be more susceptible to the infiltration of oxic freshwater, specifically because our study area is proximal to the basin margins. For example, aerobic oxidation has been hypothesized to occur at basin margins in other coalbed methane reservoirs, such as the San Juan Basin (Scott *et al.* 1994).

Because aerobic oxidation would remove ethane and higher-order hydrocarbon gases and reduce the quantity of residual thermogenic methane, the addition of a constant volume of biogenic methane to previously oxidized thermogenic gases would yield a mixture with a seemingly increasing proportion of biogenic contributions based on hydrocarbon composition. Therefore, we conclude that it is conceivable that the modern composition of the Springfield and Seelyville natural gases may appear to include a lower proportion of thermogenic natural gas today because of past episodes of aerobic oxidation. However, if aerobic oxidation did alter the hydrocarbon composition, one would also expect to observe the highest $^4He/CH_4$ values in the most oxidized, $\delta^{13}C$-CH_4-heavy endmember because oxidation would remove methane without altering helium concentrations. Instead, the samples with the lowest $^4He/CH_4$ values also display the most negative

$\delta^{13}C$-CH_4 values (Fig. 11). Thus, although we cannot rule out that aerobic oxidation altered the initial composition of an exogenous thermogenic natural gas, the ensemble of data is most easily explained by an exogenous thermogenic natural gas with an apparent range of thermal maturities that later mixed and was 'over-printed' by a helium- and ethane-poor endogenous biogenic endmember.

Mantle contributions

In addition to the data discussed above, the presence of samples with elevated helium isotopic compositions ($^3He/^4He$ reported as R/R_A) supports our interpretation for exogenous sources of natural gas in the Springfield and Seelyville coal seams (Fig. 7). An intriguing geological consideration is that ultramafic (i.e. mantle-derived) igneous dykes of Permian age (*c.* 270 Ma) have been discovered in outcrop and during mining operations in the southeastern margin of the Illinois Basin (Reynolds *et al.* 1997; Stewart *et al.* 2005; Schimmelmann *et al.* 2009). Mantle-derived fluids (e.g. ^3He and CO_2) can be degassed from an intrusive body and stored within sedimentary reservoirs, such as structural or stratigraphic hydrocarbon traps, for extended periods of geological time and can subsequently mix with or migrate into regions of locally produced fluids (Ballentine *et al.* 2001). While ^3He is a conservative tracer that preserves evidence of mantle contributions, CO_2 can readily dissolve in formation waters and form DIC, precipitate as carbonates, or be altered by microbial activity. Although there are no known magmatic intrusions observed in cuttings or cores within our study area of the Illinois Basin, we do observe helium isotopic values (R/R_A) above the crustal endmember in all samples. Observed high He/Ne values (Fig. 7) are consistent with the presence of small but quantifiable inputs of mantle-derived gases with the majority of samples plotting between the crustal endmember and 1% mantle (Hilton 1996; Darrah *et al.* 2015*b*). It is intriguing that the samples with the highest $^3He/^4He$ ratios are those with the highest proportions of ASW (maximum $^3He/^4He$ of 0.901 or *c.* 10% mantle contribution with a He/Ne ratio of 51.5), which could suggest deep-reaching conduits that act as migration pathways for magmatic fluids also facilitate downward migration of meteoric waters.

One may consider that the magmatic intrusives could produce highly variable thermal maturities within the coal seams owing to the increased heat that accompanies the intruding magmatic body. We therefore cannot completely dismiss the potential for variable heating of coals related to magmatic intrusions. However, previous studies in the southern Illinois Basin have demonstrated that the spatial influence of contact metamorphism on increased vitrinite reflectance values is limited to a proximity of about 1.2 times the dyke thickness (Stewart *et al.* 2005; Schimmelmann *et al.* 2009). Because we observe no statistically significant correlation between $^3He/^4He$ and $\delta^{13}C$-CH_4 and based on the lack of evidence for dyke intrusions locally, we conclude that the mantle-derived ^3He is not related to thermogenic natural gas production. However, ^3He does provide further evidence for the presence of an exogenous fluid that has mixed with the natural gas produced from CBM gas wells in the study area at an undetermined time in the past.

Conclusion

Although biogenic methane dominates natural gas composition in the Pennsylvanian coals in our study area located in the eastern Illinois Basin, we quantify thermogenic natural gas contributions to the majority of gases sampled in the current study. By integrating hydrocarbon and noble gas data, we identify a coherent mixture between biogenic methane and an exogenous thermogenic source. We determine that the thermogenic endmember is distinguished by a positive relationship between the occurrence of ethane, ^4He, exogenous ^{20}Ne, enriched $^{20}Ne/^{36}Ar$ and enriched $\delta^{13}C$-CH_4. In comparison, the more biogenic endmember contained more negative $\delta^{13}C$-CH_4, is significantly more enriched in gases derived from ASW (e.g. ^{36}Ar) and contains lower concentrations of ethane.

Resolving biogenic–thermogenic gas mixtures is a challenging problem, made more difficult by the lack of specific endmember identification in terms of conventional stable isotope ratios. Without the information encoded in noble gas isotope ratios, previously published mixing models from this study area did not address other local potential options for exogenous thermogenic natural gas contributions, but instead compared data with hypothetical biogenic and thermogenic endmembers. These studies concluded that thermogenic contributions ranged from only 0.1 to 0.4% by volume (Strąpoć *et al.* 2007; Mastalerz *et al.* 2017). In comparison, hydrocarbon mixing models that consider a New Albany Shale-like gas source as the thermogenic endmember, with its higher thermal maturity than the Pennsylvanian coals, yield estimates of exogenous thermogenic gas contributions that range from 4.6% to 17.1%. Similarly, mixing models that utilize conservative noble gases (i.e. $^4He/CH_4$, $^{20}Ne/^{36}Ar$) yield estimates of exogenous thermogenic gas that range from 6.3% to 19.2%, which is roughly consistent with our hydrocarbon mixing values when one considers the potential for an endogenous endmember. The noble gas ratios provide evidence that these coalbed gases are more thermogenic (or at

least more exogenous) than previously thought, but we see no indication that a significant quantity of thermogenic gas was produced from heating of the Pennsylvanian coals themselves in the study area. This interpretation is consistent with recent clumped isotopologue measurements from the Antrim Shale in the nearby Michigan Basin.

A summary of our interpretations is as follows:

(1)   Based on correlations between methane, ethane and helium-4, we suggest that a thermogenic natural gas with relatively enriched values of $\delta^{13}C$-CH_4, high $^4He/CH_4$, ^4He, $^{21}Ne^*$ and $^{40}Ar^*$, and low C_1/C_2+ (as compared with biogenic methane) is present and quantifiable in the Springfield and Seelyville coal seams.

(2)   A subset of low-thermal-maturity gases could result from endogenous thermogenic hydrocarbon generation. However, the majority of gas samples in this study require an exogenous thermogenic source from more deeply buried units to account for elevated [He]. For illustrative purposes, we used a New Albany Shale-like endmember as an example of a deeper and more thermogenic (but perhaps not purely thermogenic) gas to test the hypothesis for an exogenous source of natural gas. The presence of mantle-derived ^3He further supports our interpretation of an exogenous source of natural gas.

(3)   Deeper, higher thermal maturity stratigraphic units, such as the New Albany Shale, are required to account for the exogenous thermogenic endmember. We hypothesize that dual-phase (free gas + brine) migration of thermogenic hydrocarbon gas would fractionate natural gas composition based on solubility, increasing the C_1/C_2+, $^4He/CH_4$, and $^{20}Ne/^{36}Ar$ and decreasing $^{84}Kr/^{36}Ar$. Eventually a component of these thermogenic natural gases migrated to and charged the stratigraphic traps of the Springfield and Seelyville coal seams, where it apparently remained in the Springfield and Seelyville coal seams within the study area.

(4)   We hypothesize that the subsequent formation of a dry biogenic gas decreased the $\delta^{13}C$-CH_4 and elevated the C_1/C_2+ ratios of the resulting mixture, which diluted the geochemical signature of the migrated thermogenic methane. Unfortunately, the present dataset cannot resolve contributions from primary and secondary biogenic sources. Future work including carbon isotopes of ethane and propane should be undertaken to further test this hypothesis.

This study demonstrates that the integration of noble gas and hydrocarbon geochemistry permits a more robust evaluation of the source and migration processes of natural gas in the crust compared with hydrocarbon analyses alone. A suitable understanding of these processes is essential for understanding the genetic source of hydrocarbon gases in coalbed methane reservoirs as well as the roles of microbes and post-genetic modification processes in the subsurface. Further, these findings imply that established techniques, such as the usage of isotopic and molecular content of hydrocarbons alone, should be re-evaluated in other basins.

We acknowledge financial support from NSF EAGER (EAR-1249255) and NSF SusChem (EAR-1441497) to T.H.D, NSF Earth Sciences Postdoctoral Fellowship (EAR-1249916) to D.S.V., and salary support for M.T.M. and C.J.W. from the School of Earth Sciences at The Ohio State University. We thank Larry Neely and Jason Neely (Maverick Energy) for permission to sample their CBM wells and their unyielding assistance in sampling, sharing of information, and their knowledge about the study area. In addition, we thank Dr Maria Mastalerz (Indiana Geological Survey) for helpful discussions. We also thank Dr Robert Poreda (U. Rochester), Andrew Hunt (USGS) and an anonymous reviewer for reviews of earlier versions of the manuscript.

References

AESCHBACH-HERITG, W., PEETERS, F., BEYERLE, U. & KIPFER, R. 1999. Interpretation of dissolved atmospheric noble gases in natural waters. *Water Resources Research*, **35**, 2779–2792.

AESCHBACH-HERTIG, W., EL-GAMAL, H., WIESER, M. & PALCSU, L. 2008. Modeling excess air and degassing in groundwater by equilibrium partioning with a gas phase. *Water Resources Research*, **44**, W08449.

AHMED, A.J., JOHNSTON, S., BOYER, C., LAMBERT, S.W., BUSTOS, O.A., PASHIN, J.C. & WRAY, A. 2009. Coalbed methane: clean energy for the world. *Oilfield Review*, **21**, 4–13.

ALPERIN, M.J. & HOEHLER, T.M. 2009. Anaerobic methane oxidation by archaea/sulfate-reducing bacteria aggregates: thermodynamic and physical constraints. *American Journal of Science*, **309**, 869–957.

ALPERIN, M. & HOEHLER, T. 2010. Biogeochemistry: the ongoing mystery of sea-floor methane. *Science*, **329**, 288–289.

ANDERSON, J.S., ROMANAK, K.D., YANG, C., LU, J., HOVORKA, S.D. & YOUNG, M.H. 2017. Gas source attribution techniques for assessing leakage at geologic CO_2 storage sites: evaluating a CO_2 and CH_4 soil gas anomaly at the Cranfield CO_2-EOR site. *Chemical Geology*, **454**, 93–104.

BALLENTINE, C.J. & BURNARD, P.G. 2002. Production, release and transport of noble gases in the continental crust. *Reviews in Mineralogy and Geochemistry*, **47**, 481–538.

BALLENTINE, C.J., ONIONS, R.K., OXBURGH, E.R., HORVATH, F. & DEAK, J. 1991. Rare-gas constraints on hydrocarbon accumulation, crustal degassing, and groundwater-flow in the Pannonian Basin. *Earth and Planetary Science Letters*, **105**, 229–246.

BALLENTINE, C.J., MAZUREK, M. & GAUTSCHI, A. 1994. Thermal constraints on crustal rare-gas release and migration-evidence from Alpine fluid inclusions. *Geochimica et Cosmochimica Acta*, **58**, 4333–4348.

BALLENTINE, C.J., SCHOELL, M., COLEMAN, D. & CAIN, B.A. 2001. 300-Myr-old magmatic CO_2 in natural gas reservoirs of the west Texas Permian basin. *Nature*, **409**, 327–331.

BALLENTINE, C.J., BURGESS, R. & MARTY, B. 2002. Tracing fluid origin, transport and interaction in the crust. *Reviews in Mineralogy and Geochemistry*, **47**, 539–614.

BARRY, P., LAWSON, M., MEURER, W., DANABALAN, D., BYRNE, D., MABRY, J. & BALLENTINE, C.J. 2017. Determining fluid migration and isolation times in multiphase crustal domains using noble gases. *Geology*, **45**, 775–778, https://doi.org/10.1130/G38900.1

BARRY, P.H., LAWSON, M., MEURER, W.P., WARR, O., MABRY, J.C., BYRNE, D.J. & BALLENTINE, C.J. 2016. Noble gases solubility models of hydrocarbon charge mechanism in the Sleipner Vest gas field. *Geochimica et Cosmochimica Acta*, **194**, 291–309.

BATES, B.L., MCINTOSH, J.C., LOHSE, K.A. & BROOKS, P.D. 2011. Influence of groundwater flowpaths, residence times and nutrients on the extent of microbial methanogenesis in coal beds: Powder River Basin, USA. *Chemical Geology*, **284**, 45–61.

BERNARD, B.B., BROOKS, J.M. & SACKETT, W.M. 1976. Natural-gas seepage in Gulf of Mexico. *Earth and Planetary Science Letters*, **31**, 48–54.

BUSCHBACH, T.C. & KOLATA, D.R. 1990. Regional setting of the Illinois Basin. *In*: LEIGHTON, M.W., KOLATA, D.R., OLTZ, D.F. & EIDEL, J. (eds) *Interior Cratonic Basins*. AAPG, Memoirs, **51**, 29–55.

CHUNG, H.M., GORMLY, J.R. & SQUIRES, R.M. 1988. Origin of gaseous hydrocarbons in subsurface environments-theoretical considerations of carbon isotope distribution. *Chemical Geology*, **71**, 97–103.

CLUFF, R.M. 1980. Paleoenvironment of the New Albany Shale Group (Devonian–Mississippian) of Illinois. *Journal of Sedimentary Petrology*, **50**, 767–780.

COLEMAN, D.D., RISATTI, J.B. & SCHOELL, M. 1981. Fractionation of carbon and hydrgoen isotopes by methane-oxidizing bacteria. *Geochimica et Cosmochimica Acta*, **45**, 1033–1037.

COOK, P.G., SOLOMON, D.K., SANFORD, W.E., BUSENBERG, L.N., PLUMMER, L.N. & POREDA, R.J. 1996. Inferring shallow groundwater flow in saprolite and fractured rock using environmental tracers. *Water Resources Research*, **32**, 1501–1509.

CRAIG, H., LUPTON, J.E. & HORIBE, Y. 1978. A mantle component in circum Pacific volcanic glasses; Hakone, the Marianas, and Mt. Lassen. *In*: ALEXANDER, E.C. & OZIMA, M. (ed.) *Terrestrial Rare Gases*. Japan Science Society Press, Tokyo, 3–16.

CUOCO, E., TEDESCO, D., POREDA, R.J., WILLIAMS, J.C., DE FRANCESCO, S., BALAGIZI, C. & DARRAH, T.H. 2013. Impact of volcanic plume emissions on rain water chemistry during the January 2010 Nyamuragira eruptive event: implications for essential potable water resources. *Journal of Hazardous Materials*, **244**, 570–581.

DARRAH, T.H. & POREDA, R.J. 2012. Evaluating the accretion of meteoritic debris and interplanetary dust particles in the GPC-3 sediment core using noble gas and mineralogical tracers. *Geochimica et Cosmochimica Acta*, **84**, 329–352.

DARRAH, T.H., TEDESCO, D., TASSI, F., VASELLI, O., CUOCO, E. & POREDA, R.J. 2013. Gas chemistry of the Dallol region of the Danakil Depression in the Afar region of the northern-most East African Rift. *Chemical Geology*, **339**, 16–29.

DARRAH, T.H., VENGOSH, A., JACKSON, R.B., WARNER, N.R. & POREDA, R.J. 2014. Noble gases identify the mechanisms of fugitive gas contamination in drinking-water wells overlying the Marcellus and Barnett Shales. *Proceedings of the National Academy of Sciences of the United States of America*, **111**, 14076–14081.

DARRAH, T.H., JACKSON, R.B., VENGOSH, A., WARNER, N.R. & POREDA, R.J. 2015*a*. Noble gases: a new technique for fugitive gas investigation in groundwater. *Groundwater*, **53**, 23–28.

DARRAH, T.H., JACKSON, R.B. ET AL. 2015*b*. The evolution of Devonian hydrocarbon gases in shallow aquifers of the northern Appalachian Basin: insights from integrating noble gas and hydrocarbon geochemistry. *Geochimica et Cosmochimica Acta*, **170**, 321–355.

DOUGLAS, P.M.J., STOLPER, D.A. ET AL. 2016. Diverse origins of Arctic and Subarctic methane point source emissions identified with multiply-substituted isotopologues. *Geochimica et Cosmochimica Acta*, **188**, 163–188.

DROBNIAK, A., MASTALERZ, M., RUPP, J. & EATON, N. 2004. Evaluation of coalbed gas potential of the Seelyville Coal Member, Indiana, USA. *International Journal of Coal Geology*, **57**, 265–282.

ETIOPE, G., FEYZULLAYEV, A. & BACIU, C.L. 2009*a*. Terrestrial methane seeps and mud volcanoes: a global perspective of gas origin. *Marine and Petroleum Geology*, **26**, 333–344.

ETIOPE, G., FEYZULLAYEV, A., MILKOV, A.V., WASEDA, A., MIZOBE, K. & SUN, C.H. 2009*b*. Evidence of subsurface anaerobic biodegradation of hydrocarbons and potential secondary methanogenesis in terrestrial mud volcanoes. *Marine and Petroleum Geology*, **26**, 1692–1703.

ETIOPE, G., BACIU, C.L. & SCHOELL, M. 2011. Extreme methane deuterium, nitrogen and helium enrichment in natural gas from the Homorod seep (Romania). *Chemical Geology*, **280**, 89–96.

FABER, E., STAHL, W. & WHITICAR, M. 1992. *Distinction of Bacterial and Thermogenic Hydrocarbon Gases*. R. Vially, Paris.

FURMANN, A., SCHIMMELMANN, A., BRASSELL, S.C., MASTALERZ, M. & PICARDAL, F. 2013. Chemical compound classes supporting microbial methanogenesis in coal. *Chemical Geology*, **339**, 226–241.

GAO, L., BRASSELL, S.C., MASTALERZ, M. & SCHIMMELMANN, A. 2013. Microbial degradation of sedimentary organic matter associated with shale gas and coalbed methane in eastern Illinois Basin (Indiana), USA. *International Journal of Coal Geology*, **107**, 152–164.

GILFILLAN, S.M., BALLENTINE, C.J. ET AL. 2008. The noble gas geochemistry of natural CO_2 gas reservoirs from the Colorado Plateau and Rocky Mountain provinces, USA. *Geochimica et Cosmochimica Acta*, **72**, 1174–1198.

GILFILLAN, S.M.V., SHERWOOD LOLLAR, B. ET AL. 2009. Solubility trapping in formation water as dominant CO_2 sink in natural gas fields. *Nature*, **458**, 614–618.

GOLDING, S.D., BOREHAM, C.J. & ESTERLE, J.S. 2013. Stable isotope geochemistry of coal bed and shale gas and related production waters: a review. *International Journal of Coal Geology*, **120**, 24–40.

GREEN, T.W., WOLFE, S.R. & TEDESCO, S.A. 2003. Pilot tests in the Illinois and western interior basins to determine commercial productivity from Pennsylvanian-aged coals. Society of Petroleum Engineers SPE-84430-MS.

HARKNESS, J.S., DARRAH, T.H. ET AL. 2017. The geochemistry of naturally occurring methane and saline groundwater in an area of unconventional shale gas development. *Geochimica et Cosmochimica Acta*, **208**, 302–334.

HEAD, I.M., GRAY, N.D. & LARTER, S.R. 2014. Life in the slow lane; biogeochemistry of biodegraded petroleum containing reservoirs and implications for energy recovery and carbon management. *Frontiers in Microbiology*, **5**, 1–23.

HILTON, D.R. 1996. The helium and carbon isotope systematics of a continental geothermal system: results from monitoring studies at Long Valley caldera (California, USA). *Chemical Geology*, **127**, 269–295.

HOLLAND, G., SHERWOOD LOLLAR, B., LI, L., LACRAMPE-COULOUME, G., SLATER, G.F. & BALLENTINE, C.J. 2013. Deep fracture fluids isolated in the crust since the Precambrian era. *Nature*, **497**, 357.

HOLOCHER, J., PEETERS, F., AESCHBACH-HERTIG, W., KINZELBACH, W. & KIPFER, R. 2003. Kinetic model of gas bubble dissolution in groundwater and its implications for the dissolved gas composition. *Environmental Science and Technology*, **37**, 1337–1343.

HUNT, A.G. 2000. *Diffusional Release of Helium-4 from Mineral Phases as Indicators of Groundwater Age and Depositional History*. University of Rochester, Department of Earth and Environmental Sciences.

HUNT, A.G., DARRAH, T.H. & POREDA, R.J. 2012. Determining the source and genetic fingerprint of natural gases using noble gas geochemistry: a northern Appalachian Basin case study. *AAPG Bulletin*, **96**, 1785–1811.

JACKSON, R.B., VENGOSH, A. ET AL. 2013. Increased stray gas abundance in a subset of drinking water wells near Marcellus shale gas extraction. *Proceedings of the National Academy of Sciences of the United States of America*, **110**, 11250–11255.

JAMES, A.T. & BURNS, B.J. 1984. Microbial alteration of subsurface natural-gas accumulations. *AAPG Bulletin*, **68**, 957–960.

JENDEN, P.D., DRAZAN, D.J. & KAPLAN, I.R. 1993. Mixing of thermogenic natural gases in northern Appalachian Basin. *AAPG Bulletin*, **77**, 980–998.

KANG, M., CHRISTIAN, S. ET AL. 2016. Identification and characterization of high methane-emitting abandoned oil and gas wells. *Proceedings of the National Academy of Sciences of the United States of America*, **113**, 13636–13641.

KARACAN, C.O., DROBNIAK, A. & MASTALERZ, M. 2014. Coal bed reservoir simulation with geostatistical property realizations for simultaneous multi-well production history matching: a case study from Illinois Basin, Indiana, USA. *International Journal of Coal Geology*, **131**, 71–89.

KENNEDY, B.M., HIYAGON, H. & REYNOLDS, J.H. 1990. Crustal neon – a striking uniformity. *Earth and Planetary Science Letters*, **98**, 277–286.

KINNAMAN, F.S., VALENTINE, D.L. & TYLER, S.C. 2007. Carbon and hydrogen isotope fractionation associated with the aerobic microbial oxidation of methane, ethane, propane and butane. *Geochimica et Cosmochimica Acta*, **71**, 271–283.

MARTINI, A.M., BUDAI, J.M., WALTER, L.M. & SCHOELL, M. 1996. Microbial generation of economic accumulations of methane within a shallow organic-rich shale. *Nature*, **383**, 155–158.

MARTINI, A.M., WALTER, L.M., BUDAI, J.M., KU, T.C.W., KAISER, C.J. & SCHOELL, M. 1998. Genetic and temporal relations between formation waters and biogenic methane: Upper Devonian Antrim Shale, Michigan Basin, USA. *Geochimica et Cosmochimica Acta*, **62**, 1699–1720.

MARTINI, A.M., WALTER, L.M., KU, T.C., BUDAI, J.M., MCINTOSH, J.C. & SCHOELL, M. 2003. Microbial production and modification of gases in sedimentary basins: a geochemical case study from a Devonian shale gas play, Michigan basin. *AAPG Bulletin*, **87**, 1355–1375.

MASTALERZ, M., KVALE, E.P., STANKIEWICZ, B.A. & PORTLE, K. 1999. Organic geochemistry in Pennsylvanian tidally influenced sediments from SW Indiana. *Organic Geochemistry*, **30**, 57–73.

MASTALERZ, M., DROBNIAK, A., RUPP, J.A. & SHAFFER, N.R. 2004. *Characterization of Indiana's Coal Resource: Availability of the Reserves, Physical and Chemical Properties of Coal, and Present and Potential Uses*. Indiana Geological Survey, Indianapolis, IN.

MASTALERZ, M., SCHIMMELMANN, A., DROBNIAK, A. & CHEN, Y. 2013. Porosity of Devonian and Mississippian New Albany Shale across a maturation gradient: insights from organic petrology, gas adsorption, and mercury intrusion. *AAPG Bulletin*, **97**, 1621–1643.

MASTALERZ, M., DROBNIAK, A. & SCHIMMELMANN, A. 2017. Characteristics of microbial coalbed gas during production; example from Pennsylvanian Coals in Indiana, USA. *Geosciences*, **7**, 26–44.

MCINTOSH, J.C., WALTER, L.M. & MARTINI, A.M. 2002. Pleistocene recharge to midcontinent basins: effects on salinity structure and microbial gas generation. *Geochimica et Cosmochimica Acta*, **66**, 1681–1700.

MCINTOSH, J.C., WALTER, L.M. & MARTINI, A.M. 2004. Extensive microbial modification of formation water geochemistry: case study from a Midcontinent sedimentary basin, United States. *Geological Society of America Bulletin*, **116**, 743–759.

MCINTOSH, J.C., SCHLEGEL, M.E. & PERSON, M. 2012. Glacial impacts on hydrologic processes in sedimentary basins: evidence from natural tracer studies. *Geofluids*, **12**, 7–21.

MOORE, T.A. 2012. Coalbed methane: a review. *International Journal of Coal Geology*, **101**, 36–81.

OXBURGH, E.R. & ONIONS, R.K. 1987. Helium loss, tectonics, and the terrestrial heat-budget. *Science*, **237**, 1583–1588.

OXBURGH, E., O'NIONS, R. & HILL, R. 1986. Helium isotopes in sedimentary basins. *Nature*, **324**, 632–635.

PALLASSER, R.J. 2000. Recognising biodegradation in gas/oil accumulations through the delta C-13 compositions of gas components. *Organic Geochemistry*, **31**, 1363–1373.

PAPE, T., BAHR, A. ET AL. 2010. Molecular and isotopic partitioning of low-molecular-weight hydrocarbons during migration and gas hydrate precipitation in deposits of a high-flux seepage site. *Chemical Geology*, **269**, 350–363.

PASHIN, J.C., MCINTYRE-REDDEN, M.R., MANN, S.D., KOPASKA-MERKEL, D.C., VARONKA, M. & OREM, W. 2014. Relationships between water and gas chemistry in mature coalbed methane reservoirs of the Black Warrior Basin. *International Journal of Coal Geology*, **126**, 92–105.

PEPPER, A.S. & CORVI, P.J. 1995. Simple kinetic-models of petroleum formation. 3. Modeling an open system. *Marine and Petroleum Geology*, **12**, 417–452.

POREDA, R.J., JENDEN, P.D., KAPLAN, I.R. & CRAIG, H. 1986. Mantle helium in Sacramento Basin Natural-gas wells. *Geochimica et Cosmochimica Acta*, **50**, 2847–2853.

POREDA, R.J., CRAIG, H., ARNÓRSSON, S. & WELHAN, J.A. 1992. Helium isotopes in Icelandic geothermal systems: I. ^{3}He, gas chemistry, and ^{13}C relations. *Geochimica et Cosmochimica Acta*, **56**, 4221–4228.

REYNOLDS, R.L., GOLDHABER, M.B. & SNEE, L.W. 1997. Paleomagnetic and ^{40}Ar/^{39}Ar results from the Grant Intrusive Breccia and comparison to the Permian Downeys Bluff Sill – evidence for Permian Igneous Activity at Hicks Dome, Southern Illinois Basin. US Government Printing Office.

RITTER, D., VINSON, D. ET AL. 2015. Enhanced microbial coalbed methane generation: a review of research, commercial activity, and remaining challenges. *International Journal of Coal Geology*, **146**, 28–41.

ROWE, D. & MUEHLENBACHS, A. 1999*a*. Low-temperature thermal generation of hydrocarbon gases in shallow shales. *Nature*, **398**, 61–63.

ROWE, D. & MUEHLENBACHS, K. 1999*b*. Isotopic fingerprints of shallow gases in the Western Canadian Sedimentary Basin: tools for remediation of leaking heavy oil wells. *Organic Geochemistry*, **30**, 861–871.

RUPPERT, L.F., KIRSCHBAUM, M.A., WARWICK, P.D., FLORES, R.M., AFFOLTER, R.H. & HATCH, J.R. 2002. The US Geological Survey's national coal resource assessment: the results. *International Journal of Coal Geology*, **50**, 247–274.

SCHIMMELMANN, A., MASTALERZ, M., GAO, L., SAUER, P.E. & TOPALOV, K. 2009. Dike intrusions into bituminous coal, Illinois Basin: H, C, N, O isotopic responses to rapid and brief heating. *Geochimica et Cosmochimica Acta*, **73**, 6264–6281.

SCHLEGEL, M.E., MCINTOSH, J.C., BATES, B.L., KIRK, M.F. & MARTINI, A.M. 2011*a*. Comparison of fluid geochemistry and microbiology of multiple organic-rich reservoirs in the Illinois Basin, USA: evidence for controls on methanogenesis and microbial transport. *Geochimica et Cosmochimica Acta*, **75**, 1903–1919.

SCHLEGEL, M.E., ZHOU, Z., MCINTOSH, J.C., BALLENTINE, C.J. & PERSON, M.A. 2011*b*. Constraining the timing of microbial methane generation in an organic-rich shale using noble gases, Illinois Basin, USA. *Chemical Geology*, **287**, 27–40.

SCHLEGEL, M.E., MCINTOSH, J.C., PETSCH, S.T., OREM, W.H., JONES, E.J.P. & MARTINI, A.M. 2013. Extent and limits of biodegradation by in situ methanogenic consortia in shale and formation fluids. *Applied Geochemistry*, **28**, 172–184.

SCHOELL, M. 1980. The hydrogen and carbon isotopic composition of methane from natural gases of various origins. *Geochimica et Cosmochimica Acta*, **44**, 649–661.

SCHOELL, M. 1983. Genetic characterization of natural gases. *AAPG Bulletin*, **67**, 2225–2238.

SCHOELL, M. 1988. Multiple origins of methane in the Earth. *Chemical Geology*, **71**, 1–10.

SCOTT, A.R., KAISER, W.R. & AYERS, W.B. 1994. Thermogenic and secondary biogenic gases, San Juan Basin, Colorado and New Mexico – implications for coalbed gas producibility. *AAPG Bulletin*, **78**, 1186–1209.

SHERWOOD LOLLAR, B. & BALLENTINE, C.J. 2009. Insights into deep carbon derived from noble gases. *Nature Geoscience*, **2**, 543–547.

SMITH, S.P. & KENNEDY, B.M. 1982. The solubility of noble gases in water and in NaCl brine. *Geochimica et Cosmochimica Acta*, **47**, 503–515.

SOLOMON, D.K., HUNT, A. & POREDA, R.J. 1996. Source of radiogenic helium 4 in shallow aquifers: implications for dating young groundwater. *Water Resources Research*, **32**, 1805–1813.

STEWART, A.K., MASSEY, M., PADGETT, P.L., RIMMER, S.M. & HOWER, J.C. 2005. Influence of a basic intrusion on the vitrinite reflectance and chemistry of the Springfield (No. 5) coal, Harrisburg, Illinois. *International Journal of Coal Geology*, **63**, 58–67.

STOLPER, D.A., MARTINI, A.M. *ET AL.* 2015. Distinguishing and understanding thermogenic and biogenic sources of methane using multiply substituted isotopologues. *Geochimica et Cosmochimica Acta*, **161**, 219–247.

STRĄPOĆ, D., MASTALERZ, M., EBLE, C. & SCHIMMELMANN, A. 2007. Characterization of the origin of coalbed gases in southeastern Illinois Basin by compound-specific carbon and hydrogen stable isotope ratios. *Organic Geochemistry*, **38**, 267–287.

STRĄPOĆ, D., MASTALERZ, M., SCHIMMELMANN, A., DROBNIAK, A. & HASENMUELLER, N.R. 2010. Geochemical constraints on the origin and volume of gas in the New Albany Shale (Devonian-Mississippian), eastern Illinois Basin. *AAPG Bulletin*, **94**, 1713–1740.

STRĄPOĆ, D., MASTALERZ, M. *ET AL.* 2011. Biogeochemistry of microbial coal-bed methane. *Annual Review of Earth and Planetary Sciences*, **39**, 617–656.

STRĄPOĆ, D., MASTALERZ, M., SCHIMMELMANN, A., DROBNIAK, A. & HEDGES, S. 2008a. Variability of geochemical properties in a microbially dominated coalbed gas system from the eastern margin of the Illinois Basin, USA. *International Journal of Coal Geology*, **76**, 98–110.

STRĄPOĆ, D., PICARDAL, F.W. *ET AL.* 2008b. Methane-producing microbial community in a coal bed of the Illinois Basin. *Applied and Environmental Microbiology*, **74**, 2424–2432.

TANG, Y., PERRY, J.K., JENDEN, P.D. & SCHOELL, M. 2000. Mathematical modeling of stable carbon isotope ratios in natural gases. *Geochimica et Cosmochimica Acta*, **64**, 2673–2687.

TAYLOR, S.R. & MCLENNAN, S.M. 1995. The geochemical evolution of the continental crust. *Reviews of Geophysics*, **33**, 241–265.

TEDESCO, D., TASSI, F., VASELLI, O., POREDA, R.J., DARRAH, T., CUOCO, E. & YALIRE, M.M. 2010. Gas isotopic signatures (He, C, and Ar) in the Lake Kivu region (western branch of the East African rift system): geodynamic and volcanological implications. *Journal of Geophysical Research – Solid Earth*, **115**, B01205, https://doi.org/10.1029/2008JB006227

TILLEY, B. & MUEHLENBACHS, K. 2013. Isotope reversals and universal stages and trends of gas maturation in sealed, self-contained petroleum systems. *Chemical Geology*, **339**, 194–204.

TISSOT, B. & WELTE, D. 2012. *Petroleum Formation and Occurrence: A New Approach to Oil and Gas Exploration*. Springer Science & Business Media, Berlin.

USEIA 2013. *Annual Energy Outlook 2013*. Office of Integrated and International Energy Analysis, Washington, DC.

VALENTINE, D.L., KESSLER, J.D. *ET AL.* 2010. Propane respiration jump-starts microbial response to a deep oil spill. *Science*, **330**, 208–211.

VINSON, D.S., BLAIR, N.E., MARTINI, A.M., LARTER, S.R., OREM, W.H. & MCINTOSH, J.C. 2017. Microbial methane from in situ biodegradatoin of coal and shale: a review and reevaluation of hydrogen and carbon isotope signatures. *Chemical Geology*, **453**, 128–145.

WANG, D.T., GRUEN, D.S. *ET AL.* 2015. Nonequilibrium clumped isotope signals in microbial methane. *Science*, **348**, 428–431.

WANG, D.T., WELANDER, P.V. & ONO, S. 2016. Fractionation of the methane isotopologues (CH$_4$)-C-13, (CH3D)-C-12, and (CH3D)-C-13 during aerobic oxidation of methane by *Methylococcus capsulatus* (Bath). *Geochimica et Cosmochimica Acta*, **192**, 186–202.

WEISS, R. 1971a. Effect of salinity on the solubility of argon in water and seawater. *Deep-Sea Research*, **17**, 721.

WEISS, R. 1971b. Solubility of helium and neon in water and seawater. *Journal of Chemical and Engineering Data*, **16**, 235.

WEN, T., CASTRO, M.C., ELLIS, B.R., HALL, C.M. & LOHMANN, K.C. 2015. Assessing compositional variability and migration of natural gas in the Antrim Shale in the Michigan Basin using noble gas geochemistry. *Chemical Geology*, **417**, 356–370.

WEN, T., CASTRO, M.C., NICOT, J.P., HALL, C.M., LARSON, T., MICKLER, P.J. & DARVARI, R. 2016. Methane sources and migration mechanisms in shallow groundwaters in Parker and Hood Counties, Texas—A heavy noble gas analysis. *Environmental Science & Technology*, **50**, 12012–12021, https://doi.org/10.1021/acs.est.6b01494

WEN, T., CASTRO, M.C. *ET AL.* 2017. Characterizing the noble gas isotopic composition of the Barnett Shale and Strawn Group and constraining the source of stray gas in the trinity aquifer, North-Central Texas. *Environmental Science & Technology*, **51**, 6533–6541.

WHITEHILL, A.R., JOELSSON, L.M.T., SCHMIDT, J.A., WANG, D.T., JOHNSON, M.S. & ONO, S. 2017. Clumped isotope effects during OH and Cl oxidation of methane. *Geochimica et Cosmochimica Acta*, **196**, 307–325.

WHITICAR, M.J. 1999. Carbon and hydrogen isotope systematics of bacterial formation and oxidation of methane. *Chemical Geology*, **161**, 291–314.

WHITICAR, M., FABER, E. & SCHOELL, M. 1985. Hydrogen and carbon isotopes of C1 to C5 alkanes in natural gases: abstract. *AAPG Bulletin*, **69**, 316–316.

WHITICAR, M.J., FABER, E. & SCHOELL, M. 1986. Biogenic methane formation in marine and fresh-water environments- CO_2 reduction v. acetate fermentation isotope evidence. *Geochimica et Cosmochimica Acta,* **50,** 693–709.

WHITICAR, M.J., FABER, E., WHELAN, J.K. & SIMONEIT, B.R. 1994. Thermogenic and bacterial hydrocarbon gases (free and sorbed) in Middle Valley, Juan de Fuca Ridge, Leg 139. *Proceedings of the Ocean Drilling Progam, Scientific Results,* **139**. Ocean Drilling Program, College Station, TX.

YATSEVICH, I. & HONDA, M. 1997. Production of nucleogenic neon in the Earth from natural radioactive decay. *Journal of Geophysical Research – Solid Earth,* **102,** 10291–10298.

ZHOU, Z. & BALLENTINE, C.J. 2006. He-4 dating of groundwater associated with hydrocarbon reservoirs. *Chemical Geology,* **226,** 309–327.

ZHOU, Z., BALLENTINE, C.J., KIPFER, R., SCHOELL, M. & THIBODEAUX, S. 2005. Noble gas tracing of groundwater/coalbed methane interaction in the San Juan Basin, USA. *Geochimica et Cosmochimica Acta,* **69,** 5413–5428.

ZHOU, Z., BALLENTINE, C.J., SCHOELL, M. & STEVENS, S.H. 2012. Identifying and quantifying natural CO_2 sequestration processes over geological timescales: the Jackson Dome CO_2 Deposit, USA. *Geochimica et Cosmochimica Acta,* **86,** 257–275.

Testing clumped isotopes as a reservoir characterization tool: a comparison with fluid inclusions in a dolomitized sedimentary carbonate reservoir buried to 2–4 km

JOHN M. MacDONALD[1,2]*, CÉDRIC M. JOHN[1] & JEAN-PIERRE GIRARD[3]

[1]*Carbonate Research Group, Department Earth Science and Engineering, South Kensington Campus, Imperial College London, London SW7 2AZ, UK*

[2]*Present address: School of Geographical & Earth Sciences, University of Glasgow, Gregory Building, Glasgow G12 8QQ, UK*

[3]*TOTAL, Le Centre Scientifique et Technique Jean Féger (CSTJF), Avenue Larribau, 64018 Pau, France*

**Correspondence: John.MacDonald.3@glasgow.ac.uk*

Abstract: Constraining basin thermal history is a key part of reservoir characterization in carbonate rocks. Conventional palaeothermometric approaches cannot always be used: fluid inclusions may be reset or not present, while $\delta^{18}O$ palaeothermometry requires an assumption on the parent fluid composition. The clumped isotope palaeothermometer, however, is a promising technique for constraining the thermal history of basins. In this study, we test if clumped isotopes record temperatures of recrystallization in deeply-buried dolomitic reservoirs, through comparison with fluid-inclusion data. The studied reservoir is the Cretaceous Pinda Formation, offshore Angola, a deeply-buried dolomitized sedimentary carbonate hydrocarbon reservoir. It provides an ideal test case as samples from industry wells are available over a relatively wide burial depth range of *c.* 2000–4000 m below seafloor (mbsf) and the constituent dolomites are relatively homogeneous.

Across this depth range, fluid-inclusion homogenization temperatures for the Pinda Formation record a range of temperatures from *c.* 110 to 170°C, increasing with depth. These closely match present-day ambient well temperatures, indicating recent resetting of the fluid inclusions. Clumped isotopes, however, record temperatures significantly (*c.* 20–60°C) below fluid-inclusion and well temperatures for the seven samples analysed. The deepest five samples (*c.* 2800–3700 mbsf) record clumped isotope temperatures of around 100–120°C, interpreted to represent a deep burial recrystallization event responsible for a massive (re)dolomitization of the reservoir. The lower clumped isotope temperatures (65 and 82°C) of the shallower (2055 and 2740 mbsf) samples are interpreted to represent physical mixing of two dolomite generations due to incomplete burial recrystallization of an early shallow dolomite. Determination of temperature through clumped isotopes allows calculation of the parent fluid $\delta^{18}O$ values. In the five deepest samples, the fluid $\delta^{18}O$ values of 3.7–6.5‰ cluster around the modern-day porewater composition (5‰), suggesting that burial dolomitization occurred in the presence of evolved brine. Mineral $\delta^{18}O$ values of *c.* −7 to −4.5‰ are lower than pristine Cretaceous marine dolomite and are in accordance with burial recrystallization. Clumped isotopes are therefore interpreted to record temperatures corresponding to open-system burial recrystallization events. This study shows that clumped isotopes are a valuable tool in characterizing the thermal history of deeply-buried (>2000 m) carbonate hydrocarbon reservoirs.

Supplementary material: All standard and sample data are available at https://doi.org/10.6084/m9.figshare.c.3945184

A key aspect in understanding the history of hydrocarbon reservoirs is to characterize their thermal history. Indeed, thermal exposure is known to be a prime factor in the degradation of hydrocarbon reservoirs (e.g. Nadeau 2011). Changing temperature, for example during burial, can cause diagenetic recrystallization which may affect physical reservoir characteristics such as porosity and permeability, and impact hydrocarbon recovery rates. The thermal history also affects hydrocarbon maturation. In basins containing carbonate rocks, the two main approaches to characterizing thermal history in rocks of all ages are fluid-inclusion (e.g. McLimans 1987; Barker & Goldstein 1990; Goldstein 2001) and $\delta^{18}O$ (McCrea 1950; Epstein *et al.* 1951) palaeothermometry. These techniques are well established but have some drawbacks which may limit their application. Fluid-inclusion palaeothermometry can be used to

From: Lawson, M., Formolo, M. J. & Eiler, J. M. (eds) 2018. *From Source to Seep: Geochemical Applications in Hydrocarbon Systems*. Geological Society, London, Special Publications, **468**, 189–202. First published online December 14, 2017, https://doi.org/10.1144/SP468.7

© 2018 The Author(s). Published by The Geological Society of London. All rights reserved. For permissions: http://www.geolsoc.org.uk/permissions. Publishing disclaimer: www.geolsoc.org.uk/pub_ethics

infer burial temperatures, and hence potentially burial recrystallization, by measuring homogenization temperature (T_h). This is obtained by heating a two-phase inclusion until the vapour bubble disappears and all that remains is a single-phase liquid (e.g. Goldstein & Reynolds 1994). However, two-phase fluid inclusions are not always present or may not be large enough to reliably measure. This severely limits their use as a palaeothermometer in fine-grained carbonates. In addition, in the context of basin evolution, perhaps the most significant drawback is that they may stretch during subsequent burial, resulting in a resetting of T_h temperatures at present-day temperatures (e.g. Goldstein & Reynolds 1994). This leads to fluid inclusions recording maximum burial temperatures: that is, a temperature higher than the temperature at which the host mineral precipitated or recrystallized.

The $\delta^{18}O$ palaeothermometer is based on the relationship between mineral $\delta^{18}O$, temperature and the $\delta^{18}O$ of the parent fluid from which the mineral formed (McCrea 1950; Epstein *et al.* 1951). Measuring $\delta^{18}O$ in carbonates has been routine for many decades (e.g. Urey *et al.* 1951; Emiliani 1955; Shackleton 1967) but in order to calculate temperature, the $\delta^{18}O$ of the parent fluid must be known. In most cases, this value is unconstrained and an assumption must be made. An incorrect assumption by only 1‰ may lead to a temperature estimate greater than 10°C away from the true mineral formation temperature.

Clumped isotopes is a promising method which avoids the pitfalls in these other palaeothermometers. This technique is based on the thermodynamic relationship between carbonate mineral growth temperature and the abundance of chemical bonding ('clumping') between ^{13}C and ^{18}O isotopes (expressed as Δ_{47}) within single carbonate ions (e.g. Ghosh *et al.* 2006*a*; Schauble *et al.* 2006; Eiler 2007). This relationship has led to the application of the carbonate clumped isotope palaeothermometer to address a range of geological questions. Most studies have focused on surface and near-surface applications where the temperatures involved are in the *c.* 0–35°C range, including palaeoclimate (e.g. Affek *et al.* 2008; Passey *et al.* 2010; Thiagarajan *et al.* 2014) and palaeoaltimetry (e.g. Ghosh *et al.* 2006*b*; Huntington *et al.* 2010; Lechler *et al.* 2013; Carrapa *et al.* 2014) studies. The carbonate clumped isotope palaeothermometer can also be applied in the *c.* 50–300°C range, relevant to processes such as diagenesis and dolomitization (Ferry *et al.* 2011; Passey & Henkes 2012; MacDonald *et al.* 2013, 2015; Dale *et al.* 2014; Henkes *et al.* 2014; Sena *et al.* 2014; Vandeginste *et al.* 2014; Geske *et al.* 2015; John 2015; Kluge & John 2015; Kluge *et al.* 2015; Shenton *et al.* 2015; Stolper & Eiler 2016).

The ability of carbonate clumped isotopes to retain temperatures of original precipitation or fluid-driven recrystallization at burial temperatures of 50–300°C is not well established yet. Burial of carbonate rocks may lead to open-system recrystallization, whereby pre-existing carbonate minerals dissolve on contact with a fluid and reprecipitate. Assuming complete open-system recrystallization, the Δ_{47} value should reflect the ambient temperature during recrystallization. However, recent work (Dennis & Schrag 2010; Passey & Henkes 2012; Henkes *et al.* 2014) has suggested that temperatures attained during burial may lead to closed-system resetting of clumped isotopes, where intracrystal diffusion of O and C occurs, and Δ_{47} does not necessarily record the temperature at which open-system recrystallization occurred but is locked at a temperature between formation temperature and the maximum temperature of burial. Clumped isotope measurements of pristine carbonatites have given a wide range of temperatures, many around *c.* 150–250°C, which are interpreted as blocking temperatures during cooling (Dennis & Schrag 2010). Without recrystallization during burial diagenesis, the mechanism proposed for attaining the cooling temperatures is solid-state closed-system bond reordering of C—O bonds in the carbonate lattice (Dennis & Schrag 2010; Passey & Henkes 2012). If calcite is exposed to temperatures greater than *c.* 100°C for millions of years, bond reordering may occur (Dennis & Schrag 2010; Passey & Henkes 2012; Henkes *et al.* 2014; Stolper & Eiler 2015), bringing the clumped isotope temperatures to the ambient temperature.

The fact that bond reordering can occur in calcite held at >100°C for millions of years indicates that it is a key process to consider when investigating the utility of clumped isotopes in the characterization of deeply-buried reservoirs. Studies on bond reordering have concentrated on calcite, while in this study we focus on a dolomitic reservoir. The only available kinetics data on bond reordering in dolomite are from a published conference abstract by Bonifacie *et al.* (2013); this unpublished experimental data indicated dolomite was resistant to solid-state bond reordering up to *c.* 300°C. This suggests that bond reordering will have a lesser effect, if any, on the preservation of temperature signatures recorded by clumped isotopes in dolomite. In this study, we compare clumped isotope temperatures with an extensive suite of previously acquired fluid-inclusion and ambient well temperatures (Walgenwitz *et al.* 1990; Eichenseer *et al.* 1999) to investigate if clumped isotopes in dolomite are affected by bond reordering, or whether they are recording temperatures of geologically meaningful events such as burial recrystallization.

Sample characterization

Geological setting

The dolostone samples investigated in this study are from the Cretaceous dolomite Pinda Formation from offshore Angola (Fig. 1a). The tectonic framework of the Pinda Formation begins with the break-up of SW Gondwana during the Early Cretaceous (Reyre 1984). The initial rifting phase and associated sedimentary deposition was dominated by lacustrine sediments (Eichenseer *et al.* 1999), followed by evaporite formation (e.g. Coajou & Ribeiro 1994). As continental rifting continued northwards, marine transgression took place and marine carbonates of the Pinda Formation were deposited during Albian time (Koutsoukos *et al.* 1991). Halokinesis during the Albian resulted in turtle-back and raft structures which created hydrocarbon traps (Bremner *et al.* 1993; Eichenseer *et al.* 1999).

Modelling of the burial history of the Pinda Formation by TOTAL indicates that after deposition of the thick Pinda sequence, there was little or no burial until the Oligocene (Fig. 1b). This is supported by Walgenwitz *et al.* (1990), who indicated that diagenesis mainly took place in the late Oligocene following reactivated subsidence. Initial dolomitization of the Pinda carbonates is interpreted to have occurred during early diagenesis, producing dolomite microrhombs (Eichenseer *et al.* 1999). Early dolomitization was amplified by later shallow burial recrystallization, producing cloudy dolosparites with a clear outer rim (Eichenseer *et al.* 1999). It was proposed that this later recrystallization was triggered by mixed saline–meteoric fluid circulations at shallow depth (mixing-zone dolomitization) and resulted in a significant modification of the original geochemical signature of the dolostone (Eichenseer *et al.* 1999). Eichenseer *et al.* (1999) indicated no evidence in support of a late deep burial (high-temperature) dolomitization event, and considered that the bulk of the dolomitization was of early shallow origin. The current model of dolomitization for the Pinda Formation therefore relies solely on a mixing-zone dolomitization mechanism, which took place during early diagenetic stages in relation with sea-level fall and prior to any significant burial.

Fluid inclusions

Fluid-inclusion thermometry was conducted by TOTAL, and details of the microthermometric procedures and measurements are given in Walgenwitz *et al.* (1990). Fluid-inclusion measurements were conducted on two-phase fluid inclusions located in crystals of dolosparite and in recrystallized feldspar grains. A large suite of homogenization temperatures was obtained from about 40 samples across the Pinda Formation in the study area (13 wells).

Homogenization temperatures (T_h) range from *c.* 110 to 170°C (Fig. 2). The fluid-inclusion assemblage exhibits a roughly linear temperature–depth relationship, coincident with a suite of downhole ambient well temperature readings (Fig. 2). This was interpreted by Eichenseer *et al.* (1999) as reflecting progressive resetting of fluid-inclusion T_h values during burial down to present-day depths. However, the error bars on many of these are large and make confident interpretation difficult (see the further discussion below). Homogenization temperatures (sample averages) corresponding to the specific core plug samples selected for clumped isotope analysis range from 111 to 152°C (Table 1).

Clumped isotopes

Methodology

Carbonate clumped isotope measurements were carried out in the Qatar Stable Isotope Laboratory at Imperial College London. Samples were powdered using a dental drill. In order to remove hydrocarbons and other organic matter, the powders were treated with four–eight washes in a cold cleaning solution of 3% H_2O_2 and 0.9 wt% sodium hexametaphosphate. Subsequently, the treated powders were washed with ethanol and deionized (DI) water before being dried in an oven at 50°C to evaporate any remaining water. Where this treatment was not sufficient, sample powders were treated in a TePla 400 oxygen plasma asher at the London Centre for Nanotechnology at University College London. Powders were placed in Petri dishes and heated at *c.* 40°C for 5 min in an oxygen plasma. The method resulted in the complete combustion of all organic matter under vacuum after a 5 min treatment. Given the short treatment time, the temperature in the oxygen plasma asher only reaches *c.* 40°C, well below the temperature threshold for initiating closed-system bond reordering. Additionally, this method was tested on a Carrara Marble standard and was not found to have any effect on clumped isotope systematics. For clumped isotope analysis, 5–6 mg of sample powder was reacted under vacuum with 104% phosphoric acid at 90°C for 20 min. Water generated during the reaction was separated from the produced CO_2 by first trapping it in liquid nitrogen, then swapping the liquid nitrogen for an ethanol–liquid nitrogen mixture held at *c.* −90°C. The water remained frozen while the CO_2 was passed through a Poropak Q chromatography trap held at −35°C. Full details on the purification procedure are described in Dale *et al.* (2014).

The purified CO_2 was measured in one of two Thermo Fisher MAT 253 isotope ratio mass spectrometers (Pinta and Nina: see the Supplementary material) in dual inlet mode with a measurement

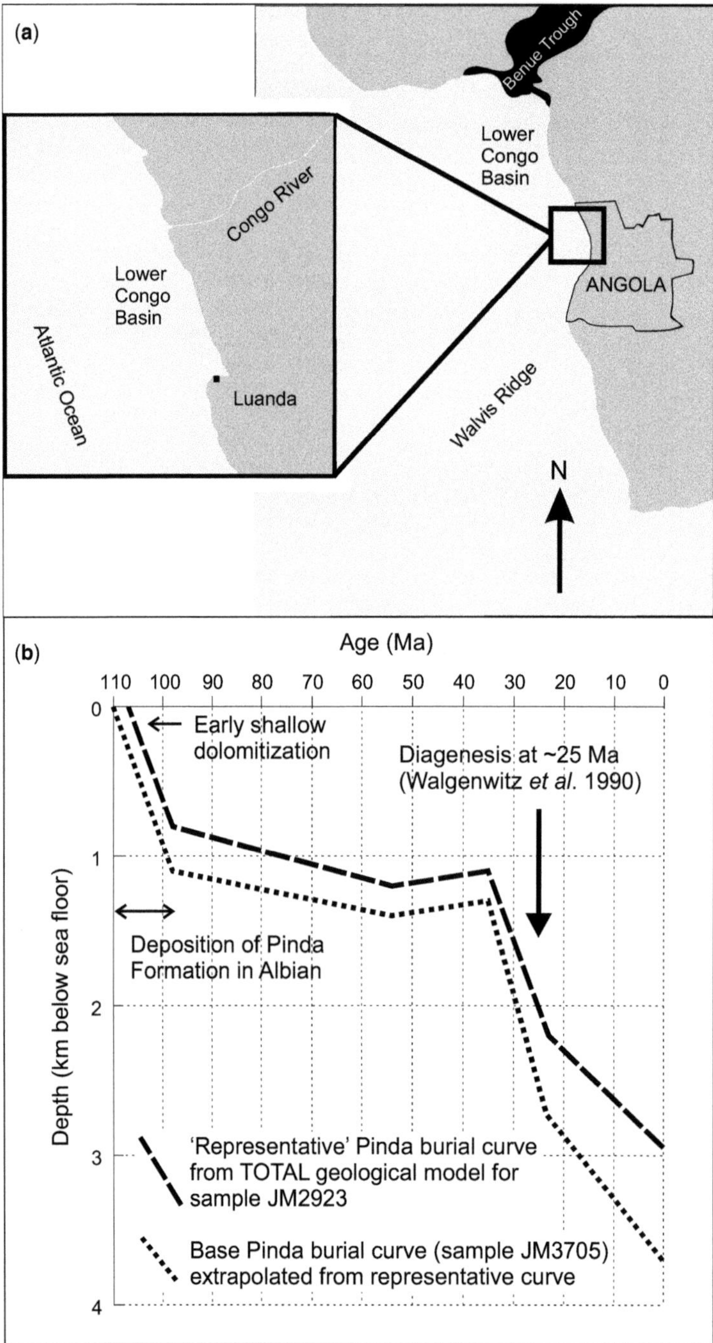

Fig. 1. (a) Map showing the general location of the studied samples, offshore Angola, after Eichenseer *et al.* (1999). (b) Burial history of the Pinda Formation.

time of *c.* 2.5 h. Masses 44–49 were measured, with 48 and 49 used to test for potential contamination as CO_2 with those molecular masses is extremely rare relative to mass 47 (e.g. Eiler 2007). Only samples with δ_{48} and Δ_{48} values that fall within 2‰ of the heated gas line (Δ_{48} offset) and had a 49 parameter

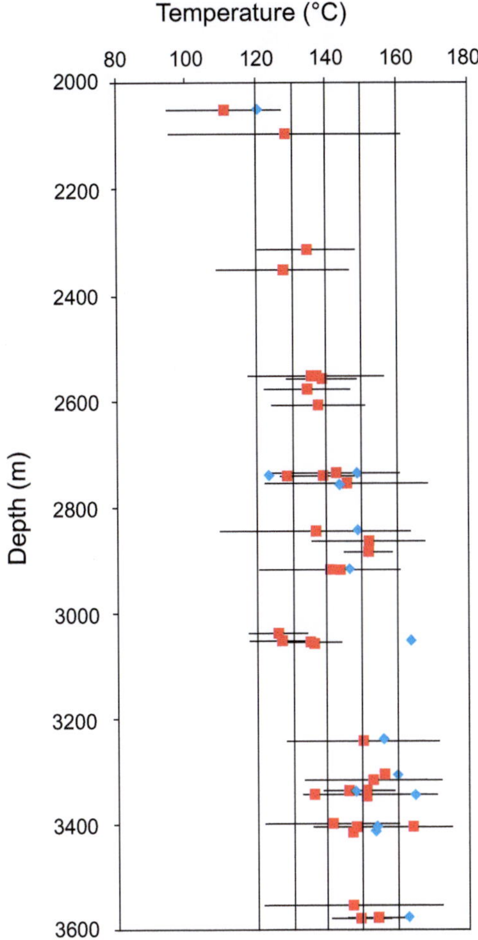

Fig. 2. Fluid-inclusion temperatures for the Pinda Formation (data reproduced from Eichenseer *et al.* 1999). Fluid-inclusion homogenization temperatures shown as red squares represent sample averages, while horizontal black bars represent the range of individual T_h values. Current ambient well temperatures are shown as blue diamonds, and are in close concordance with T_h average values (with one exception at *c.* 3 km depth).

empirical acid fractionation factor of +0.069‰ calculated in Guo *et al.* (2009) and experimentally derived by Wacker *et al.* (2013) was added to the Δ_{47} value to bring the data into the 25°C scale used in calibrations. This acid fractionation factor is consistent with internal laboratory tests at Imperial College London, and with the values used for the Imperial College calibration (Kluge *et al.* 2015). The Carrara Marble standard gave a linearity- and acid-corrected average Δ_{47} value of 0.382 ± 0.020‰ across both mass spectrometers, while the ETH-3 value was 0.700 ± 0.016‰; these compare well with published values of 0.395‰ (Dennis *et al.* 2011) and 0.705‰ (Meckler *et al.* 2014), respectively. All standard and sample data are given in the Supplementary material. For $\delta^{18}O_{\text{dolomite}}$, the acid fractionation factor of Rosenbaum & Sheppard (1986) was applied. The calibration of Kluge *et al.* (2015) was used to convert Δ_{47} into temperature, as this is the only experimental calibration that extends over the 25–250°C range, and it was derived using the same methods and instruments as in this study. $\delta^{18}O_{\text{porewater}}$ values were calculated with the carbonate–water equilibrium fractionation factors from Land (1980). The calculated $\delta^{18}O_{\text{porewater}}$ are reported in the Vienna Standard Mean Ocean Water (VSMOW) scale, whilst measured $\delta^{18}O$ and $\delta^{13}C$ values are reported in the Vienna PeeDee Belemnite (VPDB) scale.

Results

Seven dolostone bulk samples from six different wells were selected for clumped isotope analysis. The samples were taken from depths ranging from 2055 to 3705 m below seafloor (mbsf). They are generally unimodal planar-s (Sibley & Gregg 1987) dolosparites (Fig. 3a); one sample contains very minor calcite (Fig. 3b) and most have isolated quartz grains (Fig. 3a). The crystal size range is *c.* 50–500 μm and the shape is anhedral to euhedral. The smallest, most euhedral dolomite rhombs are found lining moldic pores (Fig. 3c) after dissolution of ooids – a fabric commonly preserved (Fig. 3b). Dolomitized shelly fragments are also occasionally observed (Fig. 3d).

The $\delta^{13}C$ values of the seven samples are all within a narrow range of *c.* 3–3.5‰ and do not correlate with depth (Fig. 4a; Table 1). $\delta^{18}O$ values range from *c.* −7 to −3.5‰ (Fig. 4b; Table 1) and do not correlate with depth or $\delta^{13}C$, although the shallowest sample has the least depleted $\delta^{18}O$ value. Δ_{47} values range from 0.495 to 0.592‰, which translates to temperatures of 65–122°C, with temperature errors in the range of ±1 to ±7°C. Temperatures increase with burial depth in the two shallower samples before clustering around 100–120°C

of <0.2 were accepted (Dale *et al.* 2014). Mass spectrometer non-linearity was corrected based on methods described in Huntington *et al.* (2009) using a 30 day moving average of heated gases as this provided a suitable number of heated gases and carbonate standards to correct against. All Δ_{47} values reported are in the Carbon Dioxide Equilibrated Scale (CDES) (Dennis *et al.* 2011) using a secondary transfer function based on Carrara Marble, ETH-3 (an externally verified carbonate standard (Iso C in Meckler *et al.* 2014)) and heated gases. Following correction to the absolute reference frame, the

Table 1. *Summary table of sample details, isotopic values and temperatures*

Sample	Depth (m)	No. of analyses	Mineral $\delta^{13}C$ (‰) PDB	SD (‰)	Mineral $\delta^{18}O$ (‰) PDB	SD (‰)	Δ_{47} CDES (‰)	SE (‰)	T_h (°C)	T_{well} (°C)	$T(\Delta_{47})$ (°C)	SE (°C)	Fluid $\delta^{18}O$ (‰) SMOW	SE (‰)
JM2055	2055	3	3.18	0.03	−3.57	0.10	0.592	0.001	111	110	65	1	1.7	0.2
JM2740	2740	3	3.05	0.10	−4.74	0.11	0.560	0.005	142	141	82	3	2.7	0.4
JM2848	2848	5	3.12	0.12	−5.75	0.21	0.530	0.011	136	149	99	5	3.7	0.8
JM2923	2923	3	3.26	0.01	−6.94	0.03	0.495	0.003	142	147	122	1	4.8	0.2
JM3319	3319	3	3.47	0.03	−4.72	0.11	0.503	0.010	152	161	117	6	6.5	0.8
JM3637	3637	5	3.06	0.16	−5.69	0.27	0.528	0.004	135	n/a	100	2	4.2	0.4
JM3705	3705	3	3.11	0.15	−6.09	0.52	0.502	0.008	146	n/a	118	7	5.2	0.8

CDES, Carbon Dioxide Equilibration Scale (Dennis *et al.* 2011); T_h, fluid-inclusion homogenization temperature (sample average); T_{well}, current ambient well temperature; T_{cl}, clumped isotope temperature.

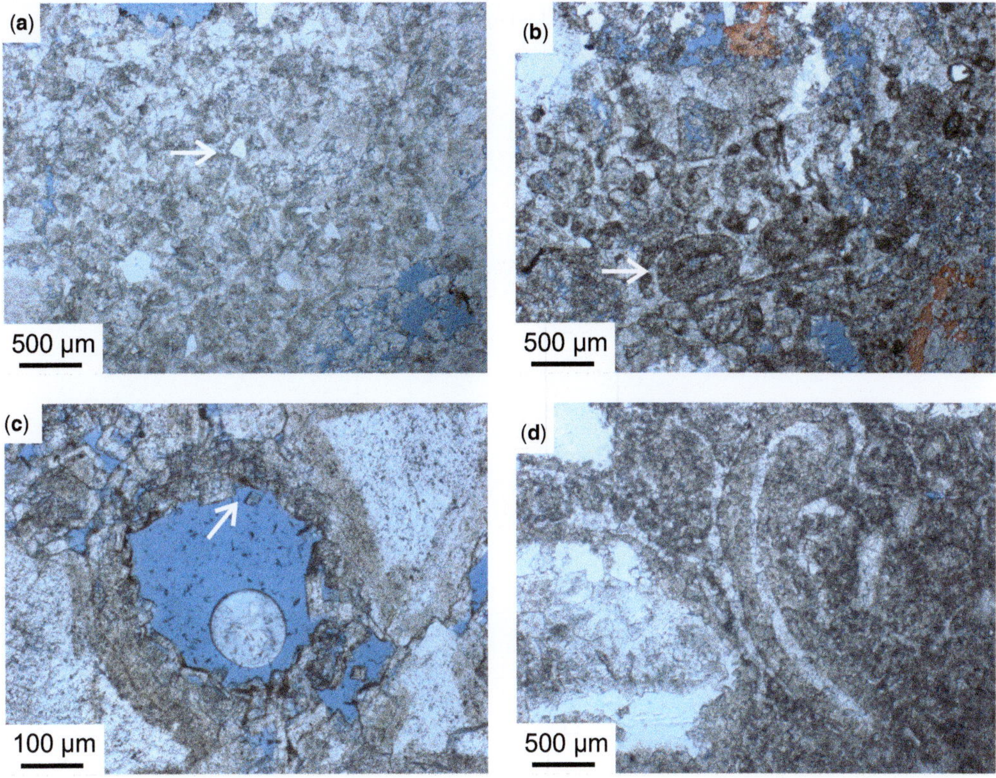

Fig. 3. Plane-polarized light photomicrographs showing typical petrographical textures of samples from the Pinda Formation: (**a**) blocky planar-s dolosparite texture typical of many of the samples, the arrow denotes an example of a sand grain; (**b**) red staining indicating a minor calcite component; the arrow denotes an example of a preserved ooid fabric typical of many of the samples; (**c**) a moldic pore after an ooid, lined with small euhedral dolomite rhombs; and (**d**) a rare preserved shell fabric.

in the five deeper samples (Fig. 4c; Table 1). The shallowest sample (JM2055) has a calculated parent fluid $\delta^{18}O$ value of 1.7‰, rising to 2.7‰ in the next deepest sample (JM2740) (Fig. 4d; Table 1). The five deeper samples, which record clumped isotope temperatures of *c.* 100–120°C (average = 111°C), exhibit calculated fluid $\delta^{18}O$ values of 3.7–6.5‰ (average = 4.9‰), with no depth trend. These five samples plot within or close to the modern-day pore-water $\delta^{18}O$ of *c.* 5‰ (Walgenwitz 1989).

Discussion

Interpretation of the fluid-inclusion homogenization temperatures is complicated by the fact that they have large error bars and record a spread of temperatures of *c.* 110–170°C, which is likely to reflect thermal re-equilibration during burial subsequent to fluid-inclusion trapping. It is well known that not all fluid inclusions are affected in the same way

by thermal resetting, some fluid inclusions being more resistant than others depending on their size and shape (Goldstein 2001). The lowest T_h values recorded in the fluid inclusions are at *c.* 110°C (Fig. 2) and would therefore represent the most conservative temperature for any burial dolomitization episode. Eichenseer *et al.* (1999) did not interpret any such late burial dolomitization. The higher T_h values recorded by fluid inclusions up to 170°C in the Pinda Formation reflect a subsequent increase in the ambient temperature and thermal resetting of fluid inclusions without dolomite recrystallization. Where ambient temperatures were available (five of the seven samples), homogenization temperatures are within 10°C of the present-day ambient temperature (Fig. 4c; Table 1).

Clumped isotope temperatures, however, fall well below present-day ambient well temperatures and average fluid-inclusion temperatures at all investigated depths by about 20–60°C (Table 1; Fig. 4c). The clumped isotope temperature of *c.* 100–120°C

J. M. MacDONALD *ET AL.*

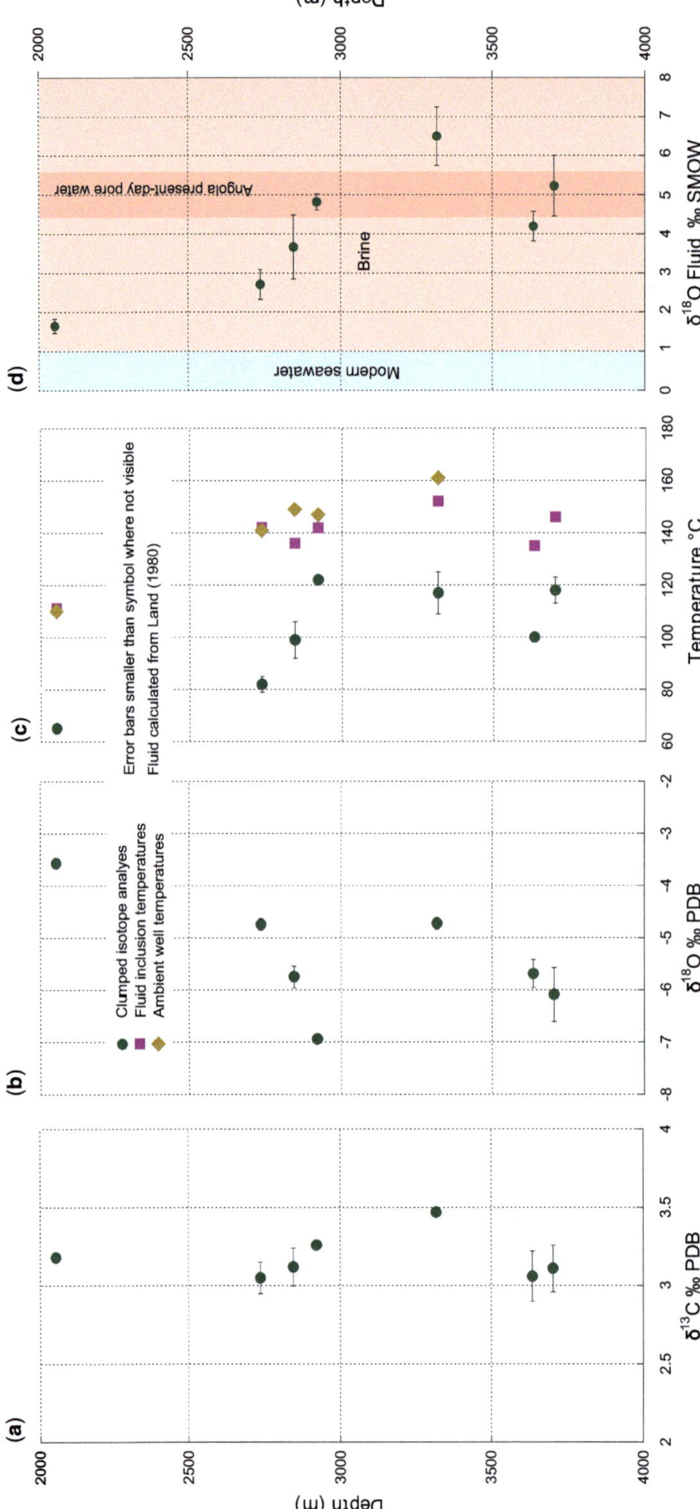

Fig. 4. Isotope data and Δ_{47}-derived temperatures for the seven samples from the Pinda Formation analysed for clumped isotopes, plotted against sample depth: (**a**) $\delta^{13}C$; (**b**) $\delta^{18}O$; (**c**) clumped isotope, fluid-inclusion and ambient well temperatures; and (**d**) parent fluid $\delta^{18}O$, shaded areas represent typical source fluid values.

recorded by the five deeper samples is consistent over a *c.* 1000 m depth range (*c.* 2800–3700 mbsf: Fig. 4c) and is close to the lower end of the range of fluid-inclusion T_h values (*c.* 110°C: Fig. 2). This strongly suggests that the clumped isotopes in these five samples are recording a single open-system recrystallization event, interpreted to represent a deep burial recrystallization episode responsible for the blocky planar-s texture observed in the studied samples. The fluid $\delta^{18}O$ values (*c.* 3.5–6.5‰) calculated for these samples are clustered around modern-day porewater composition (5‰) and suggest consistent brine $\delta^{18}O$ values with the time of burial diagenesis. The similarity between present and palaeo-fluid compositions is suggestive of open-system recrystallization (dissolution–reprecipitation). With open-system recrystallization, the $\delta^{18}O$ value of the carbonate will reflect the ambient temperature and $\delta^{18}O$ value of the fluid that enabled recrystallization to occur. Therefore, if either of these two parameters (ambient temperature or fluid $\delta^{18}O$) did vary during the open-recrystallization process, then the mineral $\delta^{18}O$ values should also vary over a depth range in a single formation/basin. This is the case of the five deeper samples of the Pinda Formation investigated in this study which have undergone complete burial recrystallization and show a 2‰ range in their bulk $\delta^{18}O$ values, reflecting a variation of *c.* 20°C in temperature and *c.* 2.8‰ in fluid $\delta^{18}O$. Eichenseer *et al.* (1999) interpreted measured bulk $\delta^{18}O$ values of these sediments to record isotopic resetting during burial of early seafloor dolomite. Indeed, the mineral $\delta^{18}O$ values reported by Eichenseer *et al.* (1999) and in our study are more depleted than typical marine carbonate values from the Cretaceous (Veizer *et al.* 1999), and cannot reflect early seafloor conditions.

After deposition of the *c.* 1–1.5 km-thick Pinda Formation in the Albian, there was little or no burial until the Oligocene (Fig. 1b) when the Pinda sediments were buried under a thick sequence of Oilgocene and Miocene sediments, with a steady subsidence continuing to the present day. The base of the Pinda Formation was buried from *c.* 1.4 km at the beginning of the Oligocene to *c.* 2.7 km at the end of the Oligocene, reaching a burial temperature of *c.* 100°C, based on the present-day geothermal gradient of 40–45°C km^{-1} (Walgenwitz *et al.* 1990). According to Walgenwitz *et al.* (1990), a major diagenetic event occurred at *c.* 25 Ma, towards the end of the episode of Oligocene subsidence. It therefore seems highly plausible that bulk and clumped isotopes record burial recrystallization of dolomite during a major diagenetic event at about 100–120°C and at *c.* 25 Ma (Fig. 1b).

The two shallowest samples (2055 and 2739 mbsf) have clumped isotope temperatures of 65 and 82°C, somewhat lower than the deeper samples. The present-day well temperatures at these shallower depths (111 and 141°C, respectively) are also lower than that of the deeper samples (147–161°C), and close to the hotter end of the temperature of the later burial dolomitization episode (*c.* 120°C). The clumped isotope temperatures of these samples are, therefore, best explained as reflecting incomplete dolomite recrystallization and physical mixing of the two generations of dolomite: that is, early shallow dolomite (*c.* 35°C) and burial dolomite (*c.* 100–120°C).

In all samples, fluid inclusions have been reset to near-present-day temperatures while clumped isotopes have not, and are instead interpreted to record open-system recrystallization. This is in accord with previous studies of subsurface carbonates which have suggested that recrystallization temperatures are recorded in clumped isotope systematics. John (2015) found that clumped isotopes were recording deposition and recrystallization in a suite of Cretaceous oysters from Oman. One well-preserved individual recorded a primary precipitation temperature of 37 ± 4°C and a fluid $\delta^{18}O$ value typical of slightly-evaporated Cretaceous seawater in a restricted platform basin. Other oysters from the suite, however, recorded clumped isotope temperatures of *c.* 60–70°C, indicating at least partial burial recrystallization. Ferry *et al.* (2011) calculated clumped isotope temperatures of *c.* 40–80°C in the Latemar Platform which they attributed to initial dolomitization, while Sena *et al.* (2014) used clumped isotopes to show that early dolomite re-equilibrated through neomorphism from protodolomite to well-ordered dolomite with fluids at 60–70°C in a shallow burial setting in a Tethyan carbonate platform in Oman.

It is, however, important to address the potential for closed-system bond reordering (Passey & Henkes 2012; Henkes *et al.* 2014; Stolper & Eiler 2015) to have occurred. Assuming open-system behaviour, the Δ_{47} value should reflect the ambient temperature during recrystallization. However, recent work has indicated that closed-system resetting of clumped isotope systematics can occur at high temperatures, thus resetting the Δ_{47} value recorded earlier during deposition or after open-system recrystallization. If calcite is exposed to temperatures of greater than *c.* 100°C for tens to hundreds of millions of years, this bond reordering may occur (Dennis & Schrag 2010; Passey & Henkes 2012; Henkes *et al.* 2014). The reordering process is interpreted to be a two-stage process, with initial rapid isotope diffusion resulting in a change of *c.* 1–40°C at ambient temperatures of *c.* 75–120°C sustained for *c.* 100 myr. Following this, slow isotope exchange reactions between adjacent carbonate groups at >*c.* 150°C sustained for >*c.* 100 myr could bring the clumped isotope temperatures to the ambient temperature (Henkes *et al.* 2014; Stolper & Eiler 2015). Such bond reordering in calcite has been documented in natural samples. Dennis & Schrag (2010) hypothesized that some form of closed-system reordering must be occurring during slow cooling of rocks as they found that carbonatites with no petrographical or geochemical evidence for diagenesis were recording clumped isotope temperatures well below those expected for carbonatite crystallization temperatures. This study led to the initial development of the bond reordering hypothesis, showing that the clumped isotope composition of calcite and phosphate did change in laboratory experiments when these minerals were exposed for several weeks at temperatures in excess of 200°C (Passey & Henkes 2012; Henkes *et al.* 2014; Stolper & Eiler 2015). Shenton *et al.* (2015) further confirmed with natural samples that different components (e.g. brachiopods, crinoids) of calcite limestones from the Palmarito Formation, Venezuela, and the Bird Springs Formation, Nevada, were affected by bond reordering during burial (to *c.* 4–5 km, 150–175°C) and exhumation.

The only available constraints on the kinetics of bond reordering in dolomite are from the published conference abstract of Bonifacie *et al.* (2013), which indicated that dolomite was resistant to solid-state bond reordering up to *c.* 300°C. Until the full results of these experiments – and, hence, the parameters required for modelling closed-system bond reordering in dolomite – are published, the exact effects of bond reordering on the Pinda Formation dolostones cannot be determined. We note that 300°C is clearly above the well temperature offshore Angola, where our samples come from, but how would the Pinda dolomite behave if we assumed dolomite had similar Arrhenius parameters to calcite? In the absence of dolomite-specific modelling parameters, we used the bond reordering model for calcite of Passey & Henkes (2012) to investigate a 'worst-case' scenario for bond reordering in the Pinda dolostones. This was conducted using the input parameters for optical calcite MGB-CC-1 (Passey & Henkes 2012) using their equation (13) (first-order kinetics), and the evolution of Δ_{47} through time is modelled in the style of the thermal history reordering models (THRMs) of Shenton *et al.* (2015).

Four scenarios were modelled. Three of these use a 'representative' Pinda burial curve, based on the dashed line in Figure 1b from thermal modelling by TOTAL for sample JM2923. These scenarios vary in their diagenetic histories: no post-deposition recrystallization (scenario 1); shallow dolomitization only (scenario 2); shallow dolomitization and burial recrystallization at *c.* 25 Ma at a burial temperature of 115°C (scenario 3) (Walgenwitz *et al.* 1990). Scenario 4 is specific to the base of the Pinda Formation

(sample JM3705), using an extrapolation of the representative Pinda curve to the present-day depth of *c.* 3.7 km and ambient temperature inferred to be 170°C from the geothermal gradient (dotted line in Fig. 1b). For all four scenarios, deposition is assumed to be at 25°C with shallow dolomitization at 35°C and burial recrystallization at 25 Ma, if applicable (Walgenwitz *et al.* 1990). For each of the (re)crystallization events (deposition, early dolomitization and recrystallization during burial), the Δ_{47} is controlled by the temperature of the respective process, regardless of whether any change on Δ_{47} due to bond reordering has occurred before, as each represents an open-system (re)precipitation. Modelling results are shown in Figure 5.

In all four scenarios, no closed-system bond reordering occurs before the end of the Oligocene burial despite the ambient temperature reaching *c.* 130°C in the base Pinda scenario. In scenario 4 (base Pinda), the burial curve indicates that the open-system burial recrystallization will occur at *c.* 130°C: that is, towards the end of the Oligocene. As the ambient temperature increases from 130 to 170°C in the 25 myr between recrystallization and the present day, the temperature would be high enough to drive the Δ_{47} values in calcite to the ambient temperature, resulting in 100% bond reordering (Fig. 5a) and a recorded temperature of 170°C. With the 'representative' Pinda burial history (sample JM2923, scenario 3), the burial curve indicates that the open-system burial recrystallization will occur at *c.* 115°C towards the end of the Oligocene (Fig. 5b). As the ambient temperature increases from 115 to 147°C (sample JM2923: Table 1; Fig. 4c) in the 25 myr between recrystallization and the present day, modelling indicates that partial reordering of the Δ_{47} value would occur in calcite to a present-day clumped isotope temperature of 120°C. Open-system burial recrystallization is important in facilitating post-recrystallization closed-system bond reordering in calcite as in the scenarios where no burial recrystallization is assumed (scenarios 1 and 2), there is very limited bond reordering. With shallow dolomitization only (scenario 2), bond reordering would modify the observed temperature from 35 to 47°C; and with no post-deposition recrystallization at all, bond reordering would modify the temperature from 25 to 37°C (Fig. 5b).

Although the temperature generated through partial bond reordering of sample JM2923 (scenario 3) is close to the recorded clumped isotope temperature, it must be emphasized that these reordering models are based on the kinetics of calcite. Dolomite is perceived to have much slower reordering kinetics (Bonifacie *et al.* 2013). This is emphasized by the fact that modelling of bond reordering using the kinetics for calcite and the burial curve for the base of the Pinda Formation indicates that sample

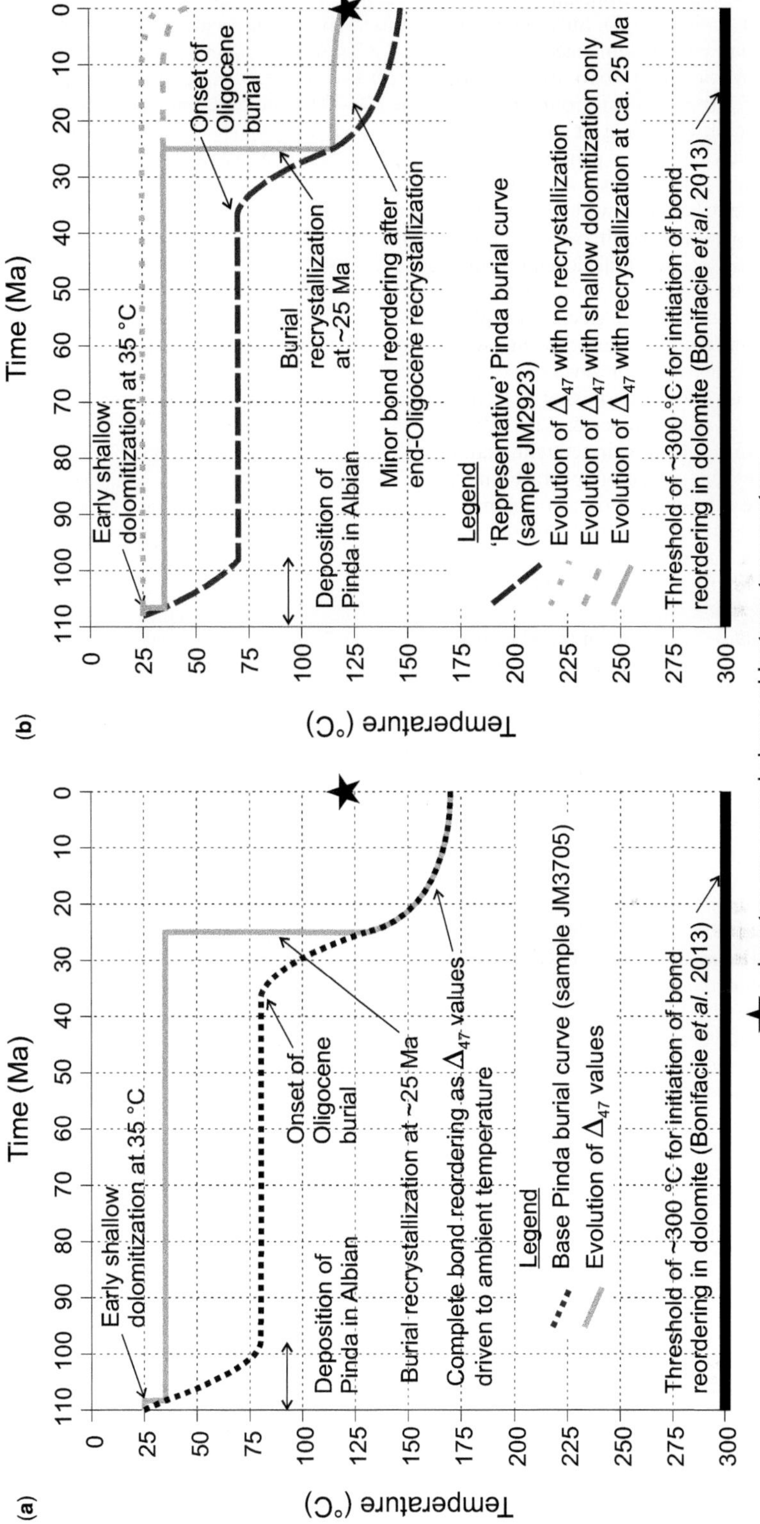

Fig. 5. 'Worst-case' scenario of post-depositional evolution of Δ_{47} values in the Pinda Formation, assuming the samples were calcite not dolomite. Modelled in the style of Shenton *et al.* (2015) using the calcite reordering kinetics of Passey & Henkes (2012) for optical calcite MGB-CC-1. (**a**) The base Pinda, based on an extrapolation of the 'representative' Pinda burial history modelled by TOTAL shown in (**b**).

JM3705 would record the present-day ambient temperature of *c.* 170°C in its clumped isotope systematics. This is not the case – the clumped isotope temperature for this sample is 118°C, well below ambient temperature (Table 1; Fig. 4c) and reflecting the interpreted recrystallization event at *c.* 110°C. We also note that we used the same Arrhenius parameters for bond reordering in the early dolomite and the recrystallized burial dolomite. This is unlikely to be the case, and recrystallized dolomite could be more resistant to bond reordering as it was equilibrated at higher temperatures. The degree of stoichiometry of (proto-)dolomite is also likely to complicate bond reordering models. We acknowledge that while the kinetics of dolomite reordering are unpublished, our interpretation that the Pinda Formation dolostones are recording the temperature of diagenetic events cannot be proven beyond doubt. However, if the conclusions of Bonifacie *et al.* (2013) are valid, the temperature conditions (*c.* 300°C) thought to be required for solid-state bond reordering in dolomite were not reached in the studied samples. This is supported by the results of our modelling simulations, which show that the temperatures recorded are compatible with a 110°C recrystallization. Therefore, our interpretation that the clumped isotopes are recording burial recrystallization is well founded.

The specificity of our study in the Pinda Formation of offshore Angola is that samples are from deep cores and are at their maximum burial depth at the present day. This enables insight into clumped isotope systematics during deep burial without the complication of any potential closed-system resetting during slow geological exhumation, as was encountered by Shenton *et al.* (2015). We suggest that in the Pinda Formation clumped isotopes are recording the palaeotemperatures of an open-system recrystallization event that resulted in massive burial dolomitization of an already dolomitic reservoir. Hence, reasonable fluid $\delta^{18}O$ values are calculated when using the clumped isotope data, while they appear rather high when using the fluid-inclusion homogenization temperatures. This illustrates that clumped isotope analysis can be a valuable tool in investigating the thermal history of deeply-buried carbonate reservoirs, and that it is of particular interest for distinguishing early shallow from late burial dolomitization processes. In the Pinda Formation dolostones, the clumped isotopes recorded open-system burial recrystallization at temperatures of *c.* 100–120°C. Our data further indicate that the clumped isotope signal was preserved during subsequent burial to conditions *c.* 1600 m deeper (*c.* 3700 mbsf) and *c.* 50°C hotter (up to 161°C). While the clumped isotope geothermometer does not yet have the spatial resolution to interrogate zonation within individual crystals or distinguish

between multiple generations of thin cement layers, this study does show that its application to bulk samples has the ability to record diagenetic information in relatively homogeneous monophasic dolostones. It has indicated that the massive blocky planar-s dolosparite in the Pinda Formation formed during late burial recrystallization at *c.* 100–120°C, a process that could not be documented from petrography and $\delta^{18}O$ (Eichenseer *et al.* 1999). Advances in methodology (e.g. Hu *et al.* 2014; Petersen & Schrag 2014) to increase spatial resolution to the millimetre- and submillimetre-scale will enable much wider usage of the clumped isotope palaeothermometer in reservoir characterization.

Conclusions

This study examined if carbonate clumped isotope palaeothermometry could be applied to characterize the thermal history of dolomitized sedimentary carbonate hydrocarbon reservoirs. It was performed on a set of seven dolostone samples from the Pinda Formation of offshore Angola presently lying at their maximum burial depth, *c.* 2000–3700 mbsf. Clumped isotope data suggest that temperatures derived from a well-established palaeothermometric technique – fluid inclusions – have been reset to current ambient well temperatures. Clumped isotope temperatures, however, record lower temperatures corresponding to open-system burial recrystallization occurring prior to present-day maximum burial. Fluid $\delta^{18}O$ values calculated for the five deepest samples overlap with present-day porewater $\delta^{18}O$, while mineral $\delta^{18}O$ supports the conclusion that clumped isotopes are recording burial recrystallization conditions. Closed-system isotopic bond reordering is not believed to have affected the Pinda Formation samples, therefore we contend that clumped isotope analysis can be a valuable tool in characterizing the thermal history of dolomite hydrocarbon reservoirs. It may be particularly useful in helping to distinguish early shallow from deep burial origins of dolomite formations.

Annabel Dale, Tobias Kluge and Simon Davis are thanked for assistance in the Qatar Stable Isotope Lab at Imperial College London. Two grants awarded to CJ from TOTAL E&P (Nos 4200059621 and 4300002831) enabled this work to be carried out. Ethan Grossman and an anonymous reviewer are thanked for their reviews, which considerably improved the manuscript, and Mike Lawson is thanked for editorial handling.

References

Affek, H.P., Bar-Matthews, M., Ayalon, A., Matthews, A. & Eiler, J.M. 2008. Glacial/interglacial temperature variations in Soreq cave speleothems as recorded by 'clumped isotope' thermometry. *Geochimica et Cosmochimica Acta*, **72**, 5351–5360.

BARKER, C.E. & GOLDSTEIN, R.H. 1990. Fluid-inclusion technique for determining maximum temperature in calcite and its comparison to the vitrinite reflectance geothermometer. *Geology*, **18**, 1003–1006.

BONIFACIE, M., CALMELS, D. & EILER, J. 2013. Clumped isotope thermometry of marbles as an indicator of the closure temperatures of calcite and dolomite with respect to solid-state reordering of C–O bonds. *Mineralogical Magazine*, **77**, 735.

BREMNER, A.N., LOMANDO, A.J. & MINCK, R.J. 1993. Successful exploration of the Outer Pinda trend of offshore Cabinda, Angola. Presented at the AAPG International Conference and Exhibition, October 17–20, 1993, The Hague, The Netherlands.

CARRAPA, B., HUNTINGTON, K.W., CLEMENTZ, M., QUADE, J., BYWATER-REYES, S., SCHOENBOHM, L.M. & CANAVAN, R.R. 2014. Uplift of the Central Andes of NW Argentina associated with upper crustal shortening, revealed by multiproxy isotopic analyses. *Tectonics*, **33**, 1039–1054.

COAJOU, A. & RIBEIRO, A. 1994. Angola Block 3 – from wildcat to intensive exploration 22 discoveries – 1 billion BBL recoverable oil. Presented at the 14th World Petroleum Congress, 29 May–1 June 1994, Stavanger, Norway.

DALE, A., JOHN, C.M., MOZLEY, P.S., SMALLEY, P.C. & MUGGERIDGE, A.H. 2014. Time-capsule concretions: unlocking burial diagenetic processes in the Mancos Shale using carbonate clumped isotopes. *Earth and Planetary Science Letters*, **394**, 30–37.

DENNIS, K.J. & SCHRAG, D.P. 2010. Clumped isotope thermometry of carbonatites as an indicator of diagenetic alteration. *Geochimica et Cosmochimica Acta*, **74**, 4110–4122.

DENNIS, K.J., AFFEK, H.P., PASSEY, B.H., SCHRAG, D.P. & EILER, J.M. 2011. Defining an absolute reference frame for 'clumped' isotope studies of CO_2. *Geochimica et Cosmochimica Acta*, **75**, 7117–7131.

EICHENSEER, H.T., WALGENWITZ, F.R. & BIONDI, P.J. 1999. Stratigraphic control on facies and diagenesis of dolomitized oolitic siliciclastic ramp sequences (Pinda group, Albian, offshore Angola). *AAPG Bulletin*, **83**, 1729–1758.

EILER, J.M. 2007. 'Clumped-isotope' geochemistry – the study of naturally-occurring, multiply-substituted isotopologues. *Earth and Planetary Science Letters*, **262**, 309–327.

EMILIANI, C. 1955. Pleistocene temperatures. *The Journal of Geology*, **63**, 538–578.

EPSTEIN, S., BUCHSBAUM, R., LOWENSTAM, H. & UREY, H. 1951. Carbonate-water isotopic temperature scale. *GSA Bulletin*, **62**, 417–426.

FERRY, J.M., PASSEY, B.H., VASCONCELOS, C. & EILER, J.M. 2011. Formation of dolomite at 40–80°C in the Latemar carbonate buildup, Dolomites, Italy, from clumped isotope thermometry. *Geology*, **39**, 571–574.

GESKE, A., GOLDSTEIN, R.H. ET AL. 2015. The magnesium isotope (δ^{26} Mg) signature of dolomites. *Geochimica et Cosmochimica Acta*, **149**, 131–151.

GHOSH, P., ADKINS, J. ET AL. 2006a. $^{13}C–^{18}O$ bonds in carbonate minerals: a new kind of paleothermometer. *Geochimica et Cosmochimica Acta*, **70**, 1439–1456.

GHOSH, P., GARZIONE, C.N. & EILER, J.M. 2006b. Rapid uplift of the Altiplano revealed through $^{13}C–^{18}O$ bonds in paleosol carbonates. *Science*, **311**, 511–515.

GOLDSTEIN, R.H. 2001. Fluid inclusions in sedimentary and diagenetic systems. *Lithos*, **55**, 159–193.

GOLDSTEIN, R.H. & REYNOLDS, T.J. 1994. *Systematics of Fluid Inclusions in Diagenetic Minerals*. Society for Sedimentary Geology (SEPM), Short Course, **31**.

GUO, W., MOSENFELDER, J.L., GODDARD, W.A. & EILER, J.M. 2009. Isotopic fractionations associated with phosphoric acid digestion of carbonate minerals: insights from first-principles theoretical modeling and clumped isotope measurements. *Geochimica et Cosmochimica Acta*, **73**, 7203–7225.

HENKES, G.A., PASSEY, B.H., GROSSMAN, E.L., SHENTON, B. J., PEREZ-HUERTA, A. & YANCEY, T.E. 2014. Temperature limits for preservation of primary calcite clumped isotope paleotemperatures. *Geochimica et Cosmochimica Acta*, **139**, 362–382.

HU, B., RADKE, J., SCHLÜTER, H.-J., HEINE, F.T., ZHOU, L. & BERNASCONI, S.M. 2014. A modified procedure for gas-source isotope ratio mass spectrometry: the long-integration dual-inlet (LIDI) methodology and implications for clumped isotope measurements. *Rapid Communications in Mass Spectrometry*, **28**, 1413–1425.

HUNTINGTON, K.W., EILER, J.M. ET AL. 2009. Methods and limitations of 'clumped' CO_2 isotope (Delta47) analysis by gas-source isotope ratio mass spectrometry. *Journal of Mass Spectrometry*, **44**, 1318–1329.

HUNTINGTON, K.W., WERNICKE, B.P. & EILER, J.M. 2010. Influence of climate change and uplift on Colorado Plateau paleotemperatures from carbonate clumped isotope thermometry. *Tectonics*, **29**, 19.

JOHN, C.M. 2015. Burial estimates constrained by clumped isotope thermometry: example of the Lower Cretaceous Qishn Formation (Haushi-Huqf High, Oman). *In*: ARMITAGE, P.J., BUTCHER, A.R. ET AL. (eds) *Reservoir Quality of Clastic and Carbonate Rocks: Analysis, Modelling and Prediction*. Geological Society, London, Special Publications, **435**. First published online November 18, 2015, https://doi.org/10.1144/SP435.5

KLUGE, T. & JOHN, C.M. 2015. Effects of brine chemistry and polymorphism on clumped isotopes revealed by laboratory precipitation of mono- and multiphase calcium carbonates. *Geochimica et Cosmochimica Acta*, **160**, 155–168.

KLUGE, T., JOHN, C.M., JOURDAN, A.L., DAVIS, S. & CRAWSHAW, J. 2015. Laboratory calibration of the calcium carbonate clumped isotope thermometer in the 25–250°C temperature range. *Geochimica et Cosmochimica Acta*, **157**, 213–227, https://doi.org/10.1016/j. gca.2015.02.028

KOUTSOUKOS, E.A.M., MELLO, M.R., FILHO, N.C.D., HART, M.B. & MAXWELL, J.R. 1991. The Upper Aptian Albian succession of the Sergipe Basin, Brazil – an integrated paleoenvironmental assessment. *AAPG Bulletin*, **75**, 479–498.

LAND, L.S. 1980. The isotopic and trace element geochemistry of dolomite: the state of the art. *In*: ZENGER, D.H., DUNHAM, D.W. & ETHINGTON, R.L. (eds) *Concepts of Models of Dolomitization*. Society of Economic Paleontologists and Mineralogists (SEPM), Special Publications, **28**, 87–110.

LECHLER, A.R., NIEMI, N.A., HREN, M.T. & LOHMANN, K.C. 2013. Paleoelevation estimates for the northern and central proto-Basin and Range from carbonate clumped isotope thermometry. *Tectonics*, **32**, 295–316.

MACDONALD, J.M., JOHN, C.M. & GIRARD, J.-P. 2013. Application of clumped isotope thermometry to subsurface dolostone samples. *Mineralogical Magazine*, **77**, 1662.

MACDONALD, J., JOHN, C. & GIRARD, J.-P. 2015. Dolomitization processes in hydrocarbon reservoirs: insight from geothermometry using clumped isotopes. *Procedia Earth and Planetary Science*, **13**, 265–268.

MCCREA, J.M. 1950. On the isotopic chemistry of carbonates and a paleotemperature scale. *The Journal of Chemical Physics*, **18**, 849–857.

MCLIMANS, R.K. 1987. The application of fluid inclusions to migration of oil and diagenesis in petroleum reservoirs. *Applied Geochemistry*, **2**, 585–603.

MECKLER, A.N., ZIEGLER, M., MILLAN, M.I., BREITENBACH, S.F. & BERNASCONI, S.M. 2014. Long-term performance of the Kiel carbonate device with a new correction scheme for clumped isotope measurements. *Rapid Communications in Mass Spectrometry*, **28**, 1705–1715.

NADEAU, P.H. 2011. Earth's energy 'Golden Zone': a synthesis from mineralogical research. *Clay Minerals*, **46**, 1–24.

PASSEY, B.H. & HENKES, G.A. 2012. Carbonate clumped isotope bond reordering and geospeedometry. *Earth and Planetary Science Letters*, **351–352**, 223–236.

PASSEY, B.H., LEVIN, N.E., CERLING, T.E., BROWN, F.H. & EILER, J.M. 2010. High-temperature environments of human evolution in East Africa based on bond ordering in paleosol carbonates. *Proceedings of the National Academy of Sciences of the United States of America*, **107**, 11 245–11 249.

PETERSEN, S.V. & SCHRAG, D.P. 2014. Clumped isotope measurements of small carbonate samples using a high-efficiency dual-reservoir technique. *Rapid Communications in Mass Spectrometry*, **28**, 2371–2381.

REYRE, D. 1984. Caracteres petroliers et evolution geologique d' une marge passive. Le cas du Bassin Bas Congo-Gabon [Petroleum characteristics and geological evolution of a passive margin: a case study of the Congo-Gabon Basin]. *Bulletin du Centre de Recherche Exploration–Production Elf-Aquitaine*, **8**, 303–332.

ROSENBAUM, J. & SHEPPARD, S.M.F. 1986. An isotopic study of siderites, dolomites and ankerites at high-temperatures. *Geochimica et Cosmochimica Acta*, **50**, 1147–1150.

SCHAUBLE, E.A., GHOSH, P. & EILER, J.M. 2006. Preferential formation of ^{13}C–^{18}O bonds in carbonate minerals, estimated using first-principles lattice dynamics. *Geochimica et Cosmochimica Acta*, **70**, 2510–2529.

SENA, C.M., JOHN, C.M., JOURDAN, A.L., VANDEGINSTE, V. & MANNING, C. 2014. Dolomitization of lower cretaceous peritidal carbonates by modified seawater: constraints from clumped isotopic paleothermometry, elemental chemistry, and strontium isotopes. *Journal of Sedimentary Research*, **84**, 552–566.

SHACKLETON, N. 1967. Oxygen isotope analyses and pleistocene temperatures re-assessed. *Nature*, **215**, 15–17.

SHENTON, B.J., GROSSMAN, E.L. *ET AL.* 2015. Clumped isotope thermometry in deeply buried sedimentary carbonates: the effects of bond reordering and recrystallisation. *GSA Bulletin*, **127**, 1036–1051.

SIBLEY, D.F. & GREGG, J.M. 1987. Classification of dolomite rock textures. *Journal of Sedimentary Petrology*, **57**, 967–975.

STOLPER, D.A. & EILER, J.M. 2015. The kinetics of solid-state isotope-exchange reactions for clumped isotopes: a study of inorganic calcites and apatites from natural and experimental samples. *American Journal of Science*, **315**, 363–411.

STOLPER, D. & EILER, J. 2016. Constraints on the formation and diagenesis of phosphorites using carbonate clumped isotopes. *Geochimica et Cosmochimica Acta*, **181**, 238–259.

THIAGARAJAN, N., SUBHAS, A.V., SOUTHON, J.R., EILER, J.M. & ADKINS, J.F. 2014. Abrupt pre-Bølling–Allerød warming and circulation changes in the deep ocean. *Nature*, **511**, 75-U409.

UREY, H.C., LOWENSTAM, H.A., EPSTEIN, S. & MCKINNEY, C. R. 1951. Measurement of paleotemperatures and temperatures of the Upper Cretaceous of England, Denmark, and the Southeastern United States. *GSA Bulletin*, **62**, 399–416.

VANDEGINSTE, V., JOHN, C.M., COSGROVE, J.W. & MANNING, C. 2014. Dimensions, texture-distribution, and geochemical heterogeneities of fracture- related dolomite geobodies hosted in Ediacaran limestones, northern Oman. *AAPG Bulletin*, **98**, 1789–1809.

VEIZER, J., ALA, D. *ET AL.* 1999. Sr-87/Sr-86, delta C-13 and delta O-18 evolution of Phanerozoic seawater. *Chemical Geology*, **161**, 59–88.

WACKER, U., FIEBIG, J. & SCHOENE, B.R. 2013. Clumped isotope analysis of carbonates: comparison of two different acid digestion techniques. *Rapid Communications in Mass Spectrometry*, **27**, 1631–1642.

WALGENWITZ, F. 1989. *CACO 2 (Angola) Etude de la diagenèse et des inclusions fluides du reservoir Pinda, carottes 1 à 3* [Study of diagenesis and fluid inclusions of the Pinda reservoir]. TOTAL Internal Report, EP/EXP/Lab, Pau 89/2055 RP. TOTAL, Pau, France.

WALGENWITZ, F., PAGEL, M., MEYER, A., MALUSKI, H. & MONIE, P. 1990. Thermochronological approach to reservoir diagenesis in the offshore Angola Basin – a fluid inclusion, ^{40}Ar–^{39}Ar and K–Ar investigation. *AAPG Bulletin*, **74**, 547–563.

Index

Page numbers in *italics* refer to Figures. Page numbers in **bold** refer to Tables.

accretionary prism, Barbados 87
accumulation into trap 133–136
activity coefficient 131
aerobic oxidation, post-migration *167*, 181–182
air contamination, noble gases 137, *143*
air-saturated water (ASW) 12
  coalbed methane reservoir 159, 172, *174*, 175–176, 180, 183
  noble gas isotopes *128*, 132, 137, *139*, 140, 170
  tritiated *169*
Alaskan lake, methane 37
algae 93, 108
alkane 56, 57, 61, 74
  cracking 63–68
  see also n-alkane
amino acid 54–55, 58, 59
ammonium 73
anaerobic enzymes 114
anaerobic oxidation, post-migration *167*, 181
anaerobic, bacteria 95
analytical technology, isotopic structure 76–77
Antrim Shale, Michigan Basin 162, 183
  methane *37*, 39–40
  propane 69
  shale gas 142
aquatic/terrestrial organic matter 108–110
aquifer contamination 143
aquifer recharge management 2
Archean basement *16*
argon ^{36}Ar ^{38}Ar ^{40}Ar 129–133, 137–143
  Ar* (radioactive decay) 153, 159
  biogenic v. thermogenic source *171–174*, 175–176, 178–180
asphaltene 118
associated/non-associated gas 28–36, 43–48
  Whiticar and Bernard plots *40–41*, 42
atmosphere
  He isotope ratio *169*
  Ne and Ar isotope ratios *167*
atmosphere-derived gas 178, 179
  noble gas inventory 132
  noble gases 130, 137–141
Austria, source rock 115–116

bacteria 95, 108
Barbados, oil field *93*, 94
  samples 85–87
  V isotope composition 95, *96–97*
Barents Sea, basin analysis *115*, 116–117
Bernard plot 24, 25, *41*, 42–43
Bighorn Basin (USA), integration case study 5–17
biodegradation 118
  V isotopes 85, 87, 92, *93*, 95
  δ^{13}C values 114
biogenic gas 1–2, *32*
  clumped isotopic equilibrium 37–38
  clumped isotopic non-equilibrium 38
  composition control 38–39

biogenic methane 183
  origin 23–24, *27, 29*
    shallow or deep 45–47
biogenic v. thermogenic gas source 3, 151–184
  Illinois Basin, study area 160–162
  integrated methodology 153–160
  numerical modelling 170–175
  results and discussion 175–184
  sample analysis 165–169
  sample collection 162, 165
biological activity, isotope fractionation 93
biomarker 73, **86**, 87, 99, 118, 120
  compound-specific isotope analysis *109*, 117
  molecular 105
  terpane *9*
  vanadium isotope 84
biomolecules 55–60, 73
  equilibrium control 55–57
  kinetic control 57–59
biosignature 74
biosynthesis 56, 108
  C and H isotopes 110
Birchtree mine, Canada 37, *38*, 48
bitumen *54*, 61
  isotopic structure 67–68
  δ^2H isotopes 113
blocking temperature 30
bond reordering 190, 197–198, *199*
Bravo Dome *135, 142–143*
Brazilian reservoirs, gas 43–45
bubble phase 171, 172
bulk and clumped-isotope composition 64
Bulk Silicate Earth (BSE) 89–90, *92*
bulk stable isotope composition value 76
Bunsen solubility constant, noble gases 170, 172, 173
burial 55, 95
  carbonate 190
  curve, offshore Angola *192*, 198, *199*
  diagenesis 197–200
  dolomitization 195
  history 7, *10–11, 13*, 31
  temperature 25, 29–30, 35
  see also catagenesis

C *see* carbon
calcite, bond reordering 190, 197–198, *199*
carbohydrates 54–55, 58–59
carbon dioxide 2
  He isotope ratio 137
  mantle derived *135, 142*
carbon isotopes
  C and H in exploration 105–121
    methodology 106–108
  $C_3/C_{2\text{-}3}$ ratio 24, 25, 37, 39, 41–43, 46–48
  methane/noble gases *171, 173, 174*
  in organic matter 108–113, 178
  in propane 69, *70*

carbon stable isotope δ^{13}C
  and clumped-isotope, temperature 43–45, *46*, **47**
  in methane *165–168*, 169
  natural gas **161–162**
carbonate clumped-isotope thermometry 4–5, 189–200
catagenesis *54*, 55, 59, 153, 171
characterization tools *see* petroleum exploration tools
charge history 115, 118–119, 120
chemical degradation, analytical technology 76
chemical kinetics and equilibrium 74
chemical purification methods
  V isotope analysis 88–89
chemistry, noble gases in fluids 131–132
chlorophyll 83, 90, 93, 95
chromium Cr 84, *91*
clean-burning fuel 151
climate archive 58–59
clumped isotope 45–47
  temperature 24–48
  see also methane clumped isotopes
clumped isotope, characterization tool 189–200
  fluid inclusions 191
  methodology 191–193
  results and discussion 193–200
  Δ_{47} C and O isotopes 190, *196*
clumped isotopologue (Δi value)
  Antrim Shale, measurements 183
  defined 25–27, 75
CO_2 *see* carbon dioxide
coal
  gas source 34
  pyrolysis experiment 31–33, 35
  seam sample analysis 169
  seam solids 176, **177**
coal-derived methane 44
coalbed methane 37, 141, *142*
  gas source 151–184
coalbed methane reservoirs
  natural gas source 178–179
  noble gases 175–176
compound-specific isotope analysis (CSIA) 105–106, 111
  application 114–120
  methodology 106–108
  n-alkanes 106
condensate-fraction compounds *67*, 68
condensates 117
contamination, noble gas 143
conventional hydrocarbon system 128–129, 135,
    142–143, 144
  gas *140*
conventional reservoir
  methane clumped-isotope measurement 28–30
  non-associated gases/methane 34
  oil-associated gases/methane 32–34
conventional thermogenic gas 41, 44
Cr *see* chromium
cracking 3, 23, 34, 55, 95
  ethane 32, 35, 71
  kerogen 113
  reactions 61–67
  secondary 69, 71, 74, *135*
crude oil, vanadium isotope study 83–99
  biodegradation 95
  maturation 95–98

sampling and methods 85–90
source and reservoir 90–95
crust, noble gas *142–143*, 175
  inventory 133
  production rate **131**
crustal fluid 159
  noble gases *135*, 153, *171–172*, 178–180
CSIA *see* compound-specific isotope analysis

D *see* deuterium
data inversion techniques 139
dating
  noble gases 130
  radiogenic ingrowth 139–141
degassing
  ^{4}He isotope 136
  deep crust 141
  solid Earth 132
density function theory (DFT) 63
depositional environment 87, 111
  C and H isotopes 110
  noble gases 127
  V isotope composition 90, 92–95, 98
  V/Ni 83–85
detrital zircon analysis *16*
deuterium ^{2}H δD 56, 57–60
  in clumped isotopologues 24–28, 32, 36
  D/H ratio 59
  in gas 38–48
  in propane 66–67, 69–71
diagenesis *54*, 55, 59
  carbonates 189–191, *192*
  dolomitization 190, 191
  Eh–pH 95
dolomite, bond reordering 190, 198–200
dry gas 153

Eagle Ford Shale, Texas 69, 70, 71
Eh–pH conditions 95, 98
  V isotope composition 90
elemental analysis, crude oil
  Ni, V, S isotope analysis 89
Elk Basin Field, USA *6*, 7, 12
  burial history *11, 13*
endogenous v. exogenous thermogenic gas 179–180
energy demand 1–2
environmental sciences 2
enzyme activity 38–39, 93, 114
enzyme, reversibility 39
equilibrium, biogenic gas 40–42
equilibrium, controls 55–57, 74
Escherichia coli, carbon isotope structure 57, 68
ethane 56, 178
  biogenic v. thermogenic 181–183
  clumped isotopes 25, 36, 37, 43, 49
  in coalbed methane reservoirs 159
  cracking reactions 61–65, 72
  noble gases 175–176
  stable isotopes *165, 167–168*
  $\Delta^{13}C_2H_6$ proxy for secondary cracking 71
ethanol 66
evaporative fractionation 113–114, 117, 120
exogenous thermogenic gas source 179–181
exploration *see* petroleum system, exploration tools

fatty acids 57–59, 68
fauna/fossil, palaeoseep mounds *14, 15*
fissiogenic production 140
   noble gases 130–131
fluid flow, noble gases 133–136, 144, 173
fluid inclusions, palaeothermometry 4, 189–200
   Elk Basin Field 12, *13*
   sample characterization 191, **194**
   temperature *193*
fluid migration 152–153, 159, 176
   compound-specific isotope analysis 105–106, *109*,
      114–120
   H isotope analysis 113
fluid residence time 179–180
food science forensics 76
fractionation 66–67
   methane/ethane/keragen 61–63
   noble gases 137
fugacity 131

gas chromatography 106, 165, 168, 169
gas composition, Illinois Basin **156, 158**
gas constituents, isotopic structure 68–72
   $^{13}C_2H_6$ in ethane 71
   site-specific ^{13}C in propane 69, *70*
   site-specific D in propane 69–71
gas generation 24, 35
   temperature 25–26, 31, 37
gas maturity, noble gas 141
gas migration 179–180
gas production 160, 162
gas release 141, 144
gas sample/analysis 162, 165–169
gas shale 129
gas stripping 170–171, 181
gas washing 113
gas wetness 71, *72*, 159
gas–oil *140*
   ratio (GOR) 28–34
gas–water ratio, proxy *170*
gas–water system 138, 176
gas, methane in natural gas 23–49
gas, noble 127–144
gas, thermogenic or biogenic source 151–184
geochemical properties, V isotope **86**
geochemical techniques 2–5, 127, 143
   compound-specific isotope analysis 114–115
   integrated hydrocarbon and noble gas data 153–160
   integration case study 5–17
geochronology **86**, 87, 94, 130
   groundwater 139–141
geothermal gradient 197–198
GGS-R groundwater gas stripping and re-dissolution
      170–171, 181
glacial meltwater 182
global energy demand 1–2
graphitization *54*, 55, 72–73
gravity, API **86**, 117–118, 131, 134, 139
greenhouse gas 2, 23
groundwater 5, 159, 170–171, 173, 182
   residence time 179–180
groundwater, noble gases 134, 136–141
   age 142
   modelling *128*, 144

Gulf of Mexico, gas 37, 70, 71
Gullfaks field, North Sea, biodegraded oil 114

H *see* hydrogen isotope
H_2S in crude oil 92, 95
halokinesis, trap structures 191
He *see* helium
heavy metal analysis 5
helium ^4He 143, 152–153, *170*
   $^3He/^4He$ isotope ratio 137, 142, *169*
   $^4He/^{20}Ne$, proxy for radiogenic He *170*
   biogenic v. thermogenic 175–183
   primordial 159
   terrestrial inventory 132–133, *135*
Henry's law *129*, 131–132, 134, 138–139, 159
hopanoids 59
humin 73
hydraulic fracturing 35–36, 129
hydrocarbon generation and source rock 2, 3
hydrocarbons, use of clumped-isotope studies 45–47
hydrogen isotope
   in organic matter 109–110, 111
   site-specific 69–71
   δ^2H natural gas **161–162**, 169
   δ^2H petroleum fluids 113
hydrogen isotopes of *n*-alkanes
   analytical procedure 106–108
   application 113, 114–120
   controls 108–114
   sample preparation 106
hydrogen sulphide, in crude oil 92, 95
hydrogenotropic source *38*, 178
hydropyrolysis 64–66, 71
hydrothermal fluids, V isotope composition 94

Illinois Basin, Pennsylvania 153, *154*
   geology 160, 162
illite, age analysis 4, 7, *10*, 16
in-reservoir processes 105, *109*, 119–121
   C and H isotopes 113–114, *115*, 117
   compound-specific isotope analysis 120–121
infrared spectroscopy 27, 77
instrument calibration, isotopic structure 73
integrated geochemical analysis 2–4, *115*, 120
   Bighorn Basin, USA 5–17
isoprenoids *116*, 118, 120
isotomomer, nomenclature 75
isotope composition, *n*-alkanes 105–121
isotope thermometry 4
isotope, reservoir characterization tool 189–200
isotopes *see also* methane clumped isotopes
   noble gas ratio **129**
   use in petroleum analysis 4–5
   vanadium in crude oil samples **86**
   Δ_{47} derived temperatures *196*
isotopic equilibrium 32
   defining 74
   $\delta^{13}C$ and δD 36–37
isotopic structures 53–54
   biomass to hydrocarbon evolution 54–55
   biomolecules 55–60
   cracking reaction/KIE 61–67
   fractions, bitumen/oil/condensate 67–68
   future work 73–75

isotopic structures (*Continued*)
  gas constituents 68–72
  graphitization 72–73
  kerogen 60–61
isotopic techniques 24–25
isotopologue 53
  methane 24, 25–27, 152
  nomenclature 75
isotopomer 75

jargon 76

K-40 potassium 133, 159
kerogen *54*, 55, 61, 73–75, 84
  C isotope properties 111
  cracking 69, 113
  graphitization 73
kinetic isotope effects (KIE) 56–64, 72–74, 95, 113
  experimental simulations 64–67
krypton ^{84}Kr 129, **131**, 133, 137–141, 144
  gas–water interaction 176, 180

Laramide Orogeny 7
lignin *54*, 55, 73
lipids *54*, 55, 59
  C and H isotopes 110–111

magnesium Mg in plant chlorophyll 83, 95
mantle fluid 144, 159
  gas 182–183
  He ratio 137
  noble gas *135*, 136, *142–143*, 175–176
  noble gas inventory 130, 133
Marcellus Shale, Pennsylvania 71, *72*
mass spectrometry 165, 166, 169
  analytical technology 76–77
  clumped isotopes 191–193
  GC-IRMS Isotope Ratio 106–108
  GC-MS, GC-FID 106
  IRMS (Isotope Ratio MS) 58–59, 69
  MC-ICP-MS (multicollector inductively coupled
      plasma MS) 84
  techniques 4–5, 12, 27, 68, 69, 70, 75, 77
  TIMS (thermal ionization MS) 84
  V isotope analysis 89–90
mass-dependent isotope fractionation 83
maturation, vanadium isotope 92, 95–99
maturity indicator 71, 118, 141
metal element extraction, crude oil 87–88
metalation 90, 95–99
metamorphism 55, 72–73
meteoric water 152, 170, 176, 178
  groundwater *128*, 159
methane clumped isotopes 23–49
  future research 48–49
  measurement 27
    biogenic gases 37–39
    mixed biogenic and thermogenic *37*, 39–40
    thermogenic gases 28–37
  molecular and isotopic techniques 24–25
  synthesis 45–47
  theory and nomenclature 25–27
  Whiticar and Bernard plots 40–43
  δ^{13}C v. clumped-isotope temperature 45–47

methane thermometry 4, 12, 45–47
methane, stable isotope composition **162**, *165–168*
methanogenesis 153
  in coalbeds 178
  rate of 48
  reversible 71, 72
methanogenic archaea 152, 160
methanogens 23, *38*
Mg magnesium 83, 95
microbial activity 71, 95, 99, 111, 114, *167*, 180
microbial gas 2, 3, 23
  methane 138, 155, 160
  noble gases 141, 178–179
  shallow or deep source 46–47
mid-ocean ridge basalt (MORB) 133
  He isotope ratio 175–176
migration 2, 170, 172
  Bighorn Basin 7–17
  noble gases 133–136, 178–179
molecular and isotopic composition 5, 152, 159
  gas ratio in coal gas 162, **163–164**

n-alkane *62*, 63–68, 74
n-alkanes, in exploration
  analytical methodology 105–108
  compound-specific isotope analysis 114–121
  controls on δ^{13}C and δ^2H values 108–114
natural gas
  constituents 68–72
  formation temperature 33
  methane 25–49
  production, Illinois Basin 160, 162
  production, USA 151
  reservoir classification 28
  source determination 153–159
    δ^{13}C and δD *26*, **47**
neon ^{21}Ne ^{2}Ne 129–132, 137–141, *142*, 167–168
  coalbed methane reservoir 159, 175–180
  Ne* (radioactive decay) 153, 159
New Albany Shale
  Illinois Basin 153, *154–155*, 160, 162, 165
  stable isotope composition 165, *166–167, 170, 173*
  thermogenic gas source 180–181, 183
nickel mine, methane 37
Niger Delta, integrated basin analysis *115*, 119
Nigerian oil *107*
NIST Crude Oil Reference Material 84, 85, 88–90, *91*, 94
nitrogen N$_2$ *165, 168*, 175, 176
noble gas 5
  definitions and overview 127–129
  fluid chemistry 131–132
  isotope ratio **129**
  migration 178–179
  in petroleum systems 133–136
    analysis 136–143
  primordial origin 129–130
  production, radioactive 130, **131**
  terrestrial inventory 127, 130, 132–133
noble gas in coalbed methane
  biogenic v. thermogenic source 151–184
  composition **157–158, 161**, 175
nomenclature, isotopic structure 75–76
non-equilibrium biogenic gas 40–42, 48
non-oil-associated gas *33*, 41, 44

North Sea, Rotliegend 34, *112*
nuclear magnetic resonance (NMR) 57–58, 68, 69, 75
  analytical technology 76
nucleogenic noble gases 159, 176, 178–179

oil source correlation *115*
oil window/generation temperature *33*, 97
oil-associated gas 41
oil–oil correlation *115*, 117
oil–water system 138
oil, isotopes 67–68, 106
Oman, carbonate platform 197
organic compounds
  diagenesis-maturation 110–111
  isotopic structure 53–75
  maturation/oil generation 111–113
  δ^{13}C and δ^2H isotopes of *n*-alkanes 108–110
organic geochemistry 105
organic matrix removal, crude oil 87–88
osmium ^{187}Os/^{188}Os ratio *14*
  Re–Os ratio 12
oxygen isotope composition 59, 61

palaeoaltimetry 190
palaeoclimate 108, 120, 190
palaeogeographical maps 7, *9–11*
palaeoseep mounds 12, *15*
palaeothermometry δ^{18}O 189
  see also thermometry
Pee Dee Belemnite scale (PDB) 69, 193
petroleum system
  defined 128
  elements and processes 128–129
  mixing models 136–139
  noble gas introduction 133–136
petroleum system, exploration tools
  clumped isotopes 189–200
  n-alkanes (C H isotopes) 105–121
  noble gases 128
  play summary chart *8*
petroleum systems analysis 1–4
  Bighorn Basin 5–17
  future challenges 4–5
pH, vanadium species 94–95
Phosphoria Formation 7, *9–10*, 12, 16, 17
photomicrographs, dolomite *195*
phytane 117, 118
Pinda Formation, offshore Angola 193–200
  geology 191, *192*
plants, hydrogen isotopes 59–60, 109–110
play summary chart, Bighorn Basin *8*
pollutants 53, 66
porewater δ^{18}O and δ^{13}C values 193, 195,
    197, 200
porphyrin, V isotopes 83, 90, 92, 95, 96, 99
post-genetic processes, natural gases 181–182
potassium ^{40}K 133, 159
Potiguar Basin, Brazil 30–31, 43, *44*, 69, 70, 71
Potwar Basin, in-reservoir processes *115*, 117
praseodymium, Pr/*n*C$_{17}$ ratio **86**, 95
precursor molecules 53, 57, 59
primary migration 134
primordial noble gases 129–130
pristane 117, 118

propane 56, 63, 72, 74, *167*
  clumped isotopes 25, 31, 36, 37, 43, 49
  site-specific ^{13}C 65–66, *67*, 69
  site-specific D 66–67, 69–71
protein 54
Prudhoe reservoir 120
pyrolysis 61, 63–69, 71
  analytical technology 76–77
  experiment *112*

radiogenic noble gases 130, 159, 176
radiogenic production 136, 137, *142–143*
Rayleigh distillation 43, 48, 71, 138
Rayleigh equation 114
Re *see* rhenium
redox index, crude oil 85
residence time 152–153
rhenium Re, Re–Os geochronology 12, *14*, 16
Rotliegend, Germany 34

San Juan, coalbed methane 141, *142*
sea water, temperature 197
sea water, vanadium isotopes 84, 93–95
secondary cracking 69, 71, 74, *135*
secondary migration 128–129, 134–135
sedimentary organic matter, evolution *54*, 55
Seelyville coal seam 153, *154–155*
  comingled gases 175–178, 180–183
  sample/analysis 160, 162, 169
sequestration, CO$_2$ 2
Setschenow coefficient **129**, 131–132
Sevier Orogeny 7
shale gas 25, 142, 151, 162, 181
Sirte Basin, N Africa, source correlation *115*, 119
solubility, noble gases 133–134, 137, 159
source rock 99, 117, 120, 127
  classification **133**
  compound-specific isotope analysis 106, 109, *115*, 119
Springfield coal seam 153, *154–155*
  comingled gases 175–178, 180–183
  sample/analysis 160, 162, 169
stable isotope analysis 77, 105–108
  methodology 106–108
  see also under natural gas and *n*-alkanes
stable isotope composition 53
  gas, water 152, 159
  methane 175, 176
  natural gas **161–162**
  vanadium 83–84
stable isotope values in methane
  carbon *165–168*
  δ^{13}C and δD *26*, **47**
  δ^{13}C and δ^2H 169, *170*
stable isotope values in *n*-alkanes
  δ^{13}C and δ^2H 108–114
starch 58
sugar 58, 72, 74
sulphur, crude oil 84
synthetic natural gas standards 169

Tarim, China *115*, 118–119
temperature
  clumped isotope v. fluid inclusion 189–200
  clumped isotopes, methane 24–48

temperature (*Continued*)
  gas formation 37–38, 152
  isotopic fractionation 83, 97
  noble gas isotopes 153
  numerical modelling 170
  oil generation *33*
  primary precipitation 197
  reservoir 28, 37, *139*, 141 179
terpane biomarker *9*
terpenes 59, 60
terrestrial noble gases 129–130
tertiary migration 136
thermal history 47–48, 200
  $\delta^{18}O$ 189
thermal history reordering model (THRM) 198
thermal maturity 3–4, 25, 155, 160
  index 95–96, *97–98*, 113, 114
  and magmatism 182–183
  noble gas 141, 178
  $\delta^{13}C$ value, methane 43, *167*
  $\delta^{2}H$ values 111, 117–119, 120
thermochemical sulphate reduction 114
thermogenic hydrocarbons 12
thermogenic methane 23–24, *27, 166*
  clumped-isotope measurement 28–32, 36
  temperature 46–47
    $\delta^{13}C$ and clumped-isotope 43–45
thermogenic v. biogenic gas *see* biogenic
      v. thermogenic gas
thermometry 4, 12, 45–47
  fluid inclusion 189–200
thorium Th 130, 131, 133, 152, 159
  coal seam solids 176, **177**, 180
titanium Ti 84, *91*
toluene 114
trace element 84, 85, **86**, 90, 93
trace gas components 172–175
trace metal 83
tracer, crustal fluids 159
tracer, noble gases 127–144
Tricyclic Index **86**, 95, *97–98*
tritiated air-saturated water *169*

U-238 *see* uranium
unconventional hydrocarbon system 129, 135, *142–143*, 144

  gas *140*
unconventional reservoir 151, 152
  associated gases 35–36
  methane clumped-isotope measurement 28–34
  non-associated gases 34–35
unconventional thermogenic gas 41, 44
uranium ^{238}U 130, 131, 133, 152, 159
  coal seam solids 176, **177**, 180

vanadium isotope analysis 5, 12
  V/(V+NI) ratio *93, 96*
  V/Ni ratio 83
  valance state 84, 94
vanadium isotope composition, crude oil 83–99
  biodegradation *93*, 95
  depositional conditions 90–95
  maturation 95–98
  method 85–90
  results *86*, 90, *91–93*
Venezuela *93*, 94
  La Luna Formation 84
  sampling for V isotope study 85, **86**
Vienna Pee Dee Belemnite scale (VPDB) 69, 193
vitrinite reflectance 4, 118, 160, 179, 183
volatiles 129–130, 132
volcanic degassing 132

water *see also* air-saturated water
  chemistry 152
  hydrocarbon generation 133–134
  meteoric, H isotope 110
  noble gas inventory 132
  resource protection 2
water washing 138
water–hydrocarbon interaction models 137–139, *140*
West Sak reservoir, N America *115*, 119–120
wet gas 71, *72*, 159
Whiticar plot 24–25, *40*, 41–42

xenon Xe 129–131, 133, 137–141, 144
  Xe* (radioactive decay) 153, 159

yeast, carbon isotope structure 57, 68

zircon analysis *16*, 17